AF360895

Physiological and Pathological Role of ROS

Physiological and Pathological Role of ROS

Benefits and Limitations of Antioxidant Treatment

Special Issue Editors

Sergio Di Meo
Paola Venditti
Gaetana Napolitano

MDPI • Basel • Beijing • Wuhan • Barcelona • Belgrade • Manchester • Tokyo • Cluj • Tianjin

Special Issue Editors

Sergio Di Meo
University of Naples Federico II
Italy

Paola Venditti
University Federico II of Naples
Italy

Gaetana Napolitano
University of Naples Parthenope
Italy

Editorial Office
MDPI
St. Alban-Anlage 66
4052 Basel, Switzerland

This is a reprint of articles from the Special Issue published online in the open access journal *International Journal of Molecular Sciences* (ISSN 1422-0067) (available at: https://www.mdpi.com/journal/ijms/special_issues/ROS).

For citation purposes, cite each article independently as indicated on the article page online and as indicated below:

LastName, A.A.; LastName, B.B.; LastName, C.C. Article Title. *Journal Name* **Year**, *Article Number, Page Range.*

ISBN 978-3-03936-282-0 (Hbk)
ISBN 978-3-03936-283-7 (PDF)

Contents

About the Special Issue Editors

Sergio Di Meo was a Professor of Physiology at the University Federico II, Naples, Italy. At the beginning of his academic career, he primarily focused on electrophysiology. He then continued his research activity by studying ROS production and cellular state redox in different experimental conditions, such as ischemia reperfusion, experimental and functional hyperthyroidism, and acute exercise and training, mainly in experimental mammal models. He currently continues to collaborate with the physiology section in the Department of Biology at Federico II University, Naples, Italy.

Paola Venditti is a Professor of Physiology at the University Federico II, Naples, Italy. She received her degree in 1990 from the University of Naples, Italy, and continued her research on the detection and measurement of mitochondrial free radical production, antioxidant activities, and oxidative-derived molecular damage. Following this, she received her Ph.D. degree in Physiology in 1996 from the University of Ferrara, Italy, and has since followed the role of oxidative stress in several conditions. In particular, she studies the role played by free radicals in tissues' functional adaptations to physio-pathological conditions, in which tissue oxidative stress develops (alteration of thyroid state, physical activity, exposition to xenobiotics, cold exposure, ischemia reperfusion, etc.) and the ROS-sensitive factors involved in the control of mitochondrial biogenesis and adaptation of the antioxidant system.

Gaetana Napolitano is a Researcher in Physiology at the Science and Technology Department (DIST) of the Parthenope University of Naples, Italy. She received her degree in 2009 from the University Federico II, Naples, Italy, and subsequently her Ph.D. in Physiology under the supervision of Professor Paola Venditti. She studies the involvement of mitochondrial reactive oxygen species and free radicals in functional adaptations associated with various physio-pathological conditions. Her main interests concern the physiology of mammals (insulin resistance, training, acute exercise, experimental and functional hyperthyroidism, antioxidant supplementation) and physiological adaptations of both marine and freshwater aquatic organisms following environmental pollution (micro- and nano- plastic pollution, nitrite pollution, food dye pollution, and exposition to xenobiotics).

International Journal of
Molecular Sciences

Editorial

Physiological and Pathological Role of ROS: Benefits and Limitations of Antioxidant Treatment

Sergio Di Meo [1], Gaetana Napolitano [2] and Paola Venditti [1,*

[1]   Dipartimento di Biologia, Università di Napoli Federico II, Complesso Universitario Monte Sant'Angelo, Via Cinthia, I-80126 Napoli, Italy; serdimeo@unina.it

[2]   Dipartimento di Scienze e Tecnologie, Università degli Studi di Napoli Parthenope, via Acton n. 38-I-80133 Napoli, Italy; gaetana.napolitano@uniparthenope.it

*   Correspondence: venditti@unina.it; Tel.: +39-081-2535080; Fax: +39-081-679233

Received: 20 September 2019; Accepted: 27 September 2019; Published: 27 September 2019

From their discovery in biological systems, reactive oxygen species (ROS) have been considered key players in tissue injury for their capacity to oxidize biological macromolecules. Aerobic organisms possess a system of biochemical defenses to neutralize the oxidative effects of ROS, but the balance between ROS generation and the antioxidant system is slightly in favor of the ROS so that a continuous low level of oxidative damage exists [1]. When the imbalance toward the ROS increases, as happens under several conditions, oxidative stress arises. This has been related to the onset of many pathological conditions including cardiovascular disease, diabetes, rheumatoid arthritis, cancer, and neurodegenerative disorders [2]. It has been proposed that if ROS are involved in many pathological conditions, the use of exogenous antioxidants can help their management. However, starting from the end of the 1970s, increasing experimental evidence has led to an opposing view about the ROS' role in biological systems. This suggests that living systems not only adapted to the coexistence with free radicals but developed methods to use them in critical physiological processes [2]. It has also been shown that, when the generation of ROS induces adaptive responses that are beneficial to the organism, the use of antioxidants can be detrimental [2].

The papers reported in this Special Issue deal with different aspects of reactive oxygen species (ROS) actions in living organisms.

Some papers consider the role of ROS in inducing cellular dysfunction

Thus, Hua et al. [3] treated human umbilical vein endothelial cells (HUVECs) with H_2O_2 to obtain a cell model of oxidative stress to study the role of sphingomyelin synthase 2 (SMS2) in endothelial disease (ED). They found that SMS2 induces the stress of the endoplasmic reticulum (ER) that leads to ED both activating the Wnt/β-catenin pathway and promoting intracellular cholesterol accumulation, both of which contribute to the induction of ER stress and finally lead to ED.

Querio et al. [4] used adult rat cardiomyocytes stressed with H_2O_2 or doxorubicin to verify if trimethylamine *N*-oxide (TMAO), an organic compound derived from dietary choline and L-carnitine, is a factor involved in the progression of atherosclerosis and other cardiovascular diseases. They show that TMAO does not affect the treatment's effect on cell viability, sarcomere length, intracellular ROS, and mitochondrial membrane potential. Therefore, they conclude that TMAO cannot be considered a direct cause or an exacerbating risk factor of cardiac damage at the cellular level in acute conditions.

Another work evaluates the role of ROS as agents able to induce cellular protection.

Lin et al. [5] demonstrated that ROS are involved in the mechanism underlying the protective action of carbon monoxide-releasing molecule 2 (CORM-2) against lipopolysaccharide (LPS)-induced inflammation in mice lung. CORM-2 induces the expression of heme oxygenase 1 (HO-1), a member of the heme oxygenase (HO) family, able to directly protect various organs from oxidative damages. This is due to the activation of protein kinase C (PKC)α and proline-rich tyrosine kinase (Pyk2), which, in turn, activate NOX-derived ROS generation. The ROS signal activates the extracellular signal-regulated kinase 1/2 (ERK1/2) that upregulates c-Fos and c-Jun, activator protein 1 (AP-1) subunits, which turn on the transcription of the HO-1 gene by regulating the HO-1 promoter.

ROS can also be involved in the therapeutic action of some antitumoral drugs.

Soltan et al. [6] evaluated the antitumoral action of a derivative of the plant extract indirubin, DKP-071, on cutaneous T-cell lymphoma (CTCL). DKP-071 activated the extrinsic apoptosis cascade via caspase-8 and caspase-3 through downregulation of the caspase antagonistic proteins c-FLIP and XIAP. In response to DPK-071 treatment, a strong increase of ROS levels was observed as an early effect. ROS turned out upstream of all other proapoptotic effects monitored. Thus, ROS appear as a highly active proapoptotic pathway in CTCL.

The antioxidant capacity to protect against oxidative stress-linked disease has been evaluated by Zhao et al. [7], who studied the protective effects of the treatment with artemisinin, an anti-malarial Chinese medicine, on SH-SY5Y and hippocampal neuronal cells treated with hydrogen peroxide (H_2O_2). Artemisinin prevents cell death at clinically relevant doses in a concentration-dependent manner. Artemisinin restored the nuclear morphology, prevented the increased intracellular ROS, and attenuated apoptosis. These data suggested that artemisinin protected neuronal cells. Similar results were obtained in primary cultured hippocampal neurons. Cumulatively, these results indicated that artemisinin protected neuronal cells from oxidative damage, at least in part through the activation of AMPK. These findings support the role of artemisinin as a potential therapeutic agent for neurodegenerative diseases.

Moreover, some reviews are presented in this Special Issue.

Lévy et al. [8] reviewed the current literature concerning the link between oxidative stress and protein aggregation processes, which are involved in the development of proteinopathies, such as Alzheimer's disease, Parkinson's disease, and prion disease.

Damiano et al. [9] examined the data concerning antioxidant supplementation associated with exercise in normal and sarcopenic subjects. In older people, malnutrition and physical inactivity can lead to sarcopenia, a process in which oxidative stress seems to be involved. The effects of exercise and antioxidant dietary supplements in limiting age-related muscle mass loss and performance reduction have been evaluated in many studies but the results are conflicting. This can be due to the dual effects of ROS in skeletal muscle, which at low levels increase muscle force and induce adaptations to exercise, but at higher levels lead to a muscle performance decline. Therefore, the controversial results obtained with antioxidant supplementation in older persons could, in part, reflect the lack of univocal effects of ROS on muscle mass and function.

Xiao and Meierhofer [10] reviewed the current knowledge about the three main renal cell carcinoma (RCC) subtypes—clear cell RCC (ccRCC), papillary RCC (pRCC), and chromophobe RCC (chRCC)—and highlight their mutual influence on GSH metabolism. Altered GSH metabolism contributes to the development and progression of the three renal carcinomas. All RCCs have a reduced oxidative phosphorylation capacity, and the respiratory chain is the main source of ROS. Raised oxidative stress levels in RCCs are counteracted by increased GSH levels that foster the survival of the malignancy. New studies have shown that combinatory therapy targeting two independent pathways of GSH synthesis and one involved in ROS metabolism is the key to improving the survival rate and eventually curing RCC.

Siauciunaite et al. [11] summarized the actual knowledge about the role of ROS as signaling molecules and key regulators of gene expression from an evolutionary point of view. They described recent work that has revealed significant species-specific differences in the gene expression response to ROS by exploring diverse organisms. This evidence supports the notion that during evolution, rather than being highly conserved, there is inherent plasticity in the molecular mechanisms responding to oxidative stress.

The review of Di Meo et al. [12] analyzed the literature dealing with sources of ROS production and the most important redox signaling pathways, including MAPKs that are involved in the responses to acute and chronic exercise in the muscle, particularly those involved in the induction of antioxidant enzymes.

Ismail et al. [13] collected and discussed studies analyzing the involvement of mitochondrial peroxiredoxins (Prdxs) in human cancers. They focused on signaling involving ROS and mitochondrial

Prdxs that is associated with cancer development and progression. An upregulated expression of Prdx3 and Prdx5 has been reported in different cancer types, such as breast, ovarian, endometrial, and lung cancers, as well as in Hodgkin's lymphoma and hepatocellular carcinoma. It is depicted that mitochondrial Prdxs are upregulated in a variety of cancer types and directly or indirectly regulated by transcription factors, microRNAs, and oncogenes.

It is our opinion that the articles included in this Special Issue, despite dealing with such different topics, represent an important contribution to the knowledge of the physiological and pathological role of ROS, and give some information on the benefits and limitations of antioxidant treatment.

Conflicts of Interest: The authors declare no conflicts of interest.

References

1. Poljsak, B.; Šuput, D.; Milisav, I. Achieving the Balance between ROS and Antioxidants: When to Use the Synthetic Antioxidants. *Oxid. Med. Cell. Longev.* **2013**, *2013*, 956792. [CrossRef] [PubMed]
2. Di Meo, S.; Reed, T.T.; Venditti, P.; Victor, V.M. Role of ROS and RNS Sources in Physiological and Pathological conditions. *Oxid. Med. Cell. Longev.* **2016**, *2016*, 1245049. [CrossRef] [PubMed]
3. Hua, L.; Wu, N.; Zhao, R.; He, X.; Liu, Q.; Li, X.; He, Z.; Yu, L.; Yan, N. Sphingomielin Synthase 2 Promotes Endothelial Dysfunction by Inducing Endoplasmic Reticulum Stress. *Int. J. Mol. Sci.* **2019**, *20*, 2861. [CrossRef] [PubMed]
4. Querio, G.; Antoniotti, S.; Levi, R.; Gallo, M.P. Trimethylamine N-Oxide Does Not Impact Viability, ROS Production, and Mitochondrial Membrane Potential of Adult Rat Cardiomyocytes. *Int. J. Mol. Sci.* **2019**, *20*, 3045. [CrossRef] [PubMed]
5. Lin, C.C.; Hsiao, L.D.; Cho, R.L.; Yang, C.M. Carbon Monoxide Releasing Molecule-2-Upregulated ROS-Dependent Heme Oxygenase-1 Axis Suppresses Lipopolysaccharide-Induced Airway Inflammation. *Int. J. Mol. Sci.* **2019**, *20*, 3157. [CrossRef] [PubMed]
6. Soltan, M.Y.; Sumarni, U.; Assaf, C.; Langer, P.; Reidel, U.; Eberle, J. Key Role of Reactive Oxygen Species (ROS) in Indirubin Derivative-Induced Cell Death in Cutaneous T-Cell Lymphoma Cells. *Int. J. Mol. Sci.* **2019**, *20*, 1158. [CrossRef] [PubMed]
7. Zhao, X.; Fang, J.; Li, S.; Gaur, U.; Xing, X.; Wang, H.; Zheng, W. Artemisinin Attenuated Hydrogen Peroxide (H_2O_2)-Induced Oxidative Injury in SH-SY5Y and Hippocampal Neurons via the Activation of AMPK Pathway. *Int. J. Mol. Sci.* **2019**, *20*, 2680. [CrossRef] [PubMed]
8. Lévy, E.; El Banna, N.; Baïlle, D.; Heneman-Masurel, A.; Truchet, S.; Rezaei, H.; Huang, M.E.; Béringue, V.; Martin, D.; Vernis, L. Causative Links between Protein Aggregation and Oxidative Stress: A Review. *Int. J. Mol. Sci.* **2019**, *20*, 3896. [CrossRef] [PubMed]
9. Damiano, S.; Muscariello, E.; La Rosa, G.; Di Maro, M.; Mondola, P.; Santillo, M. Dual Role of Reactive Oxygen Species in Muscle Function: Can Antioxidant Dietary Supplements Counteract Age-Related Sarcopenia? *Int. J. Mol. Sci.* **2019**, *20*, 3815. [CrossRef] [PubMed]
10. Xiao, Y.; Meierhofer, D. Glutathione Metabolism in Renal Cell Carcinoma Progression and Implications for Therapies. *Int. J. Mol. Sci.* **2019**, *20*, 3672. [CrossRef] [PubMed]
11. Siauciunaite, R.; Foulkes, N.S.; Calabrò, V.; Vallone, D. Evolution Shapes the Gene Expression Response to Oxidative Stress. *Int. J. Mol. Sci.* **2019**, *20*, 3040. [CrossRef] [PubMed]
12. Di Meo, S.; Napolitano, G.; Venditti, P. Mediators of Physical Activity Protection against ROS-Linked Skeletal Muscle Damage. *Int. J. Mol. Sci.* **2019**, *20*, 3024. [CrossRef] [PubMed]
13. Ismail, T.; Kim, Y.; Lee, H.; Lee, D.S.; Lee, H.S. Interplay Between Mitochondrial Peroxiredoxins and ROS in Cancer Development and Progression. *Int. J. Mol. Sci.* **2019**, *20*, 4407. [CrossRef] [PubMed]

 International Journal of
Molecular Sciences

Article

Sphingomyelin Synthase 2 Promotes Endothelial Dysfunction by Inducing Endoplasmic Reticulum Stress

Lingyue Hua [1,†], Na Wu [1,†], Ruilin Zhao [1], Xuanhong He [1], Qian Liu [1], Xiatian Li [1], Zhiqiang He [1], Lehan Yu [2] and Nianlong Yan [1,*]

[1] Department of Biochemistry and Molecular Biology, School of Basic Medical Science, Nanchang University, Nanchang 330006, Jiangxi, China; hly3288551238@163.com (L.H.); wn13907096825@163.com (N.W.); zrl953226930@163.com (R.Z.); hxhxhong@163.com (X.H.); liucandice0412@163.com (Q.L.); 15807939939@163.com (X.L.); hzq3231103954@163.com (Z.H.)

[2] School of Basic Medical Experiments Center, Nanchang University, Nanchang 330006, Jiangxi, China; yulehan@sohu.com

* Correspondence: yannianlong@163.com

† These authors contributed equally to this work.

Received: 14 April 2019; Accepted: 4 June 2019; Published: 12 June 2019

Abstract: Endothelial dysfunction (ED) is an important contributor to atherosclerotic cardiovascular disease. Our previous study demonstrated that sphingomyelin synthase 2 (SMS2) promotes ED. Moreover, endoplasmic reticulum (ER) stress can lead to ED. However, whether there is a correlation between SMS2 and ER stress is unclear. To examine their correlation and determine the detailed mechanism of this process, we constructed a human umbilical vein endothelial cell (HUVEC) model with SMS2 overexpression. These cells were treated with 4-PBA or simvastatin and with LiCl and salinomycin alone. The results showed that SMS2 can promote the phosphorylation of lipoprotein receptor-related protein 6 (LRP6) and activate the Wnt/β-catenin pathway and that activation or inhibition of the Wnt/β-catenin pathway can induce or block ER stress, respectively. However, inhibition of ER stress by 4-PBA can decrease ER stress and ED. Furthermore, when the biosynthesis of cholesterol is inhibited by simvastatin, the reduction in intracellular cholesterol coincides with a decrease in ER stress and ED. Collectively, our results demonstrate that SMS2 can activate the Wnt/β-catenin pathway and promote intracellular cholesterol accumulation, both of which can contribute to the induction of ER stress and finally lead to ED.

Keywords: atherosclerosis; sphingomyelin synthase 2; endothelial dysfunction; endoplasmic reticulum stress; β-catenin

---

1. Introduction

Angiocardiopathy is a significant cause of death in many countries. Atherosclerosis (AS), which is a major cause of angiocardiopathy, is an inflammatory disease that leads to clogged arteries [1]. Additionally, endothelial dysfunction (ED) plays a crucial role in the pathogenesis of atherosclerotic cardiovascular disease [2]. Various harmful stimuli, such as oxidative stress and inflammation, can lead to ED, and reactive oxygen species (ROS) can induce oxidative stress, which plays an essential role in ED [3,4]. Since H_2O_2 is a key ROS, in this research, human umbilical vein endothelial cells (HUVECs) were treated with H_2O_2 to establish a cell model of oxidative stress [5].

Sphingomyelin (SM) is a type of sphingolipid that is important for the composition of biological membranes and plasma lipoproteins [6,7]. The production of SM requires many enzymatic reactions, and sphingomyelin synthase (SMS), which has two isoforms (sphingomyelin synthase 1 (SMS1) and sphingomyelin synthase 2 (SMS2)), is a critical enzyme in the final step of the production of SMS [8].

Studies have shown that SM participates in AS [9–11]. The level of SM in normal arterial tissue is significantly lower than that in atherosclerotic lesions [10]. Chemical inhibition of sphingolipid biosynthesis can markedly reduce the size of AS lesions in ApoE KO (apolipoprotein E knock out) mice [11]. These studies have mainly concentrated on the impact of SMS on reverse cholesterol transport and foam cell production in the process of AS development. However, our recent study indicated that SMS2 can also promote ED by activating the Wnt/β-catenin pathway under conditions of oxidative stress [12]. The typical Wnt/β-catenin pathway plays a critical role in many physiological processes, such as tissue patterning, the specification of cell fate, and cell proliferation [13,14]. During the process of transmembrane signal transduction, Wnt combines with the transmembrane receptor frizzled (FZD) and the coreceptor low-density lipoprotein receptor-related protein 6 (LRP6) to induce the phosphorylation of LRP6, which is necessary for activating the downstream Wnt/β-catenin pathway [13,14]. Since ED plays a crucial role in the initiation of AS [2], the Wnt/β-catenin pathway also participates in AS and its development [15–20]. For example, Bhatt et al. found that Wnt5a expression in serum from atherosclerotic patients is associated with the severity of atherosclerotic lesions [17,18]. However, the detailed mechanism of SMS2 related with the Wnt/β-catenin pathway and ED (AS) is not clear.

The endoplasmic reticulum (ER) is an organelle that participates in protein folding, calcium homeostasis, and lipid biosynthesis. Many factors, including hyperlipidemia and oxidative stress, can disrupt homeostasis in the ER and the unfolded protein response (UPR) to induce ER stress [21,22]. During the process of ER stress, the chaperone GRP78 dissociates from PERK, IRE1, and ATF6, activating their downstream signaling pathways and influencing homeostasis in cells [23,24]. ER stress is strongly linked to the development of AS, and expression of GRP78, p-PERK, p-IRE1, ATF6, and CHOP is increased in ApoE knockout mice [25,26]. In addition, many atherogenic risk factors can activate ER stress during the initial stages of AS, strengthening ED and AS [27,28]. Undoubtedly, ER stress is involved in not only AS but also ED.

Importantly, SMS2, ER stress, and the Wnt/β-catenin pathway are all related to ED. Although our previous study revealed that SMS2 can lead to ED by inducing the Wnt/β-catenin pathway, the relationship between SMS2 and ER stress and the specific mechanism by which SMS2 regulates the Wnt/β-catenin pathway needs further research. Therefore, we aimed to identify the mechanism using HUVECs.

2. Results

2.1. SMS2 Can Activate ER Stress

Both ER stress and SMS2 are associated with ED; however, the mechanism involved needs further study. First, we established SMS2 overexpression in HUVECs. These results (Figure 1A) showed that the amounts of SMS2 and the ER stress marker protein GRP78 in the S group were upregulated compared with those in the C group (C, transfected with empty plasmids; S, cells overexpressing SMS2; $p < 0.001$; $n = 3$). Furthermore, an ER stress cell model was established by treating cells with tunicamycin (10 μg/mL) for 24 h. The results (Figure 1B) verified that expression of SMS2 and GRP78 was upregulated by 46.3% and 44.8% in the tunicamycin group compared with that in the C group, respectively ($p < 0.001$; $n = 3$). To rule out the possibility that endoplasmic reticulum stress was not caused by protein overload but the overexpression of SMS2, we treated the HUVECs with 20 μmol/L Dy105 (an inhibitor of SMS2). We then measured the activity of SMS2 and expression of GRP78. Based on the data presented in Figure 1C, we identified that the SMS enzyme activity was markedly decreased compared with that in the C group; this activity was decreased by 60.09% compared with that in the C group ($p < 0.001$; $n = 3$). In addition, the expression of GRP78 was decreased by 40.5% (Figure 1D; $p < 0.001$; $n = 3$). These findings demonstrate that ER stress is significantly induced by SMS2.

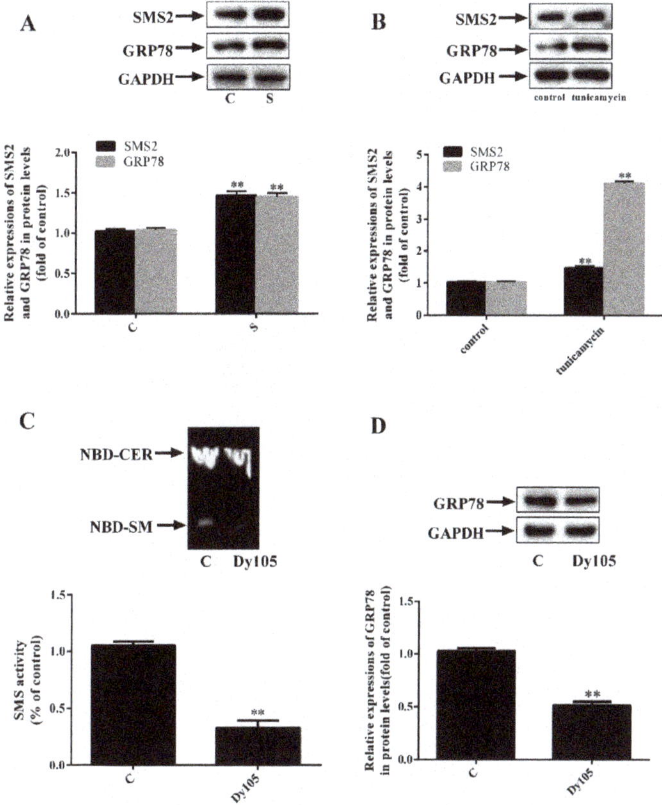

Figure 1. Sphingomyelin synthase 2 (SMS2) overexpression activates endoplasmic reticulum (ER) stress. Either a SMS2 overexpressed plasmid was used to transfect human umbilical vein endothelial cells (HUVECs) or the cells were treated with tunicamycin (10 μg/mL). (**A**) The protein levels of SMS2 and GRP78 were measured by a western blot analysis. (**B**) The protein levels of SMS2 and GRP78 were measured by a western blot analysis. (**C**) SMS activity was measured by thin-layer chromatography. (**D**) The expression of GRP78 was measured by a western blot analysis. $n = 3$, * $p < 0.05$, and ** $p < 0.001$ vs. the C group. (**A**) C, transfected with empty plasmids; S, cells overexpressing SMS2. (**C,D**) C, control group; Dy105, cells treated with Dy105. (**C**) NBD-CER, Norbornadiene -ceramide, NBD-SM, Norbornadiene-sphingomyelin.

2.2. SMS2 Can Trigger ER Stress by Provoking the Wnt/β-Catenin Pathway

To further explore the specific mechanism of SMS2-induced ER stress, LiCl (40 μmol/L) and salinomycin (5 μmol/L) were used to activate and inhibit the Wnt/β-catenin pathway, respectively. The results showed that, compared with the C group, the levels of the ER stress-related proteins GRP78, CHOP, and β-catenin were upregulated by 45.94%, 59.51%, and 94.55% in the Li group and decreased by 45.5%, 41.36%, and 28.4% in the Sal group, respectively (Figure 2A: C, control cells; Sal, salinomycin; Li, LiCl group; $p < 0.001$; $n = 3$). However, relative expression of phosphorylated β-catenin was decreased by 24.9% in the Li group compared with that in the C group and increased by 67.7% in the Sal group compared with that in the C group (Figure 2A: $p < 0.05$; $n = 3$). Additionally, we found that the expression of the total ATF6 and cleaved ATF6 (P50) were significantly increased by 96.5% and 126.3% compared with the C group, by activating the Wnt/β-catenin pathway. On the contrary, in the Sal group the expression of the total ATF6 and cleaved ATF6 were significantly decreased by 50.6% and 60.2%

compared with the C group. (Figure 2B: $p < 0.05$; $n = 3$). These results suggest that the provocation of Wnt/β-catenin can induce ER stress and that the suppression of Wnt/β-catenin can inhibit ER stress. Previous papers published by the authors have shown that SMS2 can cause dysfunction in endothelial cells by inducing the Wnt/β-catenin pathway. As shown in Figure 2C, compared with the C group, relative expression of β-catenin, phosphorylated LRP6, and LRP6 was upregulated by 101.9%, 132.9%, and 104.6% in the SMS2 group, respectively ($p < 0.001$; $n = 3$). In contrast, relative expression of phosphorylated β-catenin was reduced by 45.7%. These results suggest that SMS2 is able to trigger ER stress by inducing Wnt/β-catenin signaling.

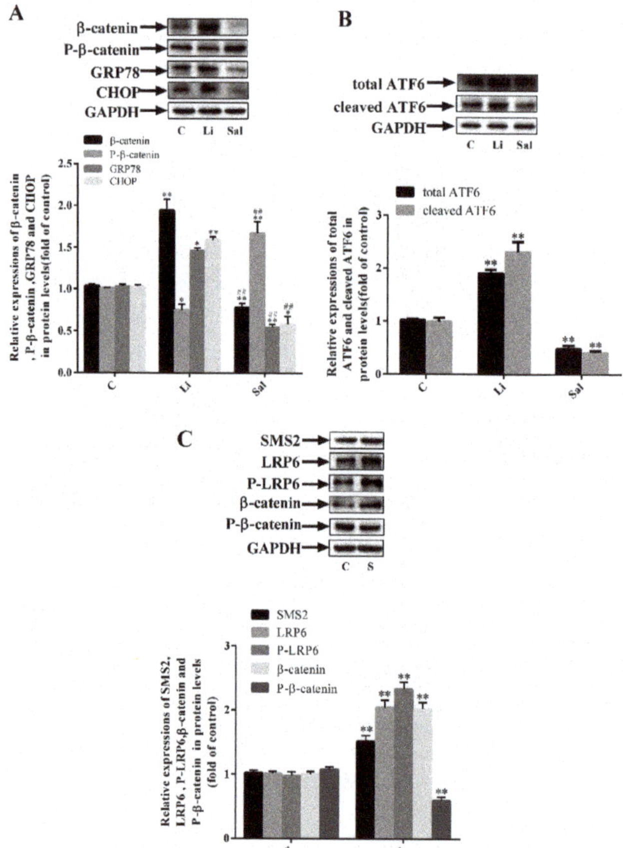

Figure 2. SMS2 can trigger ER stress by inducing the Wnt/β-catenin pathway. (**A**) Western blot analysis detected the protein expression of β-catenin, phosphorylated β-catenin, GRP78, and CHOP. (**B**) Western blotting analysis detected the protein expression of the total ATF6 and cleaved ATF6. (**C**) Western blotting analysis detected the protein expression of SMS2, β-catenin, phosphorylated β-catenin, lipoprotein receptor-related protein 6 (LRP6), and phosphorylated LRP6. $n = 3$, * $p < 0.05$ and ** $p < 0.001$ vs. the C group; ## $p < 0.001$ vs. the Li group. C, control cells; S, cells overexpressing SMS2; Li, LiCl group, control cells treated with LiCl (40 μmol/L) for 24 h; Sal, salinomycin group, control cells treated with salinomycin (5 μmol/L) for 24 h.

2.3. Inhibition of ER Stress Can Decrease SMS2-Induced ED

To prove the correlation between SMS2 and ER stress, cells were transfected with an empty plasmid or an SMS2 overexpression plasmid, treated with the ER stress inhibitor 4-PBA for 24 h, and treated with

H$_2$O$_2$ for 24 h to establish an oxidative stress model. The results indicated that the GRP78 and CHOP protein expression levels in the S group were increased by 42.8% and 32.3%, respectively, compared with those in the C group. In the PBA group, the GRP78 and CHOP protein expression levels were significantly decreased (by 21.6% and 57.4%, respectively) compared with those in the C group. Furthermore, the total ATF6 and cleaved ATF6 protein expression levels in the S group were upregulated by 210.7% and 163.3% and downregulated by 32.1% and 40.2% in the PBA group, respectively, compared with those in the C group. In particular, in the S+PBA group, the levels of GRP78, CHOP, total ATF6, and cleaved ATF6 were markedly increased compared with those in the PBA group and down-regulated compared with those in the S group (Figure 3A,B: C, cells transfected with empty plasmids; S, cells overexpressing SMS2; PBA, empty plasmids treated with 4-PBA (10 mmol/L) for 24 h; S+PBA, cells overexpressing SMS2 treated with 4-PBA (10 mmol/L) for 24 h; and all cells were treated with H$_2$O$_2$ (450 µmol/L) for 24 h. $p < 0.001$, $n = 3$). These data suggest that SMS2 can induce but that PBA inhibits ER stress.

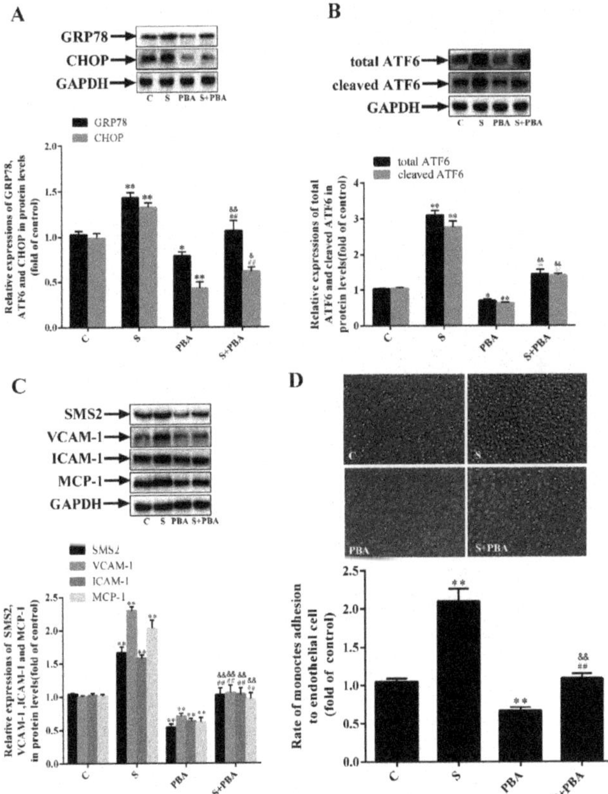

Figure 3. SMS2 can induce ER stress and endothelial dysfunction (ED) by inducing the Wnt/β-catenin pathway. (**A**) The protein levels of GRP78 and CHOP were determined by a western blot analysis. (**B**) The protein levels of the total ATF6 and cleaved ATF6 were determined by a western blot analysis. (**C**) The protein levels of VCAM-1, ICAM-1, and MCP-1 were determined by a western blot analysis. (**D**) The adhesion ratio of THP-1 cells to HUVECs (magnification 40×). $n = 3$, * $p < 0.05$, and ** $p < 0.001$ vs. the C group; ## $p < 0.001$ vs. the S group. & $p < 0.05$ and && $p < 0.001$ vs. the PBA group. C, cells transfected with empty plasmids treated with H$_2$O$_2$ (450 µmol/L) for 24 h; S, cells overexpressing SMS2 treated with H$_2$O$_2$ (450 µmol/L) for 24 h; PBA, empty plasmids treated with 4-PBA (10 mmol/L) for 24 h and then treated with H$_2$O$_2$ (450 µmol/L) for 24 h; S+PBA, cells overexpressing SMS2 treated with 4-PBA (10 mmol/L) for 24 h and then treated with H$_2$O$_2$ (450 µmol/L) for 24 h.

We then investigated the relationship among SMS2, ER stress, and ED. The results (Figure 3C) suggest that, compared with the transfection with the empty plasmid in the C group, the transfection with the SMS2 overexpression plasmid activated ER stress and increased the expression of the adhesion-related molecules ICAM-1, VCAM-1, and MCP-1 by 58.1%, 12.6%, and 103.2%, respectively. In contrast, the levels of these adhesion-related molecules were decreased by 42.8%, 29.3%, and 36.6% after the inhibition of ER stress by 4-PBA, compared with those in the C group without treatment. In addition, in the S+PBA group, the levels of ICAM-1, VCAM-1, and MCP-1 were markedly increased compared with those in the PBA group and down-regulated compared with those in the S group ($p < 0.001$, $n = 3$). Monocyte adhesion reflects the degree of cell damage that can lead to ED. As illustrated in Figure 3D, the adhesion ability in the S group was observably increased (by 100.96%) compared with that in the C group (Figure 3D: $p < 0.05$; $n = 3$), though the adhesion ability in the PBA group was significantly reduced (by 33.66%) compared with that in the C group (Figure 3D: $p < 0.05$; $n = 3$). These results demonstrate that the repression of ER stress can repress ED and that SMS2 can induce ED via ER stress.

2.4. Simvastatin Can Attenuate the ER Stress Induced by SMS2

To determine whether the intracellular accumulation of cholesterol is affected by SMS2, the following experiments were performed. HUVECs were treated with different doses of simvastatin to reduce intracellular cholesterol synthesis. The results showed that the activity of LDH (lactic dehydrogenase) and a degree of cell injury were the lowest at the 0.1 µmol/L dose; therefore, the final dose of simvastatin used was 0.1 µmol/L (Figure 4A: $p < 0.001$, $n = 3$). Subsequently, the cells were stained with filipin. The results shown in Figure 4B reveal that the intracellular cholesterol accumulation in the S group was increased by 28.8% compared with that in the C group and decreased by 20.5% in the Sim group compared with that in the C group. In addition, the cholesterol accumulation in the S+Sim group was increased by 20.2% compared with that in the Sim group and decreased by 23.1% compared with that in the S group (C, cells transfected with empty plasmids; S, cells overexpressing SMS2; Sim, empty plasmids treated with simvastatin (0.1 µmol/L) for 24 h; S+Sim, cells overexpressing SMS2 treated with simvastatin (0.1 µmol/L) for 24 h; all the cells were treated with H_2O_2 (450 µmol/L) for 24 h, $p < 0.001$; $n = 3$). These findings suggest that overexpression of SMS2 may contribute to intracellular cholesterol accumulation. Furthermore, we detected the proteins related to ER stress, and the results (Figure 4C,D) showed that the protein expression of GRP78, CHOP, SMS2, total ATF6, and cleaved ATF6, in the S group was increased by 93.6%, 160.9%, 117.6%, 235.4%, and 180.5%, respectively, compared with that in the C group. Expression levels of the GRP78, CHOP, SMS2, total ATF6, and cleaved ATF6 proteins in the Sim group were inhibited compared with those in the C group, indicating that simvastatin can inhibit ER stress. In the S+Sim group, the levels of GRP78, CHOP, total ATF6, and cleaved ATF6 were markedly increased compared with those in the Sim group but reduced compared with those in the S group ($p < 0.001$, $n = 3$). These findings demonstrate that overexpression of SMS2 can cause cholesterol accumulation, which may contribute to ER stress.

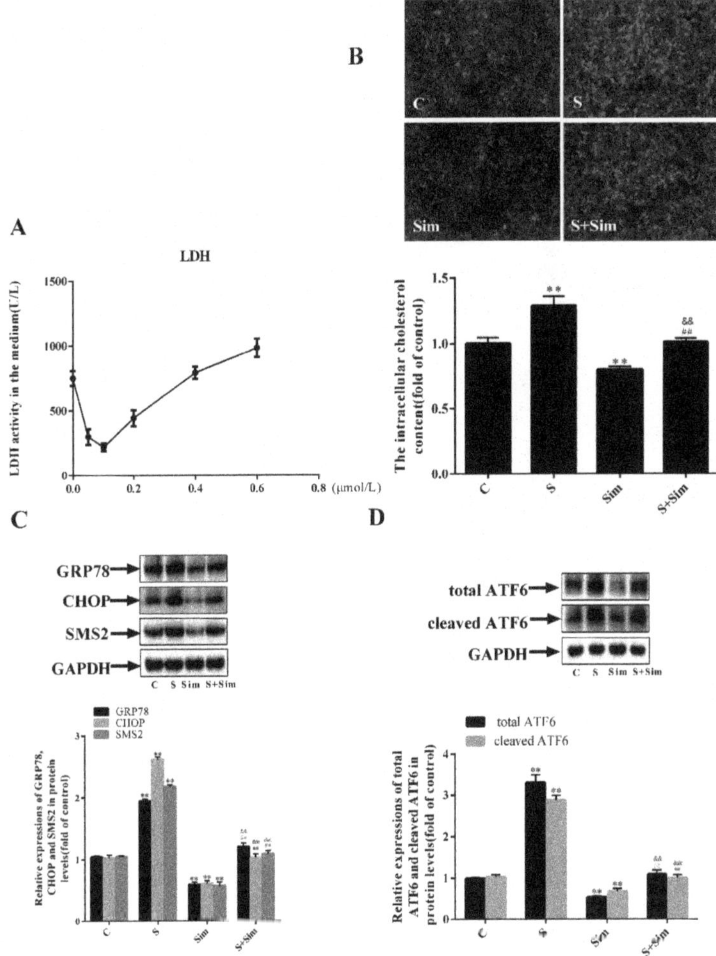

Figure 4. Overexpression of SMS2 can lead to ER stress by increasing the deposition of intracellular cholesterol. (**A**) HUVECs were treated with simvastatin at different doses (0, 0.05, 0.1, 0.2, 0.4, and 0.6 µmol/L) for 24 h, and the level of LDH in the cellular medium was detected. (**B**) The accumulation of ER cholesterol after filipin staining was visualized under a fluorescence microscope (magnification 40×). (**C**) The protein levels of GRP78 and CHOP were determined by a western blot analysis. (**D**) The protein levels of the total ATF6 and cleaved ATF6 were determined by a western blot analysis. $n = 3$, * $p < 0.05$, and ** $p < 0.001$ vs. the C group; ## $p < 0.001$ vs. the S group. && $p < 0.001$ vs. the Sim group. C, cells transfected with empty plasmids treated with H_2O_2 (450 µmol/L) for 24 h; S, cells overexpressing SMS2 treated with H_2O_2 (450 µmol/L) for 24 h; Sim, empty plasmids treated with simvastatin (0.1 µmol/L) for 24 h and then treated with H_2O_2 (450 µmol/L) for 24 h; S+Sim, cells overexpressing SMS2 treated with simvastatin (0.1 µmol/L) for 24 h and then treated with H_2O_2 (450 µmol/L) for 24 h.

2.5. Simvastatin Can Attenuate the Injury Induced by SMS2

To further elucidate the effects of cholesterol accumulation on cell injury, we measured the LDH, SOD (superoxide dismutase), and NOS (nitric oxide synthase) content. The results showed that SOD and NOS production in the HUVECs in the S group was significantly reduced compared with that in the C group; however, treatment with simvastatin increased SOD and NOS production in the Sim group compared with that in the C group. Additionally, SOD and NOS production in the S+Sim group was upregulated compared with that in the S group but decreased compared with that in the Sim group (Figure 5C,D: $p < 0.001$, $n = 3$). Conversely, LDH activity showed the opposite trend (Figure 5A: $p < 0.001$, $n = 3$). These findings indicate that SMS2 overexpression can induce HUVEC injury due to intracellular cholesterol accumulation and that simvastatin has a protective effect on cells. Furthermore, compared with the C group, the results showed that, in the S group, the level of the pro-apoptotic gene Bax was increased by 76.5%, while the level of the anti-apoptotic gene Bcl-2 was reduced by 35.9%. This finding contrasts the results observed after the simvastatin treatment. In the Sim group, the level of the pro-apoptotic gene Bax was decreased by 51.3% compared with that in the C group, while the level of the anti-apoptotic gene Bcl-2 was increased by 57.2% compared with that in the C group (Figure 5B: $p < 0.001$, $n = 3$). These data indicate that overexpression of SMS2 can lead to ER stress and ED, due to cholesterol accumulation.

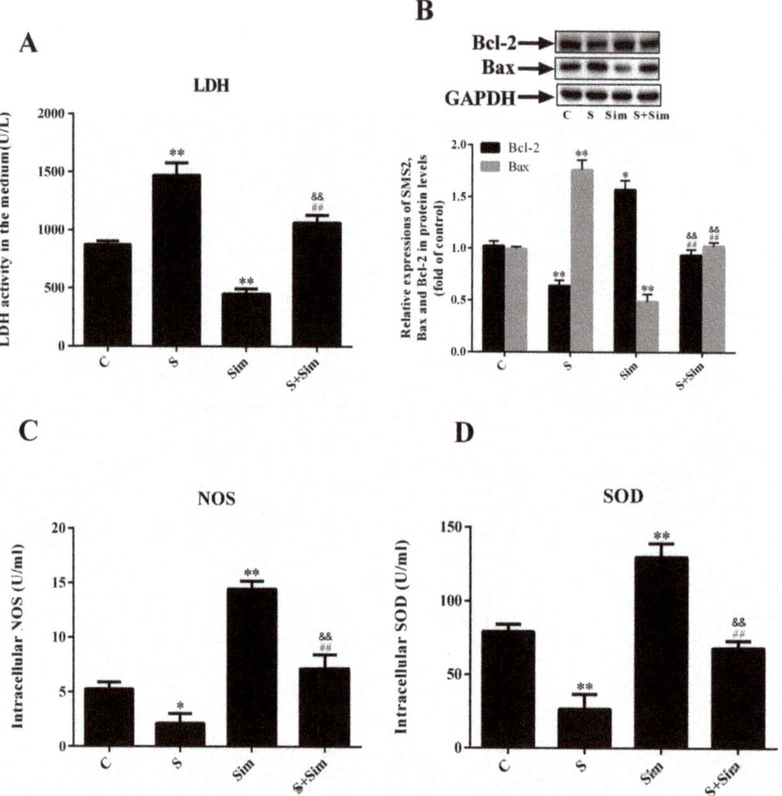

Figure 5. Overexpression of SMS2 can promote endothelial cell injury by increasing the deposition of intracellular cholesterol. (**A**) LDH, (**C**) NOS, and (**D**) SOD levels were measured with assay kits. (**B**) Western blot analysis detected the protein levels of SMS2, Bax, and Bcl-2. $n = 3$, * $p < 0.05$, and ** $p < 0.001$ vs. the C group; ## $p < 0.001$ vs. the S group. && $p < 0.001$ vs. the Sim group. C, cells transfected with empty plasmids treated with H_2O_2 (450 μmol/L) for 24 h; S, cells overexpressing SMS2 treated with H_2O_2 (450 μmol/L) for 24 h; Sim, empty plasmids treated with simvastatin (0.1 μmol/L) for 24 h and then treated with H_2O_2 (450 μmol/L) for 24 h; S+Sim, cells overexpressing SMS2 treated with simvastatin (0.1 μmol/L) for 24 h and then treated with H_2O_2 (450 μmol/L) for 24 h.

2.6. Simvastatin Can Attenuate the Adhesion Capacity Induced by SMS2

We next analyzed the adhesion capacity of HUVECs and THP-1 cells to demonstrate the effects of simvastatin. Figure 6A shows that adhesion capacity in the S group was markedly increased (by 100.5%) compared with that in the C group (Figure 6A: $p < 0.05$; $n = 3$), adhesion capacity in the Sim group was evidently reduced (by 32.9%) compared with that in the C group (Figure 6A: $p < 0.05$; $n = 3$), and adhesion capacity in the S+Sim group was higher than that in the Sim group and lower than that in the S group. These findings suggest that simvastatin reduces cholesterol deposition, thus decreasing the cell adhesion capacity. Moreover, the results showed that, compared with transfection with the empty plasmid in the C group, transfection with the SMS2 overexpression plasmid increased cholesterol accumulation and expression of the adhesion molecules VCAM-1, ICAM-1, and MCP-1 by 88.1%, 83.2%, and 44.2%, respectively. Meanwhile, compared with the C group without treatment, the level of these adhesion-related molecules was decreased by 39.6%, 28.6%, and 41.8% after treatment with simvastatin (Figure 6B: $p < 0.05$; $n = 3$). Furthermore, S+Sim decreased expression of the adhesion-related molecules compared with simvastatin treatment in the S group (Figure 6B: $p < 0.001$; $n = 3$). These data suggest that simvastatin can attenuate the adhesion capacity induced by SMS2.

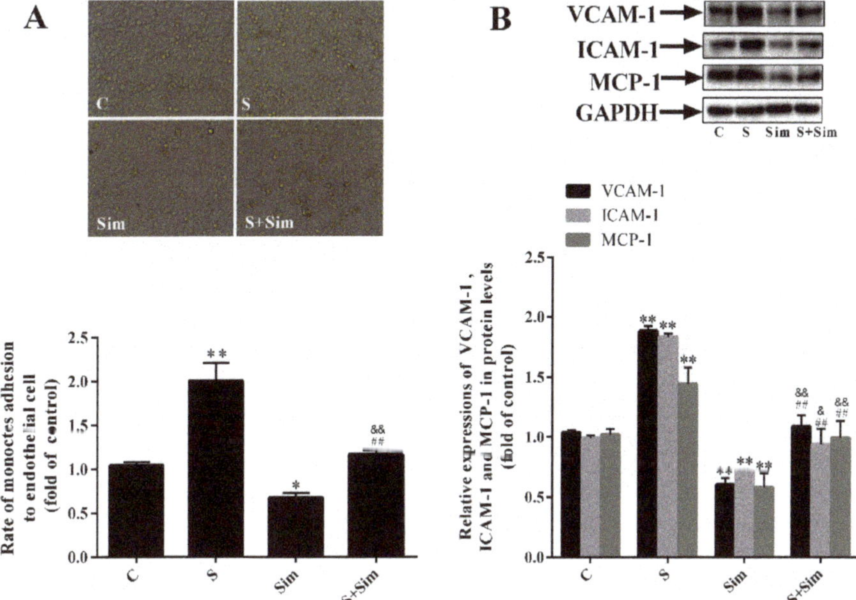

Figure 6. Overexpressed SMS2 increases the adhesion activity of HUVECs and THP-1 cells by increasing the deposition of intracellular cholesterol. (**A**) The adhesion ratio of THP-1 cells to HUVECs (magnification 40×). (**B**) Western blot analysis detected the protein level of VCAM-1, ICAM-1, and MCP-1. $n = 3$, * $p < 0.05$ and ** $p < 0.001$ vs. the C group; ## $p < 0.001$ vs. the S group. && $p < 0.001$ vs. the Sim group. C, cells treated with empty plasmids treated with H_2O_2 (450 µmol/L) for 24 h; S, cells overexpressing SMS2 treated with H_2O_2 (450 µmol/L) for 24 h; Sim, empty plasmids treated with simvastatin (0.1 µmol/L) for 24 h and then treated with H_2O_2 (450 µmol/L) for 24 h; S+Sim, cells overexpressing SMS2 treated with simvastatin (0.1 µmol/L) for 24 h and then treated with H_2O_2 (450 µmol/L) for 24 h.

3. Discussion

Our previous study demonstrated that SMS2 can activate the Wnt/β-catenin pathway [12], but the detailed mechanism has remained unclear. Both SM and cholesterol are the main components of lipid rafts [29]. Therefore, changing the expression of SMS may affect the SM content in lipid rafts and, thus, influence transmembrane signal transduction. For example, Ding et al. found that overexpression of SMS can lead to the deposition of SM in cells and lipid rafts [30]. Lipid rafts play an essential role in physiological and biochemical processes as signaling "platforms", such as the LPS receptor (lipopolysaccharides), TLR4, which must be recruited to lipid rafts to transduce extracellular signals to intracellular downstream signaling molecules [31]. During the process of Wnt/β-catenin signal transduction, the coreceptor LRP6 needs to be phosphorylated to disinhibit DKK1 (Dickkopf related protein 1) and form the FZD transmembrane protein receptor complex to promote signal transduction, and the binding between the LRP6 and FZD is affected by lipid rafts [32,33]. In this study, overexpressing SMS2 and both expressions of LRP6 and the phosphorylation of LRP6 were increased in HUVECs (Figure 2C). These results indicate that SMS2 may increase the phosphorylation of LRP6 in lipid rafts and decrease the degradation of LRP6 outside lipid rafts by promoting LRP6 endocytosis to lipid rafts [34,35]. Thereafter, the increase in expression and phosphorylation of LRP6 can lead to the provocation of the Wnt/β-catenin pathway, which contributes to ED (Figures 2 and 6).

Previously, many studies have suggested that SMS2 is involved in AS by affecting reverse cholesterol transport [9–11]. However, we recently demonstrated that SMS2 also participates in ED by stimulating the Wnt/β-catenin signal pathway [19]. Meanwhile, some reports believe that ER stress can lead to ED [36,37], but the detailed correlation between these factors remains unexplored. Therefore, we transfected HUVECs with SMS2 overexpression plasmids. The results revealed that overexpression of SMS2 can increase expression of the ER stress marker protein, GRP78 (Figure 1). Interestingly, when ER stress was induced by tunicamycin, expression of SMS2 was also significantly increased (Figure 1). These results suggest that SMS2 can promote ER stress and that ER stress may regulate SMS2 expression.

Since SMS2 can stimulate the Wnt/β-catenin pathway, we investigated whether SMS2 can induce ER stress, and subsequently ED, by activating the Wnt/β-catenin pathway in HUVECs. In this study, the Wnt/β-catenin pathway was activated or blocked by LiCl or salinomycin, respectively. The results revealed that activating Wnt/β-catenin signaling could decrease ER stress in the HUVECs and vice versa (Figure 2A,B). Mechanically, Zhang et al. suggested that the Wnt/β-catenin pathway blockage and β-catenin degradation result in the inhibition of the effect of LEF1 on ATF6 and activate ATF6-related ER stress [38]. These findings confirm that the activation of the Wnt/β-catenin pathway can promote ER stress. In fact, other studies have also shown that the Wnt/β-catenin pathway is negative with the ER stress in cancer cells, but our studies contradict these results [39,40]. These conflicting results suggest that the Wnt/β-catenin pathway has diverse functions in different types of cells. Previously, we proved that SMS2 can activate Wnt/β-catenin signaling; therefore, after the treatment with 4-PBA, compared with the simvastatin treatment in the S group, expression of ER stress-related proteins (GRP78, ATF6, and CHOP) and adhesion molecules (ICAM-1, VCAM-1, and MCP-1) and the adhesion activity were significantly decreased (Figure 3), suggesting that 4-PBA reverses the effects of SMS2 on ER stress and ED. These results further indicate that SMS2 can activate ER stress and promote oxidative stress-induced ED (Figures 3 and 7).

The ER is not only involved in protein folding and modification but is also inextricably linked to the metabolism of lipids, such as cholesterol; therefore, lipid metabolism disorders can also trigger ER stress [41,42]. For example, recent animal and human studies have identified cholesterol deposition and ER stress activation as key players in the progression of many metabolic diseases [43]. Nonetheless, cholesterol deposition causes dysfunction in β-cells and promotes autophagy by activating ER stress [44,45]. In fact, SM and cholesterol can affect each other in cells and serum. For example, patients suffering from the Niemann–Pick disease (NPD-B) cannot synthesize SM due to defective SMase (sphingomyelinase), leading to the accumulation of SM and cholesterol in the liver [46]. Furthermore, we previously confirmed that overexpression of SMS in Huh7 cells markedly enhances the levels of intracellular sphingomyelin and cholesterol [47]. To investigate whether SMS2 expression can lead to the deposition of cholesterol in HUVECs, we measured intracellular cholesterol by filipin staining. The results suggested that in HUVECs, SMS2 overexpression can increase the intracellular cholesterol levels (Figure 4). The inhibition of cholesterol synthesis by simvastatin can reverse the deposition of cholesterol induced by SMS2, and expression of the ER stress-associated proteins, GRP78, CHOP, and ATF6, in the S+Sim group was also noticeably down-regulated compared with that in the S group (Figure 4). Finally, ED was found to be significantly attenuated (Figures 4 and 5). SMS2 was shown to induce ER stress and ED by promoting intracellular cholesterol accumulation (Figures 6 and 7).

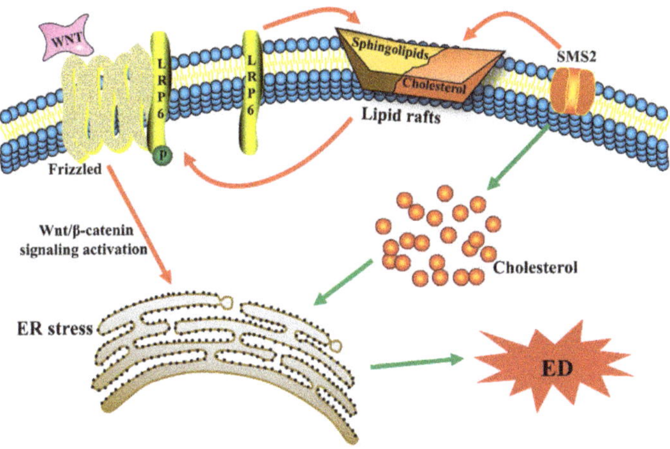

Figure 7. The possible mechanism by which endoplasmic reticulum stress is induced by SMS2. ER stress, endoplasmic reticulum stress; ED, endothelial dysfunction.

In conclusion, our results show that SMS2 (1) triggers the Wnt/β-catenin pathway and (2) promotes intracellular cholesterol accumulation, both of which contribute to the induction of ER stress and finally lead to ED. Although the mechanism of ED is very complex, we hope that our studies help to further elucidate the related mechanism.

4. Materials and Methods

4.1. Cell Culture and Reagents

HUVECs were obtained from the Cell Bank of Type Culture Collection of the Chinese Academy of Sciences (Shanghai, China) and cultured in Dulbecco's modified Eagle's medium (DMEM; cat. no. 12100-500; Beijing Solarbio Bioscience & Technology co., Ltd., Beijing, China) containing penicillin and streptomycin (100 U/mL and 0.1 mg/mL, respectively) and 10% certified fetal bovine serum (FBS; Biological Industries Israel Beit Haemek, KibbutzBeit Haemek, Israel) at 37 °C containing 5% CO2. In addition, the THP-1 cells (Cell Bank of Type Culture Collection of the Chinese Academy of Sciences, Shanghai, China) were grown in RPMI-1640 (cat. no. 31800; Beijing Solarbio Bioscience & Technology co., Ltd.) containing 10% FBS and incubated at 37 °C in a humidified atmosphere containing 5% CO2. Simvastin (cat. no. MB1222-S; Meilunbiotech Co., Ltd., Dalian, China), sodium 4-phenylbutyrate (4-PBA; cat. no. C1029659; Macklinbio co., Ltd., Shanghai, China), tunicamycin (cat. no. B7417; APExBIO co., Ltd., Shanghai, China), filipin (cat. no. B6034; APExBIO co., Ltd., Shanghai, China), and salinomycin (cat. no. HY-15597; Medchem Express co., Ltd., New Jersey, USA) were dissolved in DMSO (dimethyl sulfoxide; cat. no. 302A036; Beijing Solarbio Bioscience & Technology co., Ltd.).

4.2. Transfection with an SMS2 Overexpression Plasmid

The SMS2-overexpression plasmid was provided by Dr. Tingbo Ding (School of Pharmacy, Fudan University, Shanghai, China) and used to transfect the HUVECs. In brief, the cells were seeded in culture flasks (Haote Technologies co., Ltd., Guangzhou, China), and at the time of transfection, the cells were grown to 90–95% confluency; the medium was replaced with the antibiotic-free DMEM medium

(Biological Industries Israel Beit Haemek Ltd.). The SMS2 plasmid (4 µg; S group) or an empty control plasmid (4 µg; C group) was diluted with 400 µL DMEM (FBS-free and antibiotic-free medium), and 24 µL of Hieff Trans™ Liposomal Transfection Reagent was also diluted with 400 µL DMEM. After 5 min, the dilutions were gently mixed together and incubated at 37 °C for 20 min, and the mixture was added to each culture flask. Before adding the drugs, the medium was replaced with fresh DMEM (containing 10% FBS and antibiotics). The control and SMS2 groups were divided into two groups after 24 h, resulting in four small groups as follows: C (empty plasmid), S (SMS2), Sim (control + simvastatin), and S+Sim (SMS2 + simvastatin). Both the Sim and S+Sim groups were treated with simvastatin (0.1 µmol/L). The following set of groups were established: C (empty plasmid), S (SMS2), PBA (control + 4-PBA), and S+PBA (SMS2 + 4-PBA); both the PBA and S+PBA groups were treated with PBA (10 mmol/L). After 24 h, all groups were treated with H_2O_2 (500 µmol/L). Finally, after 24 h, the HUVECs were centrifuged at 1520× g for 5 min (at room temperature) as appropriate to collect all cells.

4.3. Measurement of the Degree of Oxidative Stress

HUVECs were cultured at a density of 1×10^5/well in 6-well plates and incubated overnight at 37 °C. After the transfection, the cells were treated with simvastatin (0.1 µmol/L) for 24 h, and then the cells were treated with H_2O_2 (500 µmol/L) for 24 h. To collect the supernatant, the HUVECs were centrifuged (1520× g, 5 min, room temperature) and harvested, followed by digestion with trypsin, according to the protocol of the manufacturer. The levels of LDH (cat. no. A020-1; Nanjing Jiancheng Bioengineering Institute, Nanjing, China) and the levels of NOS (cat. no. A014-2-1; Nanjing Jiancheng Bioengineering Institute) and SOD (cat. no. A001-1-1; Nanjing Jiancheng Bioengineering Institute) activity were measured using a microplate reader (Thermo Fisher Scientific, Inc., Waltham, MA, USA). LDH was measured at a wavelength of 450 nm, and SOD and NOS were measured at a wavelength of 560 nm.

4.4. Filipin Staining

Cells were seeded in 24-well plates (Beaver Nano-Technologies co., Ltd., Suzhou, China) at a density of 2×10^4 cells/well. Following the above transfection steps described in the HUVECs, the cells were treated with simvastatin (0.1 µmol/L) for 24 h. Subsequently, all cells were treated with H_2O_2 (500 µmol/L) for 24 h and washed with PBS (phosphate buffer saline) three times. Then, 10% paraformaldehyde was used to fix the cells at room temperature for 10 min; subsequently, the cells were washed with PBS three times again. To eliminate the paraformaldehyde, the cells were washed with PBS containing glycine (1.5 mg/mL), and then filipin was added in a dark room for 1 h. The cells were washed with PBS three times. Finally, we used UV excitation at 405 nm and confocal microscopy to observe the cholesterol aggregation.

4.5. Cell Adhesion Assay

Cells were seeded in 24-well plates in three replicates per group to obtain the average number of adhesive monocytes/well. Following the above transfection steps described in the HUVECs, THP-1 cells at a density of 5×10^3/well were added and incubated for 2 h at 37 °C. The medium was discarded, and the non-adherent THP-1 cells were removed by washing with PBS three times. The adherent THP-1 cells were counted in a single field under a phase contrast inverted microscope (Magnification, 20×; Olympus IX71; Olympus corporation, Tokyo, Japan).

4.6. Western Blot Analysis

The total proteins were extracted from all groups by a radioimmunoprecipitation assay buffer (cat. no. ROO20; Beijing Solarbio Bioscience & Technology co., Ltd.), and the protein content was measured using a BCA assay kit (Bradford Protein Assay kit; cat. no. cW0014; Beijing Kangwei century Biotechnology co., Ltd., Beijing, China). Equal amounts of protein (~50 µg) were separated by 8–10% SDS-PAGE and transferred onto polyvinylidene fluoride membranes (Immobilon-P; EMd Millipore, Billerica, MA, USA). An equal transfer was examined by staining with Ponceau red (cat. no.

CW0057S; Beijing Kangwei century Biotechnology co., Ltd.). The membranes were blocked with 5% skimmed milk or 5% BSA (bovine serum albumin) in TBS (phosphate buffer saline + Tween) for 1 h at room temperature and incubated with primary antibodies overnight at 4 °C in TBST containing 0.05% Tween 20 and 2% bovine serum albumin (cat. no. A8020; Beijing Solarbio Science & Technology co., Ltd.). Subsequently, the membranes were incubated with a 1:8,000-dilution of a horseradish peroxidase-conjugated secondary antibody for 1 h at room temperature. The peroxidase activity was visualized using an ECL kit (Bio-Rad Laboratories Inc., USA). GAPDH was used as a loading control. Anti-SMS2 (cat. no SA100531AA; 1:1,000) antibody was purchased from Abgent Biotech. (Suzhou co., Ltd., Suzhou, China). Anti-apoptosis-associated proteins B-cell lymphoma 2 (Bcl-2; cat. no. 60178-1-Ig; 1:1000), anti-Bcl-2-associated X protein (Bax; cat. no. 505992-2-Ig; 1:2000), anti-adhesion-associated proteins intracellular adhesion molecule-1 (ICAM-1; cat. no. 10831-1-AP; 1:1000), anti-Wnt/β-catenin signal pathway-associated proteins (β-catenin cat. no. 51067-2-AP; 1:2000), anti-glucose-regulated protein 78 (GRP78; cat. no. 66574-1-Ig; 1:5000), and anti-GAPDH (cat. no. HRP-60004; 1:8000) antibodies were pauchased from ProteinTech Group, Inc. (chicago, IL, USA). Anti-phosphorylated β-catenin (cat. no. DF2989; 1:1,000) antibody was pauchased from Affnity Biosciences. (Cincinnati, OH, USA). Anti-vascular cell adhesion molecule-1 (VCAM-1; cat. no. WL02474; 1:500), anti-monocyte chemoattractant protein-1 (MCP-1; cat. no. WL01755; 1:1000) anti-activating transcription factor 6 (ATF6; cat. no. Wl02407; 1:800), and anti-C/EBP homologus protein (CHOP; cat. no. WL00880; 1:800) antibody was pauchased from Wanleibio. (Shenyang co., Ltd., China). Anti-low density lipoprotein receptor-related protein 6 (LRP6: cat. no. A13325; 1:1000) antibody was pauchased from ABclonal. (Wuhan co., Ltd., Wuhan, China). Anti-phosphorylated LRP6 (cat. no. abs140173) antibody was pauchased from Absin. (Shanghai co., Ltd., shanghai, China). Anti-mouse (cat. no. SA00001-1; 1:8000) secondary antibody was pauchased from ProteinTech Group, Inc. Horseradish peroxidase-conjugated anti-rabbit (cat. no. BA1054; 1:8000) was pauchased from Boster Biological Technology, Inc. (Wuhan, China).

4.7. LiCl or Salinomycin Treatment of HUVECs

HUVECs were plated in culture flasks. After reaching 70–80% confluency, the cells were treated with lithium chloride (LiCl, 40 µmol/L) or salinomycin (5 µmol/L) for 24 h and harvested. The following three experimental groups were established: C (untreated control cells), Sal (salinomycin-treated cells), and Li (LiCl-treated cells).

4.8. Sphingomyelin Synthase Enzyme Activity Assay

The HUVECs were treated with H_2O_2 for 24 h as previously described. The HUVECs were treated with 20 µmol/L Dy105 (provided by Dr. Deyong Ye, School of Pharmacy, Fudan University) for 24 h. The treated cells were incubated with NBD-ceramide (0.1 µg/µL, cat. no. 62527; Cayman chemical company, Ann Arbor, MI, USA) at 37 °C to synthesize sphingomyelin in vitro. After 3 h of incubation, the cells were harvested by 1520× *g* for 5 min, and the medium was collected at room temperature. According to the protein content in each group, the protein levels were adjusted to the volume of the reaction system (700 µL) to ensure consistency in the amount of total protein and medium added. The lipids were extracted in chloroform: Methanol (2:1) dried under N_2 gas and separated by thin layer chromatography using chloroform:MeOH:NH4OH (14:6:1) at room temperature for 10 min [12]. The chromatography film was scanned after 10 min with an autoradiography system (Chemiluminescence Imaging System, Clinx Science Instruments Co., Ltd., Shanghai, China), and the intensity of each band was measured using Image-Pro Plus version 6.0 software (Media Cybernetics, Inc., Rockville, MD, USA) [12].

4.9. Statistical Analysis

The data were analyzed by GraphPad Prism 6.0 (San Diego, CA, USA). *t*-Tests were used for comparisons between two groups, and a one-way analysis of variance (ANOVA) was used for comparisons among multiple groups. All results were reduplicated at least thrice. $p < 0.05$ was considered statistically significant.

5. Conclusions

SMS2 1) triggers the Wnt/β-catenin pathway by increasing the phosphorylation of LRP6 and 2) promotes intracellular cholesterol accumulation, which contributes to the induction of ER stress and causes ED.

Author Contributions: L.H. and N.W.; data curation, investigation, methodology, and writing of the original draft. R.Z.; investigation, writing review, and editing. X.H.; formal analysis and software. Q.L.; validation, writing review, and editing. X.L.; resources and software. Z.H.; resources. L.Y.; methodology. N.Y.; methodology, project administration, supervision, and validation.

Funding: The present study was supported by grants from the National Natural Science Foundation of China (grant no. 81560151) and the Jiangxi Provincial Department of Science and Technology (grant no. 20181BAB205022).

Acknowledgments: We thank Shuang Liu for assistance with Figure 7.

Conflicts of Interest: The authors declare no conflict of interest.

References

1. Ross, R. Atherosclerosis—An Inflammatory Disease—NEJM. *N. Engl. J. Med.* **1999**, *340*, 115. [CrossRef] [PubMed]

2. Gimbrone, M.A.; García-Cardeña, G. Endothelial Cell Dysfunction and the Pathobiology of Atherosclerosis. *Circ. Res.* **2016**, *118*, 620–636. [CrossRef] [PubMed]

3. El Assar, M.; Angulo, J.; Rodríguez-Maas, L. Oxidative stress and vascular inflammation in aging. *Free Radic. Biol. Med.* **2013**, *65*, 380–401. [CrossRef] [PubMed]

4. Zhang, M.; Pan, H.; Xu, Y.; Wang, X.; Qiu, Z.; Jiang, L. Allicin decreases lipopolysaccharide-induced oxidative stress and infammation in human umbilical vein endothelial cells through suppression of mitochondrial dysfunction and activation of Nrf2. *Cell Physiol. Biochem.* **2017**, *41*, 2255–2267. [CrossRef] [PubMed]

5. Del Río, L.A. ROS and RNS in plant physiology: An overview. *J. Exp. Bot.* **2015**, *66*, 2827–2837. [CrossRef] [PubMed]

6. Adada, M.; Luberto, C.; Canals, D. Inhibitors of the sphingomyelin cycle: Sphingomyelin synthases and sphingomyelinases. *Chem. Phys. Lipids* **2016**, *197*, 45–59. [CrossRef]

7. Fessler, M.B.; Parks, J.S. Intracellular Lipid Flux and Membrane Microdomains as Organizing Principles in Inflammatory Cell Signaling. *J. Immunol.* **2011**, *187*, 1529–1535. [CrossRef]

8. Yamaoka, S.; Miyaji, M.; Kitano, T.; Umehara, H.; Okazaki, T. Expression Cloning of a Human cDNA Restoring Sphingomyelin Synthesis and Cell growth in Sphingomyelin Synthase-defective Lymphoid Cells. *J. Biol. Chem.* **2004**, *279*, 18688–18693. [CrossRef]

9. Jiang, X.C.; Liu, J. Sphingolipid metabolism and atherosclerosis. *Handb. Exp. Pharmacol.* **2013**, *216*, 133.

10. Zilversmit, D.B.; Mccandless, E.L.; Jordan, P.H.; Henly, W.S.; Ackerman, R.F. The Synthesis of Phospholipids in Human Atheromatous Lesions. *Circulation* **1961**, *23*, 370. [CrossRef]

11. Park, T.S.; Panek, R.L.; Mueller, S.B.; Hanselman, J.C.; Rosebury, W.S.; Robertson, A.W.; Kindt, E.K.; Homan, R.; Karathanasis, S.K.; Rekhter, M.D. Inhibition of Sphingomyelin Synthesis Reduces Atherogenesis in Apolipoprotein E-Knockout Mice. *Circulation* **2004**, *110*, 3465–3471. [CrossRef] [PubMed]

12. Zhang, P.; Hua, L.; Hou, H.; Du, X.; He, Z.; Liu, M.; Hu, X.; Yan, N. Sphingomyelin synthase 2 promotes H2O2-induced endothelial dysfunction by activating the Wnt/β-catenin signaling pathway. *Int. J. Mol. Med.* **2018**, *42*, 3344–3354. [CrossRef] [PubMed]

13. De Jaime-Soguero, A.; Abreu de Oliveira, W.A.; Lluis, F. The pleiotropic effects of the canonical Wnt pathway in early development and pluripotency. *Genes* **2018**, *9*, 93. [CrossRef] [PubMed]

14. Nusse, R.; Clevers, H. Wnt/β-catenin signaling, disease, and emerging therapeutic modalities. *Cell* **2017**, *169*, 985–999. [CrossRef] [PubMed]

15. Matthijs, B.W.; Hermans, K.C. Wnt signaling in atherosclerosis. *Eur. J. Pharmacol.* **2015**, *763*, 122–130. [CrossRef] [PubMed]

16. Ueland, T.; Otterdal, K.; Lekva, T.; Halvorsen, B.; Gabrielsen, A.; Sandberg, W.J.; Paulsson-Berne, G.; Pedersen, T.M.; Folkersen, L.; Gullestad, L.; et al. Dickkopf-1 Enhances Inflammatory Interaction Between Platelets and Endothelial Cells and Shows Increased Expression in Atherosclerosis. *Arterioscler. Thromb. Vasc. Biol.* **2009**, *29*, 1228–1234. [CrossRef]

17. Bhatt, P.M.; Lewis, C.J.; House, D.L. Increased Wnt5a mRNA Expression in Advanced Atherosclerotic Lesions, and Oxidized LDL Treated Human Monocyte-Derived Macrophages. *Open Circ. Vasc. J.* **2012**, *5*, 1–7. [CrossRef]

18. Malgor, R.; Bhatt, P.M.; Connolly, B.A.; Jacoby, D.L.; Feldmann, K.J.; Silver, M.J.; Nakazawa, M.; McCall, K.D.; Goetz, D.J. Wnt5a, TLR2 and TLR4 are elevated in advanced human atherosclerotic lesions. *Inflamm. Res.* **2014**, *63*, 277–285. [CrossRef]

19. Kim, J.; Kim, J.; Kim, D.W.; Ha, Y.; Ihm, M.H.; Kim, H.; Song, K.; Lee, I. Wnt5a Induces Endothelial Inflammation via β-Catenin-Independent Signaling. *J. Immunol.* **2010**, *185*, 1274–1282. [CrossRef]

20. Vikram, A.; Kim, Y.R.; Kumar, S.; Naqvi, A.; Hoffman, T.A.; Kumar, A.; Miller, F.J., Jr.; Kim, C.S.; Irani, K. Canonical Wnt Signaling Induces Vascular endothelial dysfunction via p66Shc-Regulated Reactive Oxygen Species Significance. *Arterioscler. Thromb. Vasc. Biol.* **2014**, *34*, 2301–2309. [CrossRef]

21. Tabas, I.; Ron, D. Integrating the mechanisms of apoptosis induced by endoplasmic reticulum stress. *Nat. Cell Biol.* **2011**, *13*, 184–190. [CrossRef] [PubMed]

22. Tabas, I. The role of endoplasmic reticulum stress in the progression of atherosclerosis. *Circ. Res.* **2010**, *107*, 839–850. [CrossRef] [PubMed]

23. Sozen, E.; Karademir, B.; Ozer, N.K. Basic mechanisms in endoplasmic reticulum stress and relation to cardiovascular diseases. *Free Radic. Biol. Med.* **2015**, *78*, 30–41. [CrossRef] [PubMed]

24. Huang, A.; Patel, S.; McAlpine, C.S.; Werstuck, G.H. The Role of Endoplasmic Reticulum Stress-Glycogen Synthase Kinase-3 Signaling in Atherogenesis. *Int. J. Mol. Sci.* **2018**, *19*, 1607. [CrossRef] [PubMed]

25. Shinozaki, S.; Chiba, T.; Kokame, K.; Miyata, T.; Kaneko, E.; Shimokado, K. A deficiency of Herp, an endoplasmic reticulum stress protein, suppresses atherosclerosis in ApoE knockout mice by attenuating inflammatory responses. *PLoS ONE* **2013**, *8*, e75249. [CrossRef] [PubMed]

26. Halleskog, C.; Mulder, J.; Dahlström, J.; Mackie, K.; Hortobágyi, T.; Tanila, H.; Kumar Puli, L.; Färber, K.; Harkany, T.; Schulte, G. WNT signaling in activated microglia is proinflammatory. *Glia* **2011**, *59*, 119–131. [CrossRef] [PubMed]

27. Amodio, G.; Moltedo, O.; Faraonio, R.; Remondelli, P. Targeting the Endoplasmic Reticulum Unfolded Protein Response to Counteract the Oxidative Stress-Induced endothelial dysfunction. *Oxid. Med. Cell. Longev.* **2018**, *2018*, 4946289. [CrossRef] [PubMed]

28. Battson, M.L.; Lee, D.M.; Gentile, C.L. Endoplasmic reticulum stress and the development of endothelial dysfunction. *Am. J. Physiol. Heart Circ. Physiol.* **2017**, *312*, H355–H367. [CrossRef] [PubMed]

29. Dong, L.; Watanabe, K.; Itoh, M.; Huan, C.-R.; Tong, X.-P.; Nakamura, T.; Miki, M.; Iwao, H.; Nakajima, A.; Kawanami, T.S.T.; et al. CD4+ T-cell dysfunctions through the imp aired lipid rafts ameliorate concanavalin A-induced hepatitis in sphingomyelin synthas1 knockout mice. *Int. Immunol.* **2012**, *24*, 327–337. [CrossRef] [PubMed]

30. Ding, T.; Li, Z.; Hailemariam, T.; Mukherjee, S.; Maxfield, F.R.; Wu, M.P.; Jiang, X.C. SMS overexpression and knockdown: Impact on cellular sphingomyelin and diacylglycerol metabolism, and cell apoptosis. *J. Lipid Res.* **2008**, *49*, 376–385. [CrossRef]

31. Li, Y.; Guan, J.; Wang, W.; Hou, C.; Zhou, L.; Ma, J.; Cheng, Y.; Jiao, S.; Zhou, Z. TRAF3-interacting JNK-activating modulator promotes inflammation by stimulating translocation of Toll-like receptor 4 to lipid rafts. *J. Biol. Chem.* **2018**, *294*, 2744–2756. [CrossRef] [PubMed]

32. Yamamoto, H.; Komekado, H.; Kikuchi, A. Caveolin is necessary for Wnt-3a-dependent internalization of lrp6 and accumulation of b-catenin. *Dev. Cell* **2006**, *11*, 213–223. [CrossRef] [PubMed]

33. Bedel, A.; Nègre-Salvayre, A.; Heeneman, S.; Grazide, M.H.; Thiers, J.C.; Salvayre, R.; Maupas-Schwalm, F. E-cadherin/β-catenin/T-cell factor pathway is involved in smooth muscle cell proliferation elicited by oxidized low-density lipoprotein. *Circ. Res.* **2008**, *103*, 694–701. [CrossRef] [PubMed]

34. Liu, C.C.; Kanekiyo, T.; Roth, B.; Bu, G. Tyrosine-based signal mediates LRP6 receptor endocytosis and desensitization of Wnt/β-catenin pathway signaling. *J. Biol. Chem.* **2014**, *289*, 27562–27570. [CrossRef] [PubMed]

35. Ma, S.; Yao, S.; Tian, H.; Jiao, P.; Yang, N.; Zhu, P.; Qin, S. Pigment epithelium-derived factor alleviates endothelial injury by inhibiting Wnt/β-catenin pathway. *Lipids Health Dis.* **2017**, *16*, 31. [CrossRef] [PubMed]

36. Banerjee, K.; Keasey, M.P.; Razskazovskiy, V.; Visavadiya, N.P.; Jia, C.; Hagg, T. Reduced FAK-STAT3 signaling contributes to ER stress-induced mitochondrial dysfunction and death in endothelial cells. *Cell. Signal.* **2017**, *36*, 154–162. [CrossRef]

37. SchrDer, M.; Kaufman, R.J. The mammalian unfolded protein response. *Annu. Rev. Biochem.* **2005**, *74*, 739–789. [CrossRef] [PubMed]

38. Zhang, Z.; Wu, S.; Muhammad, S.; Ren, Q.; Sun, C. miR-103/107 promote ER stress-mediated apoptosis via targeting the Wnt3a/β-catenin/ATF6 pathway in preadipocytes. *Lipid Res.* **2018**, *59*, 843–853. [CrossRef]

39. Cao, L.; Lei, H.; Chang, M.Z.; Liu, Z.Q.; Bie, X.H. Down-regulation of 14-3-3 β exerts anti-cancer effects through inducing ER stress in human glioma U87 cells: Involvement of CHOP-Wnt pathway. *Biochem. Biophys. Res. Commun.* **2015**, *462*, 389–395. [CrossRef]

40. Jia, X.; Chen, Y.; Zhao, X.; Lv, C.; Yan, J. Oncolytic vaccinia virus inhibits human epatocellular carcinoma MHCC97-H cell proliferation via endoplasmic reticulum stress, autophagy and Wnt pathways. *J. Gene Med.* **2016**, *18*, 211–219. [CrossRef]

41. Chaube, R.; Kallakunta, V.M.; Espey, M.G.; McLarty, R.; Faccenda, A.; Ananvoranich, S.; Mutus, B. Endoplasmic reticulum stress-mediated inhibition of NSMase2 elevates plasma membrane cholesterol and attenuates NO production in endothelial cells. *Biochim. Biophys. Acta* **2012**, *1821*, 313–323. [CrossRef] [PubMed]

42. Liu, T.; Duan, W.; Nizigiyimana, P.; Gao, L.; Liao, Z.; Xu, B.; Liu, L.; Lei, M. α-Mangostin attenuates diabetic nephropathy in association with suppression of acid sphingomyelianse and endoplasmic reticulum stress. *Biochem. Biophys. Res. Commun.* **2018**, *496*, 394–400. [CrossRef] [PubMed]

43. Sozen, E.; Ozer, N.K. Impact of high cholesterol and endoplasmic reticulum stress on metabolic diseases: An updated mini-review. *Redox. Biol.* **2017**, *12*, 456–461. [CrossRef] [PubMed]

44. Röhrl, C.; Stangl, H. Cholesterol metabolism-physiological regulation and pathophysiological deregulation by the endoplasmic reticulum. *Wien. Med. Wochenschr.* **2018**, *168*, 280–285. [CrossRef] [PubMed]

45. Mou, D.; Yang, H.; Qu, C.; Chen, J.; Zhang, C. Pharmacological Activation of Peroxisome Proliferator-Activated Receptor Increases Sphingomyelin Synthase Activity in THP-1 Macrophage-Derived Foam Cell. *Inflammation* **2016**, *39*, 1538–1546. [CrossRef] [PubMed]

46. Lee, C.Y.; Lesimple, A.; Denis, M.; Vincent, J.; Larsen, A.; Mamer, O.; Krimbou, L.; Genest, J.; Marcil, M. Increased sphingomyelin content impairs HDL biogenesis and maturation in human Niemann-Pick disease type B. *J. Lipid Res.* **2006**, *47*, 622–632. [CrossRef]

47. Yan, N.; Ding, T.; Dong, J.; Li, Y.; Wu, M. Sphingomyelin synthase overexpression increases cholesterol accumulation and decreases cholesterol secretion in liver cells. *Lipids Health Dis.* **2011**, *10*, 46. [CrossRef]

 International Journal of
Molecular Sciences

Article

Trimethylamine N-Oxide Does Not Impact Viability, ROS Production, and Mitochondrial Membrane Potential of Adult Rat Cardiomyocytes

Giulia Querio, Susanna Antoniotti, Renzo Levi and Maria Pia Gallo *

Department of Life Sciences and Systems Biology, University of Turin, Via Accademia Albertina 13,
10123 Turin, Italy; giulia.querio@unito.it (G.Q.); susanna.antoniotti@unito.it (S.A.); renzo.levi@unito.it (R.L.)
* Correspondence: mariapia.gallo@unito.it; Tel.: +39-011-670-4671

Received: 17 May 2019; Accepted: 19 June 2019; Published: 21 June 2019

Abstract: Trimethylamine N-oxide (TMAO) is an organic compound derived from dietary choline and L-carnitine. It behaves as an osmolyte, a protein stabilizer, and an electron acceptor, showing different biological functions in different animals. Recent works point out that, in humans, high circulating levels of TMAO are related to the progression of atherosclerosis and other cardiovascular diseases. However, studies on a direct role of TMAO in cardiomyocyte parameters are still limited. The purpose of this work is to study the effects of TMAO on isolated adult rat cardiomyocytes. TMAO in both 100 µM and 10 mM concentrations, from 1 to 24 h of treatment, does not affect cell viability, sarcomere length, intracellular ROS, and mitochondrial membrane potential. Furthermore, the simultaneous treatment with TMAO and known cardiac insults, such as H_2O_2 or doxorubicin, does not affect the treatment's effect. In conclusion, TMAO cannot be considered a direct cause or an exacerbating risk factor of cardiac damage at the cellular level in acute conditions.

Keywords: trimethylamine N-oxide; cardiomyocytes; cardiotoxicity; ROS; mitochondrial membrane potential

---

1. Introduction

Trimethylamine N-oxide (TMAO) is an amine oxide directly introduced through diet or synthetized from its precursors, primarily L-carnitine and choline, that are transformed into trimethylamine (TMA) by the gut microbiota. Once absorbed, TMA in most mammals is oxidized by hepatic FMOs to form TMAO which enters the systemic circulation. Several studies illustrate different biological functions of TMAO in other animals. In elasmobranchs and deep-sea fishes, it acts as an osmolyte able to counteract either osmotic or hydrostatic pressure. It is a protein stabilizer preserving protein folding and it also acts as an electron acceptor balancing oxidative stress [1,2]. TMAO is also reduced to TMA in the anaerobic metabolism of a number of bacteria. Although TMAO is involved in several reactions within cells, recent studies highlight its detrimental role when present in high plasmatic concentrations in some mammals. In fact, TMAO seems to be involved in accelerating endothelial cell senescence, enhancing vascular inflammation and oxidative stress [3,4]; it also could be involved in the stimulation of platelet hyperreactivity and in the onset of thrombosis, exacerbating atherosclerotic lesions [5]. Several studies also underline the role of TMAO in the pathogenesis of type 2 diabetes mellitus [6]. There are limited data on its function in mediating direct cardiac injuries, and those studies are mainly focused on its role in the impairment of mitochondrial metabolism [7] and calcium handling [8]. On the contrary, recent papers are re-evaluating the role of TMAO, underlining the emerging debate on its direct effect in causing or exacerbating cardiovascular diseases (CVD) [9,10]. First criticisms point out that populations having diets with high concentrations of TMAO, like those rich in fish products, when compared to Western diets rich in its precursors, have reduced risks of CVD or diseases assumed to be related to high TMAO plasma levels [11]. Another study demonstrates that TMAO does not affect macrophage foam cell formation and lesion progression in ApoE$^{-/-}$ mice expressing human cholesteryl ester transfer protein, suggesting that the molecule does not worsen atherosclerosis [12]. Furthermore, administration of TMAO seems to improve symptoms related to streptozotocin-induced diabetes in rats and mice, highlighting no direct contribution of the molecule in exacerbating this condition [13]. Finally, data about TMAO plasma concentrations in healthy and pathological subjects are not clear: the lack of plasma concentration ranges of the molecule highlights the difficulties in referring to TMAO as a protective or a damaging factor in CVD. Starting from these conflicting considerations, the aim of this work was to evaluate for the first time the effect of TMAO in an in vitro model of adult rat cardiomyocytes exposed to different concentrations of the compound from 1 h to 24 h of treatment. To show whether TMAO exacerbates or reduces induced cell stress, cardiomyocytes were simultaneously treated with TMAO and H_2O_2 or doxorubicin (DOX). Investigations were focused on cell viability after TMAO or TMAO and stressors co-treatment, assessing cell morphology and functionality with α-actinin staining and specific probes that measure oxidative stress status and mitochondrial membrane potential.

2. Results

2.1. TMAO and Cell Viability

In order to investigate the effect of TMAO on cell viability, cardiomyocytes were treated with TMAO 100 µM, TMAO 10 mM, H_2O_2 50 µM, and H_2O_2 50 µM + TMAO 100 µM. After 1 h or 24 h of treatment, cardiomyocytes were labeled with propidium iodide (PI) and marked nuclei of suffering cells were detected by confocal microscopy at 568 nm. Concentrations used were taken from the literature: TMAO 100 µM is recognized as a marker of cardiovascular risk, TMAO 10 mM is over the physiological range and was tested here to detect any effect induced by high concentrations of the compound [14]. As shown in Figure 1a, there is no effect of TMAO 100 µM or TMAO 10 mM at either time of treatment, whereas H_2O_2, used here as a positive control, had effects only after 24 h. Simultaneous treatment with H_2O_2 and TMAO did not improve or worsen the effect of the stressor on cell viability (1 h—CTRL: 83.95 ± 6.59, $n = 3$, 52 cells; TMAO 100 µM: 85.52 ± 7.01, $n = 5$, 81 cells; TMAO 10 mM: 84.08 ± 5.84, $n = 5$, 92 cells; H_2O_2: 52.92 ± 16.46, $n = 3$, 56 cells; H_2O_2 + TMAO: 50.93 ± 18.50, $n = 3$, 58 cells. 24 h—CTRL: 83.42 ± 2.29, $n = 3$, 101 cells; TMAO 100 µM: 85.66 ± 6.48, $n = 3$, 91 cells; TMAO 10 mM: 62.22 ± 10.47, $n = 3$, 119 cells; H_2O_2: 2.38 ± 2.38, $n = 3$, 82 cells (*** $p < 0.001$); H_2O_2 + TMAO: 1.33 ± 1.33, $n = 3$, 41 cells (*** $p < 0.001$)). Figure 1b displays confocal images of cardiomyocytes from a representative experiment of PI staining after 24 h of treatment. White arrows point out PI-stained, damaged cardiomyocytes.

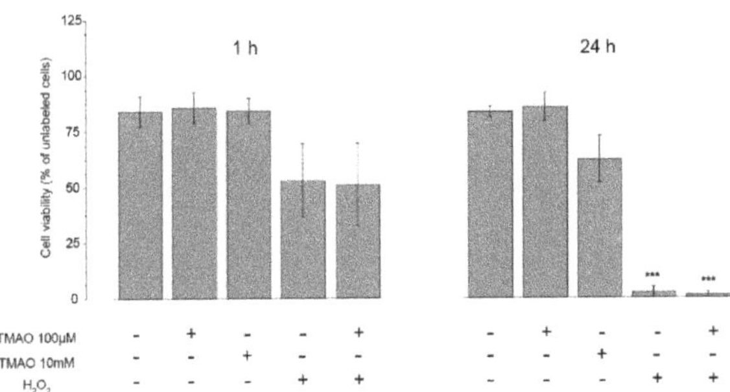

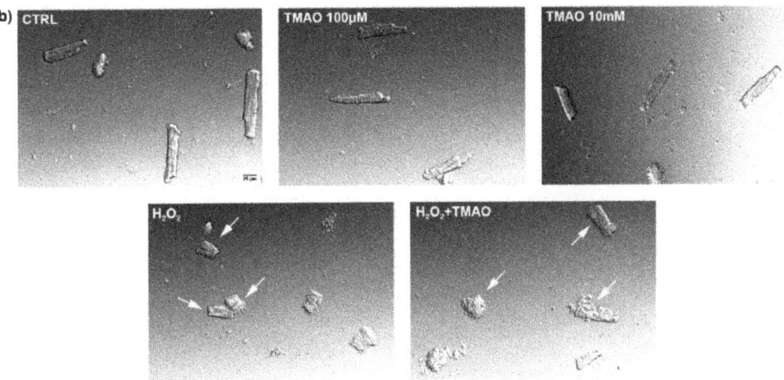

Figure 1. Cell viability after trimethylamine N-oxide (TMAO) exposure. (**a**) Bar graph of cell viability after 1 h and 24 h of treatment. Cell viability results reduced only after H_2O_2 treatment for 24 h, condition not improved or worsened by the simultaneous treatment with TMAO (refer to the main text for numerical values). (**b**) Merged images in bright field and fluorescence of cells treated for 24 h with TMAO 100 µM and TMAO 10 mM, H_2O_2, and H_2O_2+TMAO and labeled with propidium iodide (PI) (20× magnification). White arrows point out PI-stained, damaged cardiomyocytes.

2.2. TMAO and Sarcomere Length

To evaluate if TMAO was able to alter sarcomere structures after 24 h of treatment, sarcomere length was measured in α-actinin-stained cardiomyocytes. As shown in Figure 2, no changes in sarcomere length were observed in cells treated with TMAO, while H_2O_2 50 µM used as a positive control caused cardiomyocyte shrinkage, a condition that was not improved or worsened by the simultaneous treatment with TMAO. In cardiomyocytes treated with DOX 1 µM for 24 h, no sarcomere length variations were observed, because the DOX treatment in our model was designed to induce mild damage preceding cell shortening. Even so, TMAO 100 µM did not modify DOX-treated cardiomyocytes (sarcomere length in µm—CTRL: 1.69 ± 0.01, $n = 7$, 42 cells; TMAO 100 µM: 1.69 ± 0.01, $n = 6$, 34 cells; TMAO 10 mM: 1.67 ± 0.02, $n = 3$, 19 cells; H_2O_2: 1.22 ± 0.04, $n = 5$, 33 cells (*** $p < 0.001$); H_2O_2 + TMAO: 1.28 ± 0.03, $n = 3$, 24 cells (*** $p < 0.001$); DOX: 1.62 ± 0.02, $n = 3$, 16 cells; DOX + TMAO: 1.65 ± 0.01, $n = 3$, 15 cells).

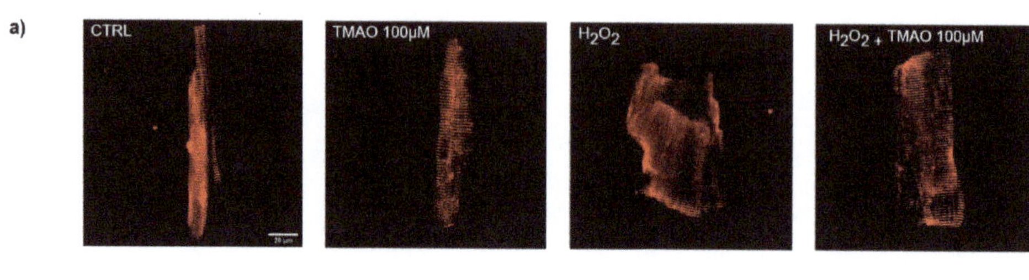

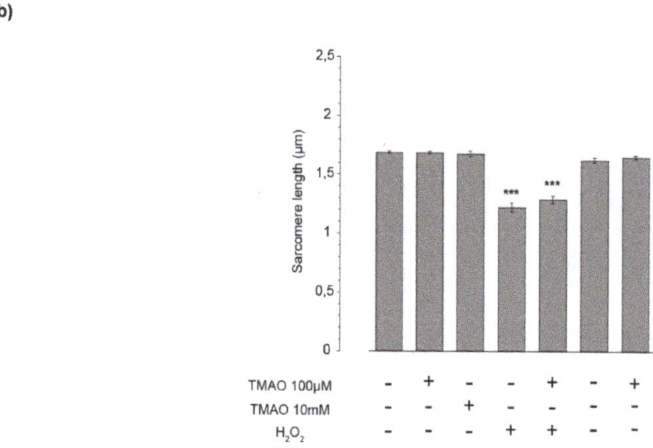

Figure 2. Sarcomere length after TMAO treatment. (**a**) Confocal microscopy images of fixed cells labeled for α-actinin protein (40× magnification). (**b**) Bar graph showing sarcomere length after 24 h of treatment with TMAO and other stressors: no cell shrinkage is measured when cells are exposed to different TMAO concentrations (refer to the main text for numerical values).

2.3. TMAO and Intracellular Reactive Oxygen Species (ROS)

In order to determine a variation in total ROS produced after treatment with TMAO for 1 h, 3 h, or 24 h, cells were labeled with the DCF-DA probe and its fluorescence was quantified and related to the control. As shown in Figure 3 (1 h, 3 h) and 4 (24 h), no fluorescence variations after TMAO treatment were detected at any concentration and time used. As a positive control, we employed DOX 1 µM for 24 h [15]; this drug caused a significant variation in ROS production with respect to the control condition. TMAO 100 µM did not modify ROS production in the DOX-treated cardiomyocytes (Figure 4) (1 h—TMAO 100 µM: 1.30 ± 0.21, $n = 3$, 34 cells; TMAO 10 mM: 1.32 ± 0.23, $n = 3$, 40 cells, vs. CTRL; 3 h—TMAO 100 µM: 0.96 ± 0.05, $n = 3$, 45 cells; TMAO 10 mM: 1.15 ± 0.09, $n = 3$, 52 cells, vs. CTRL; 24 h—TMAO 100 µM: 1.18 ± 0.04, $n = 5$, 40 cells; TMAO 10 mM: 1.24 ± 0.16, $n = 3$, 52 cells; DOX: 1.33 ± 0.11, $n = 4$, 21 cells (** $p < 0.01$); DOX+TMAO: 1.27 ± 0.01, $n = 6$, 31 cells (*** $p < 0.001$), vs. CTRL).

a)

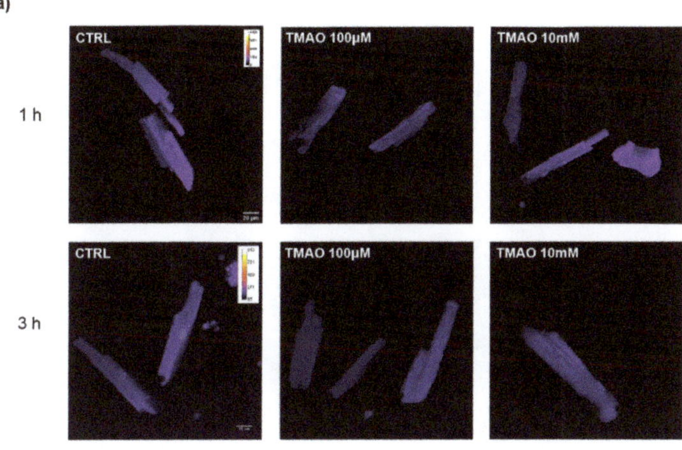

b)

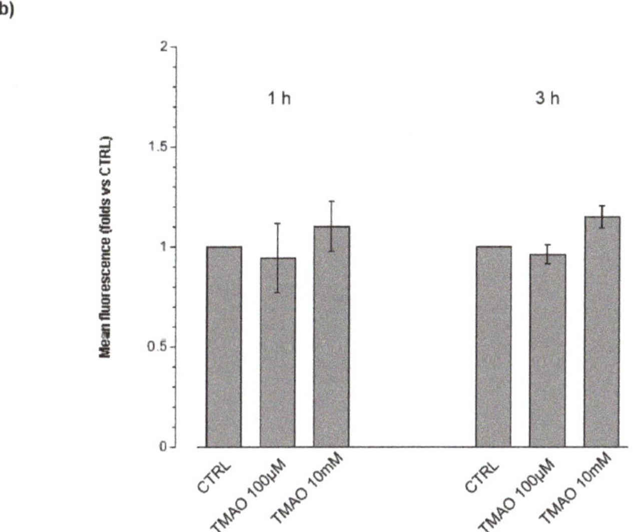

Figure 3. ROS production after 1 h and 3 h of treatment. (**a**) Confocal microscopy images of cells treated with TMAO 100 μM and TMAO 10 mM for 1 h and 3 h and labeled with the DCF-DA probe (60× magnification). (**b**) Bar graph showing mean fluorescence after 1 h and 3 h of treatment, no variations or ROS produced are detectable (refer to the main text for numerical values).

a)

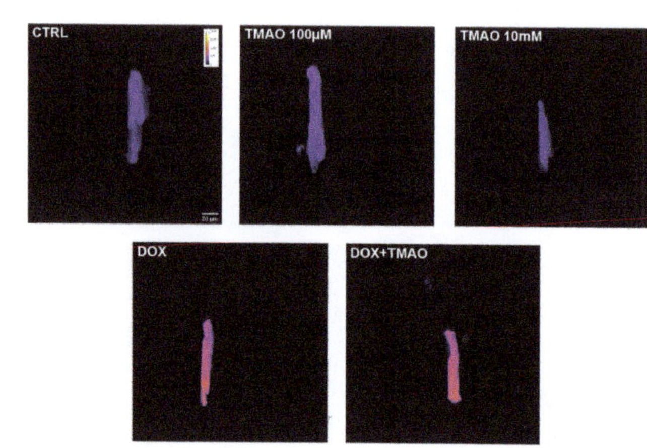

b)

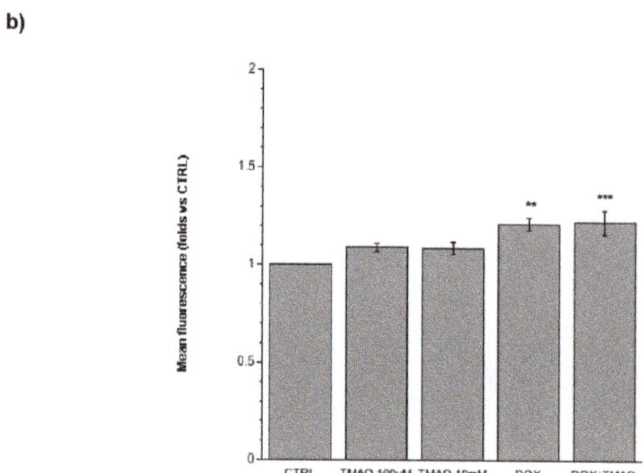

Figure 4. ROS production after 24 h of treatment. (**a**) Confocal microscopy images of cells treated with TMAO 100 µM and TMAO 10 mM for 24 h and labeled with the DCF-DA probe. In these experiments, doxorubicin (DOX) is used as a positive control (60× magnification). (**b**) Bar graph showing mean fluorescence after 24 h of treatment. No variations of ROS produced are detectable after TMAO treatment, and a small but significant increase is visible after DOX treatment (used here as a positive control); this increase is not changed by a simultaneous treatment with TMAO (refer to the main text for numerical values).

2.4. TMAO and Mitochondrial Membrane Potential

To investigate the potential metabolic damage induced by TMAO, cardiomyocytes treated with TMAO 100 µM and 10 mM for 1 h, 3 h, or 24 h were labeled with the JC-1 probe. Figure 5 (1 h, 3 h) and 6 (24 h) show variations in mitochondrial membrane potential (red/green fluorescence ratio) detected by confocal microscopy in living cells. TMAO treatment from 1 to 24 h did not cause any difference with respect to the control, indicating no mitochondrial effect of the molecule, whereas, as expected, DOX caused a depolarization of mitochondrial membrane potential after 24 h of treatment. TMAO 100 µM did not modify mitochondrial membrane potential in DOX-treated cardiomyocytes (Figure 6)

(1 h—TMAO 100 μM: 1.09 ± 0.08, $n = 5$, 65 cells; TMAO 10 mM: 1.11 ± 0.11, $n = 3$, 49 cells, vs. CTRL. 3 h—TMAO 100 μM: 1.02 ± 0.06, $n = 3$, 26 cells; TMAO 10 mM: 0.88 ± 0.05, $n = 4$, 39 cells, vs. CTRL; 24 h—TMAO 100 μM: 1.08 ± 0.11, $n = 3$, 21 cells; TMAO 10 mM: 0.95 ± 0.15, $n = 3$, 54 cells; DOX: 0.73 ± 0.02, $n = 3$, 24 cells (** $p < 0.01$); DOX + TMAO: 0.69 ± 0.05, $n = 3$, 22 cells (** $p < 0.01$), vs. CTRL).

a)

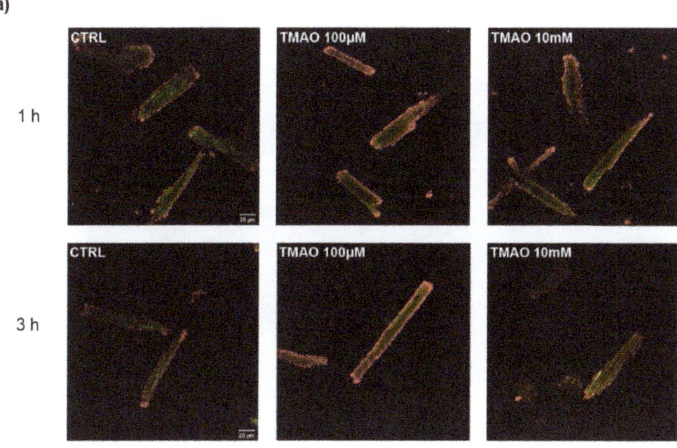

b)

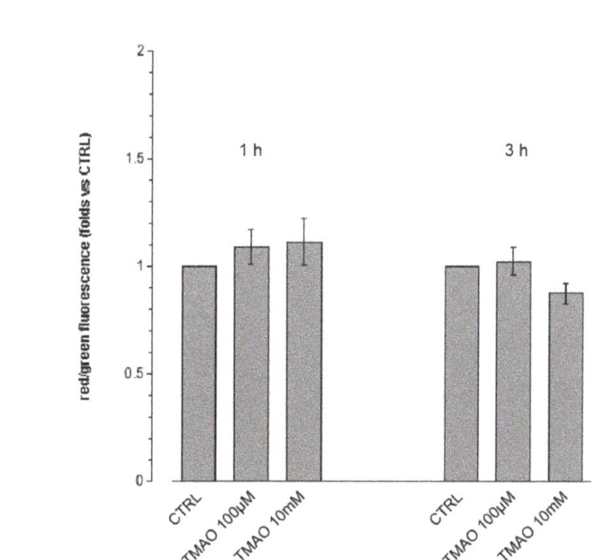

Figure 5. Mitochondrial membrane potential variation following 1 h and 3 h of treatment. (**a**) Confocal microscopy images of cells treated with TMAO 100 μM and TMAO 10 mM for 1 h and 3 h and labeled with the JC-1 probe (60× magnification). (**b**) Bar graph showing red/green fluorescence after 1 h and 3 h of treatment, no variations of mitochondrial membrane potential are detected (refer to the main text for numerical values).

a)

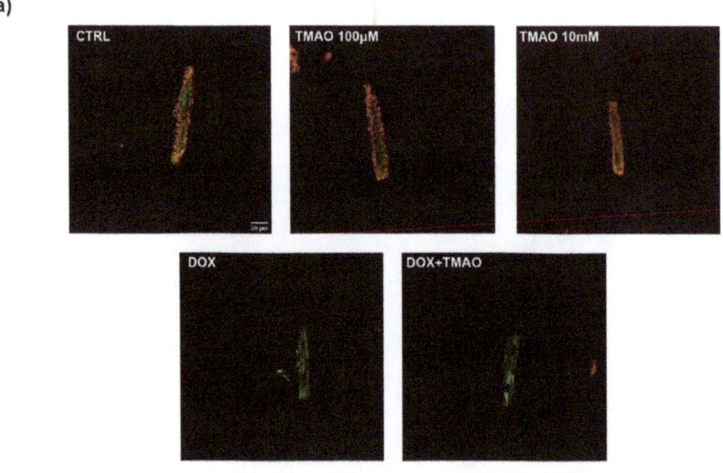

b)

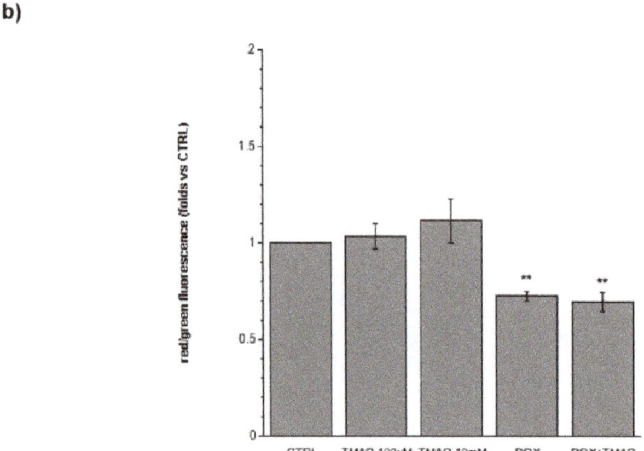

Figure 6. Mitochondrial membrane potential variation following 24 h treatment. (**a**) Confocal microscopy images of cells treated with TMAO 100 µM and TMAO 10 mM for 24 h and labeled with the JC-1 probe. In these experiments, doxorubicin (DOX) is used as a positive control (60× magnification). (**b**) Bar graph showing red/green fluorescence after 24 h of treatment, no variations of mitochondrial membrane potential are detected, a little but significant reduction of the ratio is visible after DOX treatment used here as a positive control, condition not changed in a simultaneous treatment with TMAO (refer to the main text for numerical values).

3. Discussion

This study provides novel insights into the physiological role of TMAO in isolated adult rat cardiomyocytes. Our findings do not show effects of TMAO on cell viability, sarcomere length, ROS production, and mitochondrial membrane potential within the range of concentration used. Moreover, we demonstrate that TMAO does not exacerbate or counteract the effect of known insults, such as H_2O_2 or doxorubicin, tested for up to 24 h of treatment. Taken together, these results suggest that

TMAO should not be considered a primary cause of acute cardiac damage and that the molecule could not revert or worsen existing risk factors of cardiac damage.

In the last few years, many studies suggest a strong relationship between diet, gut microbiota, and cardiovascular diseases [16]. In particular, some attention has been pointed to either TMAO directly coming from diet (fish), or produced from L-carnitine and choline conversion by gut microbiota into TMA and oxidized in the liver by FMO3 enzymes [14,17].

Experiments have mainly now focused on the role of TMAO in damaging endothelial cells. It has been described as upregulating cellular senescence, thereby reducing cell proliferation, increasing the expression of senescence markers, such as p53 and p21, and impairing cell migration [3]. TMAO also increases the oxidative stress of endothelial cells through a down-regulation of SIRT-1 and impairs NO production that causes endothelial dysfunction [4]. Hypertensive effects of TMAO have been evaluated by Ufnal and colleagues who demonstrated that TMAO has a role in stabilizing the action of Ang II and in prolonging its hypertensive effect, underlining the role of TMAO in stabilizing protein conformation and no direct role of the molecule in mediating hypertension [18]. Koeth and colleagues underlined the strong relationship between the high consumption of TMAO precursors, high TMAO plasma concentrations, and the development of atherosclerosis [19], while another study underlined the effect of the metabolite in enhancing platelet hyperreactivity and thrombosis risk in subjects with high TMAO plasma concentrations [5]. With respect to the cardiovascular effects of TMAO, Dambrova and collaborators showed that high plasma concentrations of the molecule are linked with increased body weight and insulin resistance and that it directly correlates with an augmented risk of diabetes [20].

Only a few studies are centered on the direct effect of the molecule on cardiac cells; in particular, they focus on the impairment of mitochondrial metabolism in the heart and underline TMAO as an agent that increases the severity of cardiovascular events or that enhances the progression of cardiovascular diseases [7]. Savi and colleagues showed a damaging effect of TMAO in cardiomyocytes because it worsens intracellular calcium handling with a reduced efficiency in the intracellular calcium removal and consequent loss in functionality of cardiac cells; furthermore, TMAO seems to alter energetic metabolism and to facilitate protein oxidative damage [8].

This scenario presents TMAO as either a marker or a direct agent involved in vascular and cardiac outcomes, but recent papers seem to oppose this point of view, highlighting uncertainty about the causative relation between TMAO and CVD [9]. The function of TMAO is still being debated, for example, the controversy surrounding fish-rich diets, because of the higher bioavailability of the compound in seafood products and their well-known role in lowering risk of CVD. Additionally, TMAO does not enhance atherosclerosis development because it seems not to be involved in foam cell formation even at higher concentrations than physiological ones [12], and there is no direct correlation between high plasma TMAO concentrations and coronary heart diseases [21,22]. Finally, findings by Huc et al. have also underlined a protective role of TMAO in reducing diastolic dysfunction and fibrosis in the pressure-overloaded heart [23].

The present study fits into this debate and the results presented agree with other works supporting TMAO as a non-damaging factor. In fact, it is well known that the loss of vital cardiac cells is a damaging condition that hampers primarily the functionality of the heart and has several aggravating responses also in peripheral tissues. Our first investigations underline no toxic effect in cardiomyocytes exposed to high concentrations of TMAO, highlighting the result that the molecule is not involved in inducing cardiac tissue cell loss, and no alterations of cardiac structure emerge from the evaluation of sarcomere length and cytoskeletal organization. Oxidative stress could be considered one of the causative factors of senescence in cells and one of the promoters of cardiometabolic reorganization in response to injury. With respect to TMAO as a possible inducer of ROS rising, both in a cytoplasmic and a mitochondrial environment, we show no variation in ROS production even after 24 h of treatment and we detect no depolarization of mitochondrial membrane potential, underlining the result of no direct influence by the molecule in inducing cardiac cell senescence.

4. Materials and Methods

4.1. Animal Care and Sacrifice

Experiments were performed on female adult rats which were allowed ad libitum access to tap water and standard rodent diet. The animals received human care in compliance with the Guide for the Care and Use of Laboratory Animals published by the US National Institutes of Health (NIH Publication No. 85-23, revised 1996), and in accordance with Italian law (DL-116, Jan. 27, 1992). The scientific project was supervised and approved by the Italian Ministry of Health, Rome, and by the ethical committee of the University of Torino (approval code 116/2017-PR, 3/2/2017). Rats were anaesthetized by i.p. injection of tiletamine (Zoletil 100, Virbac, Carros, France) and sacrificed by stunning and cervical dislocation.

4.2. Solutions and Drugs

Tyrode standard solution containing (in mM): 154 NaCl, 4 KCl, 1 $MgCl_2$, 5.5 D-glucose, 5 HEPES, 2 $CaCl_2$, pH adjusted to 7.34 with NaOH. Ca^{2+} free Tyrode solution containing (in mM): 154 NaCl, 4 KCl, 1 $MgCl_2$, 5.5 D-glucose, 5 HEPES, 10 2,3-Butanedione monoxime, 5 taurine, pH 7.34. All drug-containing solutions were prepared fresh before the experiments and the Tyrode solutions were oxygenated (O_2 100%) before each experiment. Unless otherwise specified, all reagents for cell isolation and experiments were purchased from Sigma-Aldrich (St. Louis, MO, USA).

4.3. Adult Rat Ventricular Cell Isolation

Isolated cardiomyocytes were obtained from the hearts of adult rats (200–300 g body weight) according to the previously described method [24]. Briefly, after sacrifice, rat heart was explanted, washed in Ca^{2+} free Tyrode solution, and cannulated via the aorta. All the following operations were carried on under a laminar flow hood. The heart was perfused at a constant flow rate of 10 mL/min with Ca^{2+} free Tyrode solution (37 °C) with a peristaltic pump for approximately 5 min to wash away the blood and then with 10 mL of Ca^{2+} free Tyrode supplemented with collagenase (0.3 mg/mL) and protease (0.02 mg/mL). Hearts were then perfused and enzymatically dissociated with 20 mL of Ca^{2+} free Tyrode containing 50 μM $CaCl_2$ and the same enzymatic concentration as before. Atria and ventricles were then separated and the ventricles were cut in small pieces and shaken for 10 min in 20 mL of Ca^{2+} free Tyrode solution in the presence of 50 μM $CaCl_2$, collagenase, and protease. Calcium ion concentration was slowly increased to 0.8 mM. Cardiomyocytes were then plated on glass cover slips or glass bottom dishes (Ibidi, Martinsried, Germany), both treated with laminin to allow cell adhesion.

4.4. Cell Viability

Cell viability was evaluated by propidium iodide (PI) staining on glass bottom dishes for adherent cells. At the end of the treatments, cells were incubated with PI (10 μg/mL, Invitrogen, Carlsbad, CA, USA) for 5 min in the dark. Nuclei of suffering cells were detected with confocal microscopy using an Olympus Fluoview 200 microscope (Shinjuku, Tokyo, Japan) at 568 nm (magnification 20×). Merged images were created with ImageJ (U.S. National Institutes of Health, Bethesda, MD, USA, https://imagej.nih.gov/ij/) and cell viability was calculated as percentage of (total cells-labeled cells)/total cells.

4.5. Evaluation of Sarcomere Length

Cardiomyocytes on glass coverslips were stimulated with TMAO and with H_2O_2 as a positive control. Cells were treated for 24 h with TMAO, at 100 μM and 10 mM, then the sarcomere protein α-actinin, localized in the Z lines, was detected using confocal microscopy. Subsequently, cells were fixed in 4% PFA for 40 min. After two washes with PBS, cells were incubated for 20 min with 0.3% Triton and 1% bovine serum albumin (BSA) in PBS and stained for 24 h at +4 °C with a mouse monoclonal anti-α-actinin primary antibody (Sigma-Aldrich, 1:800). Cover slides were washed twice with PBS and incubated 1 h at room temperature with the secondary antibody (1:2000, anti-mouse Alexa Fluor 568, Thermo Fisher Scientific, Waltham, MA, USA). After two washes in PBS, coverslips were mounted on standard slides with DABCO and observed after 24 h under a confocal microscope. Confocal fluorimetric measurements were acquired using a Leica SP2 laser scanning confocal system (Wetzlar, Germany), equipped with a 40× water-immersion objective. Image processing and analysis were performed with ImageJ software. Sarcomere length was evaluated measuring the distance between Z lanes in n = 10 sarcomeres/cell.

4.6. Intracellular Reactive Oxygen Species (ROS) Measurement

Production of ROS was evaluated by fluorescence microscopy using a 2′-7′-dichlorofluorescein diacetate probe (DCF-DA). After adhesion on glass bottom dishes, DCF-DA solution (5 μg/mL) was added to each dish 30 min prior to the end of the treatment, then the cells were washed with standard Tyrode solution. Fluorescence images at 488 nm were acquired using an Olympus Fluoview 200 microscope (magnification 60×). Fluorescence variations were calculated with the definition and measurement of regions of interest (ROIs) using ImageJ software and expressed as relative Medium Fluorescence Index (MFI) compared to control, fixed at 1.

4.7. Mitochondrial Membrane Potential Measurement

Mitochondrial membrane potential was evaluated by staining cardiomyocytes with the dye 5,5′,6,6′-tetrachloro-1,1′,3,3′-tetraethyl-imidacarbocyanine iodide (JC-1). JC-1 solution (10 μM) was added to each dish 30 min prior to the end of the treatment, then the cells were washed with standard Tyrode solution. Fluorescence images at 488 nm and 568 nm were acquired using an Olympus Fluoview 200 microscope (magnification 60×). Amounts of the monomeric form of the dye were quantified using the red/green fluorescence ratio in the ROIs using ImageJ software and expressed as folds towards control, fixed at 1.

4.8. Statistical Analysis

All data are expressed as mean ± standard error of the mean. For differences between mean values, Bonferroni's multiple comparisons test was performed. Differences with $p < 0.05$ were regarded as statistically significant.

5. Conclusions

In summary, this study demonstrates that TMAO is not directly involved in causing or exacerbating cardiac damage in an acute stress model (Figure 7). However, this study has some limitations: a very wide range of plasmatic TMAO concentrations have been presented in literature, within different mammals, and also between different sexes of the same species; so, several orders of magnitude can be considered physiological [25,26]. Therefore, to test the direct effect of the molecule, high concentrations were used, even higher than human physiological ones. Another weakness of the study could be linked to the time of treatment, because the evaluations were no longer than 24 h and they only represented an acute exposure to TMAO. Moreover, in order to better evaluate the mechanism involved in TMAO-mediated responses, it may be necessary to treat cells for a longer period to assess a chronic stress compatible with the development of CVD. Furthermore, we only studied the TMAO effect on female ventricular cardiomyocytes and it may be interesting to extend the analysis also to male cardiomyocytes as gender differences have been observed in cardioprotective mechanisms [27] and TMAO-induced intracellular calcium imbalance has been described in male cardiomyocytes [8]. Finally, our findings provide new insights into the cardiac effect of TMAO, exploring the direct treatment of isolated cardiomyocytes.

Int. J. Mol. Sci. **2019**, *20*, 3045

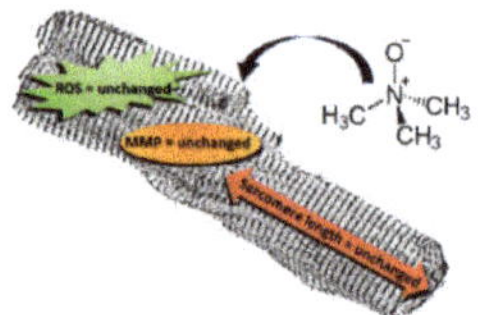

Figure 7. No effects of TMAO are detectable on isolated adult rat cardiomyocyte viability, sarcomere length, ROS production, and mitochondrial membrane potential.

Author Contributions: M.P.G. and R.L. conceived the study, assisted in its design, and revised the manuscript for important intellectual content. G.Q. and S.A. carried out all the experiments and statistical analysis, and all the authors interpreted the results. G.Q. wrote the manuscript. All authors read and approved the final manuscript.

Funding: This research received no external funding; it was supported by the local funding of the University of Turin (Maria Pia Gallo and Renzo Levi). The APC was funded by Maria Pia Gallo and Renzo Levi.

Acknowledgments: The authors acknowledge Nunzio Dibiase for his technical support.

Conflicts of Interest: The authors declare no conflict of interest.

Abbreviations

TMAO	Trimethylamine N-oxide
PI	Propidium iodide
ROS	Reactive oxygen species
DCF-DA	2′-7′-Dichlorofluorescein diacetate
JC-1	5,5′,6,6′-Tetrachloro-1,1′,3,3′-tetraethyl-imidacarbocyanine iodide

References

1. Ufnal, M.; Zadlo, A.; Ostaszewski, R. TMAO: A small molecule of great expectations. *Nutrition* **2015**, *31*, 1317–1323. [CrossRef] [PubMed]
2. Yancey, P.H. Organic osmolytes as compatible, metabolic and counteracting cytoprotectants in high osmolarity and other stresses. *J. Exp. Biol.* **2005**, *208*, 2819–2830. [CrossRef] [PubMed]
3. Ke, Y.; Li, D.; Zhao, M.; Liu, C.; Liu, J.; Zeng, A.; Shi, X.; Cheng, S.; Pan, B.; Zheng, L.; et al. Gut flora-dependent metabolite Trimethylamine-N-oxide accelerates endothelial cell senescence and vascular aging through oxidative stress. *Free Radic. Biol. Med.* **2018**, *116*, 88–100. [CrossRef] [PubMed]
4. Li, T.; Chen, Y.; Gua, C.; Li, X. Elevated Circulating Trimethylamine N-Oxide Levels Contribute to Endothelial Dysfunction in Aged Rats through Vascular Inflammation and Oxidative Stress. *Front. Physiol.* **2017**, *8*, 350. [CrossRef] [PubMed]
5. Zhu, W.; Gregory, J.C.; Org, E.; Buffa, J.A.; Gupta, N.; Wang, Z.; Li, L.; Fu, X.; Wu, Y.; Mehrabian, M.; et al. Gut Microbial Metabolite TMAO Enhances Platelet Hyperreactivity and Thrombosis Risk. *Cell* **2016**, *165*, 111–124. [CrossRef] [PubMed]
6. Shan, Z.; Sun, T.; Huang, H.; Chen, S.; Chen, L.; Luo, C.; Yang, W.; Yang, X.; Yoa, P.; Cheng, J.; et al. Association between microbiota-dependent metabolite trimethylamine-N-oxide and type 2 diabetes. *Am. J. Clin. Nutr.* **2017**, *106*, 888–894. [CrossRef] [PubMed]
7. Makrecka-Kuka, M.; Volska, K.; Antone, U.; Vilskersts, R.; Grinberga, S.; Bandere, D.; Liepinsh, E.; Dambrova, M. Trimethylamine N-oxide impairs pyruvate and fatty acid oxidation in cardiac mitochondria. *Toxicol. Lett.* **2017**, *267*, 32–38. [CrossRef] [PubMed]
8. Savi, M.; Bocchi, L.; Bresciani, L.; Falco, A.; Quaini, F.; Mena, P.; Brighenti, F.; Crozier, A.; Stilli, D.; Del Rio, D. Trimethylamine-N-Oxide (TMAO)-Induced Impairment of Cardiomyocyte Function and the Protective Role of Urolithin B-Glucuronide. *Molecules* **2018**, *23*, 549. [CrossRef]
9. Arduini, A.; Zammit, V.A.; Bonomini, M. Identification of trimethylamine N-oxide (TMAO)-producer phenotype is interesting, but is it helpful? *Gut* **2019**. [CrossRef]
10. Nowiński, A.; Ufnal, M. Trimethylamine N-oxide: A harmful, protective or diagnostic marker in lifestyle diseases? *Nutrition* **2018**, *46*, 7–12. [CrossRef]

11. Dumas, M.-E.; Maibaum, E.C.; Teague, C.; Ueshima, H.; Zhou, B.; Lindon, J.C.; Nicholson, J.K.; Stamler, J.; Elliott, P.; Chan, Q.; et al. Assessment of analytical reproducibility of 1H NMR spectroscopy based metabonomics for large-scale epidemiological research: The INTERMAP Study. *Anal. Chem.* **2006**, *78*, 2199–2208. [CrossRef] [PubMed]

12. Collins, H.L.; Drazul-Schrader, D.; Sulpizio, A.C.; Koster, P.D.; Williamson, Y.; Adelman, S.J.; Owen, K.; Sanli, T.; Bellamine, A. L-Carnitine intake and high trimethylamine N-oxide plasma levels correlate with low aortic lesions in ApoE−/− transgenic mice expressing CETP. *Atherosclerosis* **2016**, *244*, 9–37. [CrossRef] [PubMed]

13. Lupachyk, S.; Watcho, P.; Stavniichuk, R.; Shevalye, H.; Obrosova, I.G. Endoplasmic Reticulum Stress Plays a Key Role in the Pathogenesis of Diabetic Peripheral Neuropathy. *Diabetes* **2013**, *62*, 944–952. [CrossRef] [PubMed]

14. Zeisel, S.H.; Warrier, M. Trimethylamine N-Oxide, the Microbiome, and Heart and Kidney Disease. *Annu. Rev. Nutr.* **2017**, *37*, 157–181. [CrossRef] [PubMed]

15. Sarvazyan, N. Visualization of doxorubicin-induced oxidative stress in isolated cardiac myocytes. *Am. J. Physiol.* **1996**, *271*, H2079–H2085. [CrossRef]

16. Zununi Vahed, S.; Barzegari, A.; Zuluaga, M.; Letoumeur, D.; Pavon-Djavid, G. Myocardial infarction and gut microbiota: An incidental connection. *Pharmacol. Res.* **2018**, *129*, 308–317. [CrossRef]

17. Cho, C.E.; Taesuwan, S.; Malysheva, O.V.; Bender, E.; Tulchinsky, N.F.; Yan, J.; Sutter, J.L.; Caudill, M.A. Trimethylamine-N-oxide (TMAO) response to animal source foods varies among healthy young men and is influenced by their gut microbiota composition: A randomized controlled trial. *Mol. Nutr. Food Res.* **2017**, *61*, 1600324. [CrossRef]

18. Ufnal, M.; Jazwiec, R.; Dadlez, M.; Drapala, A.; Sikora, M.; Skrzypecki, J. Trimethylamine-N-oxide: A carnitine-derived metabolite that prolongs the hypertensive effect of angiotensin II in rats. *Can. J. Cardiol.* **2014**, *30*, 1700–1705. [CrossRef]

19. Koeth, R.A.; Wang, Z.; Levison, B.S.; Buffa, J.A.; Org, E.; Sheehy, B.T.; Britt, E.B.; Fu, X.; Wu, Y.; Li, L.; et al. Intestinal microbiota metabolism of L-carnitine, a nutrient in red meat, promotes atherosclerosis. *Nat. Med.* **2013**, *19*, 576–585. [CrossRef]

20. Dambrova, M.; Latkovskis, G.; Kuka, J.; Strele, I.; Konrade, I.; Grinberga, S.; Hartmane, D.; Pugovics, O.; Erglis, A.; Liepinsh, E. Diabetes is Associated with Higher Trimethylamine N-oxide Plasma Levels. *Exp. Clin. Endocrinol. Diabetes* **2016**, *124*, 251–256. [CrossRef]

21. Mueller, D.M.; Allenspach, M.; Othman, A.; Saely, C.H.; Muendlein, A.; Vonbank, A.; Drexel, H.; von Eckardstein, A. Plasma levels of trimethylamine-N-oxide are confounded by impaired kidney function and poor metabolic control. *Atherosclerosis* **2015**, *243*, 638–644. [CrossRef] [PubMed]

22. Yin, J.; Liao, S.-X.; He, Y.; Wang, S.; Xia, G.-H.; Liu, F.-T.; Zhu, J.-J.; You, C.; Chen, Q.; Zhou, L.; et al. Dysbiosis of Gut Microbiota with Reduced Trimethylamine-N-Oxide Level in Patients With Large-Artery Atherosclerotic Stroke or Transient Ischemic Attack. *J. Am. Heart Assoc.* **2015**, *4*, e002699. [CrossRef] [PubMed]

23. Huc, T.; Drapala, A.; Gawrys, M.; Konop, M.; Bielinska, K.; Zaorska, E.; Samborowska, E.; Wyczalkowska-Tomasik, A.; Paczek, L.; Dadlez, M.; et al. Chronic, low-dose TMAO treatment reduces diastolic dysfunction and heart fibrosis in hypertensive rats. *Am. J. Physiol. Heart Circ. Physiol.* **2018**, *315*, H1805–H1820. [CrossRef] [PubMed]

24. Gallo, M.P.; Femminò, S.; Antoniotti, S.; Querio, G.; Alloatti, G.; Levi, R. Catestatin induces glucose uptake and GLUT4 trafficking in adult rat cardiomyocytes. *Biomed. Res. Int.* **2018**. [CrossRef] [PubMed]

25. Lenky, C.C.; McEntyre, C.J.; Lever, M. Measurement of marine osmolytes in mammalian serum by liquid chromatography-tandem mass spectrometry. *Anal. Biochem.* **2012**, *420*, 7–12. [CrossRef] [PubMed]

26. Somero, G.N. From dogfish to dogs: Trimethylamines protect proteins from urea. *Am. J. Physiol.* **1986**, *1*, 9–12. [CrossRef]

27. Lagranha, C.L.; Deschamps, A.; Aponte, A.; Steenbergen, C.; Murphy, E. Sex differences in the phosphorylation of mitochondrial proteins result in reduced production of reactive oxygen species and cardioprotection on females. *Circ. Res.* **2010**, *106*, 1981–1991. [CrossRef] [PubMed]

International Journal of
Molecular Sciences

Article

Carbon Monoxide Releasing Molecule-2-Upregulated ROS-Dependent Heme Oxygenase-1 Axis Suppresses Lipopolysaccharide-Induced Airway Inflammation

Chih-Chung Lin [1], Li-Der Hsiao [1], Rou-Ling Cho [2] and Chuen-Mao Yang [1,2,3,*]

[1] Department of Anesthetics, Chang Gung Memorial Hospital at Linkuo, and College of Medicine, Chang Gung University, Kwei-San, Tao-Yuan 33302, Taiwan

[2] Department of Physiology and Pharmacology and Health Aging Research Center, College of Medicine, Chang Gung University, 259 Wen-Hwa 1 Road, Kwei-San, Tao-Yuan 33302, Taiwan

[3] Research Center for Chinese Herbal Medicine and Research Center for Food and Cosmetic Safety, College of Human Ecology, Chang Gung University of Science and Technology, Tao-Yuan 33302, Taiwan

[*] Correspondence: chuenmao@mail.cgu.edu.tw or chuenmao@gmail.com; Tel.: +886-3-211-8800 (ext. 5123); Fax:+886-3-211-8365

Received: 3 June 2019; Accepted: 26 June 2019; Published: 28 June 2019

Abstract: The up-regulation of heme oxygenase-1 (HO-1) is mediated through nicotinamaide adenine dinucleotide phosphate (NADPH) oxidases (Nox) and reactive oxygen species (ROS) generation, which could provide cytoprotection against inflammation. However, the molecular mechanisms of carbon monoxide-releasing molecule (CORM)-2-induced HO-1 expression in human tracheal smooth muscle cells (HTSMCs) remain unknown. Here, we found that pretreatment with CORM-2 attenuated the lipopolysaccharide (LPS)-induced intercellular adhesion molecule (ICAM-1) expression and leukocyte count through the up-regulation of HO-1 in mice, which was revealed by immunohistochemistrical staining, Western blot, real-time PCR, and cell count. The inhibitory effects of HO-1 by CORM-2 were reversed by transfection with HO-1 siRNA. Next, Western blot, real-time PCR, and promoter activity assay were performed to examine the HO-1 induction in HTSMCs. We found that CORM-2 induced HO-1 expression via the activation of protein kinase C (PKC)α and proline-rich tyrosine kinase (Pyk2), which was mediated through Nox-derived ROS generation using pharmacological inhibitors or small interfering ribonucleic acids (siRNAs). CORM-2-induced HO-1 expression was mediated through Nox-(1, 2, 4) or p47^{phox}, which was confirmed by transfection with their own siRNAs. The Nox-derived ROS signals promoted the activities of extracellular signal-regulated kinase 1/2 (ERK1/2). Subsequently, c-Fos and c-Jun—activator protein-1 (AP-1) subunits—were up-regulated by activated ERK1/2, which turned on transcription of the HO-1 gene by regulating the HO-1 promoter. These results suggested that in HTSMCs, CORM-2 activates PKCα/Pyk2-dependent Nox/ROS/ERK1/2/AP-1, leading to HO-1 up-regulation, which suppresses the lipopolysaccharide (LPS)-induced airway inflammation.

Keywords: CORM-2; NADPH oxidase; ROS; AP-1; HO-1

---

1. Introduction

Heme oxygenase (HO), a rate-limiting enzyme, metabolizes heme to biliverdin-IXα, ferrous iron, and carbon monoxide (CO). These secondary products are involved in the regulation of different physiological processes. HO-1, a member of the HO family [1,2], is inducible and directly protects various organs from oxidative damages [1,3]. Biliverdin-IXα is converted to the endogenous radical scavenger bilirubin-IXα and has anti-inflammatory properties [4,5]. The ferrous iron is rapidly sequestered with ferritin to function additional antioxidant and anti-apoptotic effects [5,6]. Moreover,

CO has been shown to exert anti-apoptotic and anti-inflammatory effects that are mediated via HO-1 up-regulation [4,7,8]. However, several pro-inflammatory cytokines and oxidative stresses can also trigger HO-1 expression [9–11]. For example, HO-1 is also induced by various factors in the airway cells of asthmatic patients [12]. Accumulating evidence concerning HO-1/CO-dependent cytoprotection elicits the mechanisms involved in the modulation of the inflammatory responses, including the down-regulation of pro-inflammatory mediators, atherosclerosis, ischemia-reperfusion systems, and airway disorders [8,13,14]

Nicotinamaide adenine dinucleotide phosphate (NADPH) oxidase (Nox)-derived reactive oxygen species (ROS) generation has been approved to regulate either the expression of inflammatory or anti-inflammatory mediators in the airway and pulmonary diseases [15]. Excessive ROS production can regulate the expression of various inflammatory genes during airway disorders [15,16]. In contrast, low levels of ROS contribute to maintain cellular redox homeostasis under physiological conditions [15]. Several studies indicate that the exogenous application of CO and HO-1 can protect against oxidative stress and hyperoxic injury in the various organs [17,18]. It has also been reported that the up-regulation of HO-1 via the Nox/ROS formation is induced by lipopolysaccharide (LPS) or cytokines [8,19]. In addition, our previous reports have indicated that the Nox/ROS system is a key player for HO-1 expression induced by lipotechoic acid (LTA) and cigarette smoke particle extract (CSPE) in human tracheal smooth muscle cells (HTSMCs) [5,20]. Our previous studies and others have demonstrated that the carbon monoxide-releasing molecule (CORM)-2 mediates the Nox-dependent ROS generation in astrocytes [21,22] and human bronchial smooth muscle cells [23]. CORMs have been confirmed to provide the exogenous CO source and induce HO-1 expression in several cell types [8,9,24]. However, the roles of Nox/ROS involved in CORM-2-induced HO-1 expression were still unknown. HO-1 expression is regulated by various intracellular signaling pathways, such as ROS, growth factor receptors, non-receptor tyrosine kinases such as Pyk2, or mitogen-activated protein kinases (MAPKs) [25,26]. In our previous study, CORM-2 has been shown to induce HO-1 expression via c-Src/epidermal growth factor receptor (EGFR)/phosphoinositide 3-kinase (PI3K)/Akt/c-Jun N-terminal kinase 1/2 (JNK1/2) and p38 MAPK pathways in HTSMCs [27]. We also noticed that many stress-activated response elements on the upstream region of the HO-1 promoter, such as nuclear factor erythroid 2-related factor 2 (Nrf2) and activator protein 1 (AP-1), are involved in the expression of HO-1 in response to oxidative stresses [28,29]. Therefore, whether an alternative Nox/ROS-mediated pathway involved in HO-1 expression induced by CORM-2 has yet to be investigated in HTSMCs.

Besides, several reports have shown that administration with low concentrations of CO or pharmacological application of CORMs can also confer protective effects in the models of inflammatory responses and tissue injury [26,30,31]. Accumulating evidence has indicated that CORM-2-liberated CO reduces inflammatory responses in sepsis by interfering with nuclear factor (NF)-κB activation [32]. Moreover, the overexpression of HO-1 by cobalt protoporphyrin (CoPPIX) can reduce tumor necrosis factor (TNF) α-induced oxidative stress and airway inflammation [7]. Therefore, CORM-2-induced HO-1 gene expression could prevent inflammatory responses. However, the detailed mechanisms by which CORM-2 induced HO-1 expression in HTSMCs are still unclear. Our results demonstrated that CORM-2-induced HO-1 expression is mediated via protein kinase C (PKC) α/proline-rich tyrosine kinase 2 (Pyk2)-dependent Nox/ROS generation linking to the ERK1/2-mediated activation of AP-1 in HTSMCs, which could protect against the lipopolysaccharide (LPS)-induced airway inflammatory diseases.

2. Results

2.1. CORM-2 Inhibits LPS-Induced Lung Inflammation in Mice

CORM-2 has been shown to protect against inflammatory responses induced by various insults [4,7,8]. First, we investigated the anti-inflammatory effects of CORM-2 on LPS-induced lung inflammation. We observed that LPS markedly induced intercellular adhesion molecule-1

(ICAM-1) expression, which was attenuated by CORM-2 via HO-1 expression, as determined by immunohistochemistry staining (Figure 1A). Bronchoalveolar lavage (BAL) fluid was collected to determine the number of leukocytes, and airway tissues were harvested to study the levels of protein and mRNA expression. As shown in Figure 1B,D, LPS significantly enhanced ICAM-1 protein expression and leukocytes count in BAL fluid. LPS also significantly induced ICAM-1 messenger ribonucleic acid (mRNA) expression (Figure 1C). In addition, pretreatment with CORM-2 also inhibited the LPS-induced ICAM-1 protein expression in HTSMCs (Figure 1E). Further, we confirmed the inhibitory effects of HO-1 induction by CORM-2 on the LPS-up-regulated ICAM-1 expression by transfection with HO-1 siRNA. We found that CORM-2 attenuated the LPS-induced ICAM-1 expression, which was partially reversed by transfection with HO-1 siRNA (Figure 1F). All of these LPS-mediated responses were attenuated by CORM-2 via the up-regulation of HO-1 in the airway tissues of mice (Figure 1). These results suggested that the up-regulation of HO-1 by CORM-2 protects airway tissues against the LPS-mediated inflammatory responses.

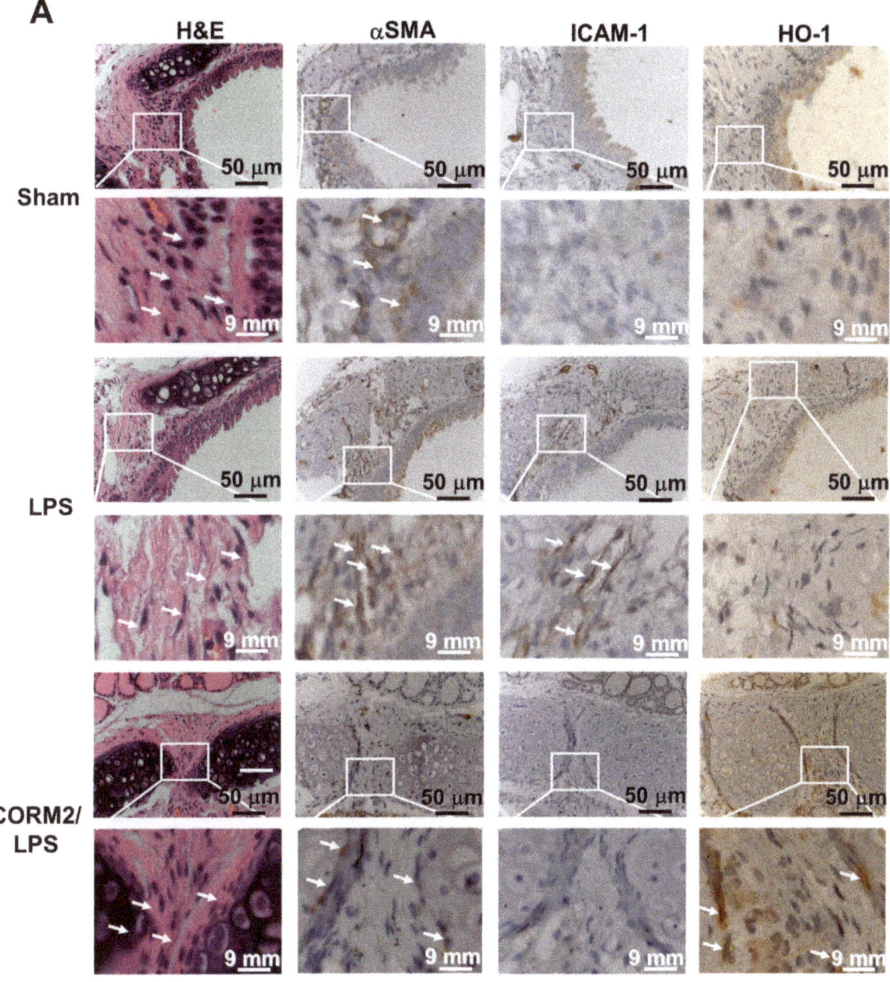

Figure 1. *Cont.*

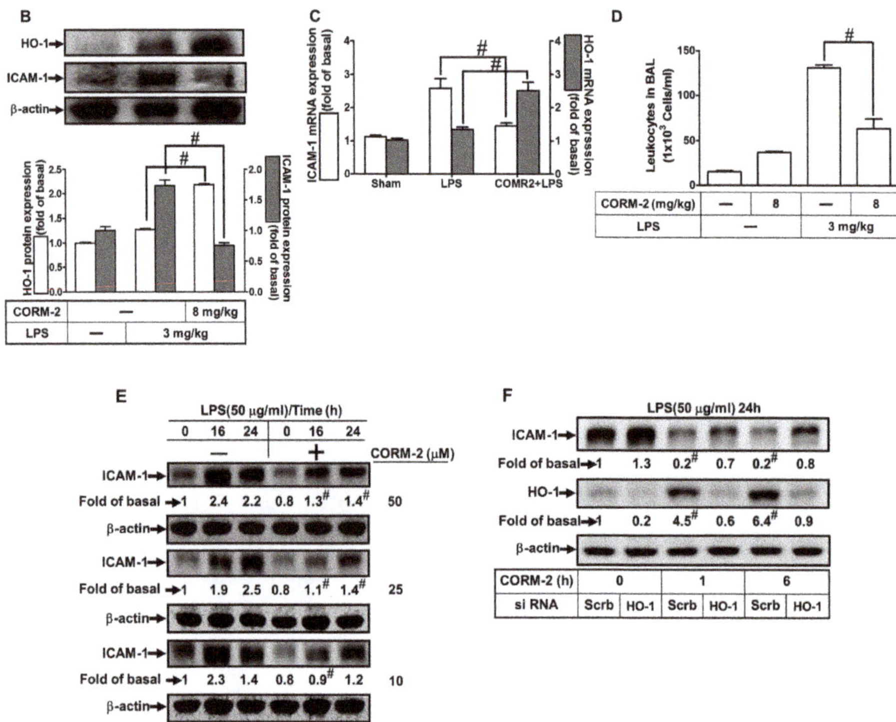

Figure 1. Induction of heme oxygenase (HO-1) by carbon monoxide-releasing molecule (CORM-2) suppresses intercellular adhesion molecule-1 (ICAM-1) expression and leukocyte infiltration in HTSMCs and mice. Institute of Cancer Research (ICR) mice were pretreated with CORM-2 (8 mg/kg of body weight) for 24 h, and then treated with lipopolysaccharide (LPS) (3 mg/kg of body weight). (**A**) H (hematoxylin) & E (Eosin) and immunohistochemical staining for α-smooth muscle actin (α-SMA), ICAM-1, and HO-1 in serial section of the airway tissues from sham [0.1 mL of DMSO–PBS (phosphate-buffered saline) (1:100) with 0.1% (*w/v*) BSA (bovine serum albumin) treated mice], LPS (LPS-injected mice) and CORM-2 + LPS mice. The arrows indicate tracheal smooth muscle cells displayed with ICAM-1 and HO-1 expression. (**B,C**) Airway tissues were homogenized to extract proteins and mRNAs (messenger ribonucleic acids), and analyzed by (**B**) Western blot and (**C**) real-time PCR to determine the levels of HO-1, ICAM-1, and β-actin (served as an internal control) protein and mRNA expression, respectively. (**D**) BAL fluid was collected to count the number of leukocytes infiltration. (**E**) Human tracheal smooth muscle cells (HTSMCs) were pretreated without or with various concentrations of CORM-2 for 1 h and then incubated with LPS (50 μg/mL) for the indicated time periods. The levels of ICAM-1 and β-actin were determined by Western blot. (**F**) Cells were transfected with scrambled (Scrb) or HO-1 siRNA (small interfering ribonucleic acid), incubated with CORM-2 (50 μM) for 1 or 6 h, and then stimulated with LPS (50 μg/mL) for 16 h. The levels of ICAM-1, HO-1, and β-actin protein were determined by western blot. Data are expressed as mean ± SEM of five independent experiments (*n* = 5). [#] *p* < 0.05, as compared with the mice exposed to the indicated reagents. Data analysis and processing are described in the section "Statistical Analysis of Data".

2.2. ROS Participate in CORM-2-Induced HO-1 Expression

Low levels of ROS have been shown to contribute to maintain cellular redox homeostasis and protect cells against oxidative stress through the up-regulation of HO-1 [8,19]. To examine whether ROS participate in HO-1 induction in HTSMCs, a ROS scavenger N-acetyl cysteine (NAC) was used for this purpose. We found that pretreatment with NAC concentration-dependently attenuated the

CORM-2-induced both of HO-1 protein (Figure 2A) and mRNA expression (Figure 2B), suggesting the involvement of ROS in the CORM-2-induced HO-1 expression. Next, we evaluated whether CORM-2 stimulated ROS generation and the scavenging efficacy of NAC. Our results indicated that CORM-2 stimulated ROS generation in a time-dependent manner with a maximal response within 4 h (Figure 2C), which was markedly attenuated by pretreatment with 10 mM of NAC (Figure 2D), indicating that NAC can efficiently scavenge ROS in HTSMCs. These results were further supported by the data of 2′,7′-chloromethyl 2′,7′-dichloro fluorescein diacetate (CMH2DCF-DA; for H_2O_2) and dihydroethidium (DHE; for O_2^-) fluorescence images observed under a fluorescent microscope (Figure 2E). Pretreatment of HTSMCs with NAC (10 mM) significantly reduced CORM-2-stimulated H_2O_2 and O_2^- generation. In addition, treatment with an inactive form of CORM-2 [iCORM-2, ruthenium (III) chloride, RuCl3] [33] failed to induce HO-1 expression (Figure 2F). These results concluded that ROS generation stimulated by CORM-2 contributes to up-regulation of HO-1 in HTSMCs.

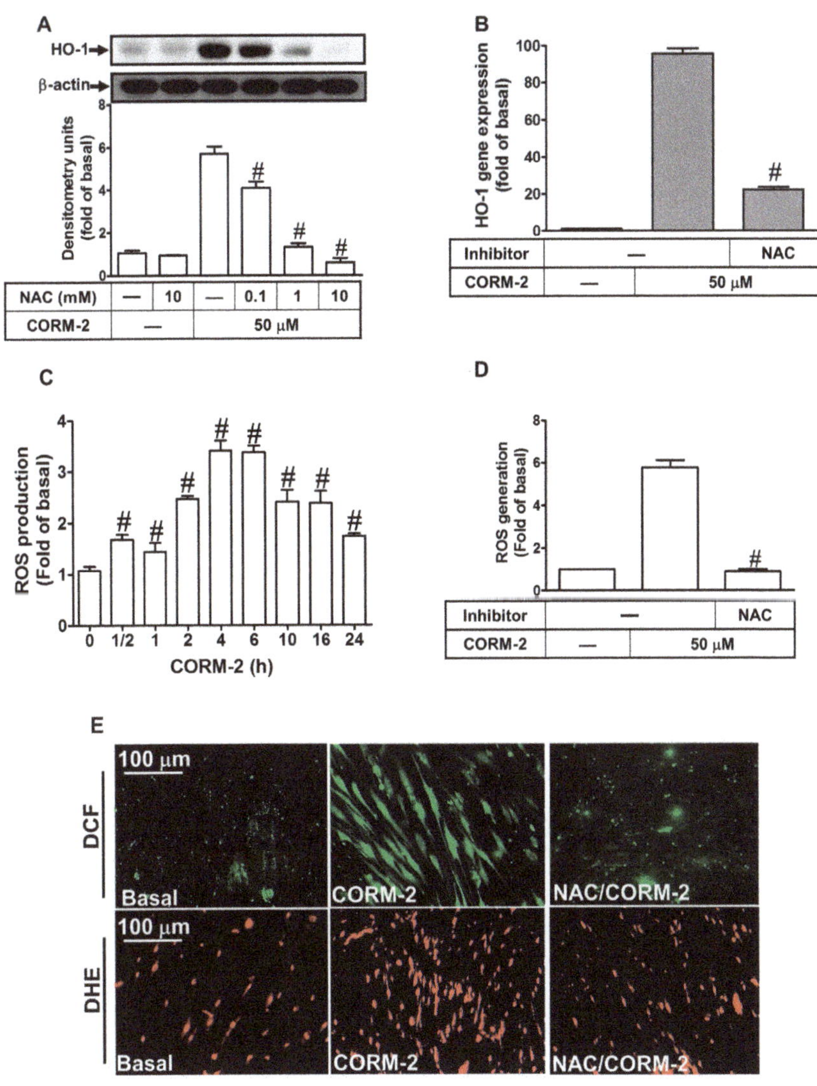

Figure 2. *Cont.*

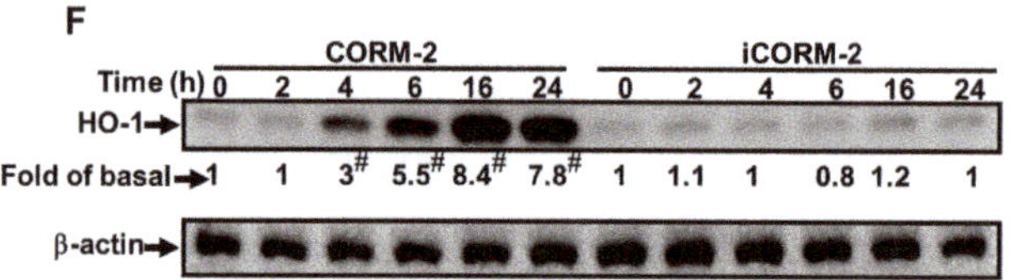

Figure 2. Reactive oxygen species (ROS) are required for CORM-2-induced HO-1 expression in HTSMCs. (**A**) Cells were pretreated with N-acetyl cysteine (NAC) for 1 h and then incubated with CORM-2 for 24 h. The protein levels of HO-1 and β-actin (served as an internal control) were determined by Western blot. (**B**) Cells were pretreated with NAC (10 mM) for 1 h, and then incubated with CORM-2 for 6 h. The mRNA expression of HO-1 was determined by real-time PCR. (**C**) Cells were incubated with the 2′,7′-dichlorofluorescin diacetate (DCF-DA) (5 µM) for 45 min, followed by stimulation with 50 µM of CORM-2 for the indicated time intervals. (**D**) Cells were pretreated without or with NAC (10 mM) for 1 h before exposure to CORM-2 for 6 h. (**C,D**) The fluorescence intensity of cells was determined. (**E**) DCF-DA, and dihydroethidium (DHE) staining, cells were treated with CORM-2 for 6 h in the absence or presence of NAC (10 mM). The fluorescence images were observed by a fluorescence microscope. Image of fluorescence microscope, 400×. (**F**) Cells were treated with 50 µM of CORM-2 or iCORM-2 for the indicated time intervals. The protein levels of HO-1 were determined by Western blot. Data are expressed as mean ± SEM of five independent experiments ($n = 5$). $^{\#} p < 0.05$, as compared with the cells exposed to vehicle (**C,F**) or CORM-2 (**A,B,D**) alone. Data analysis and processing are described in the section "Statistical Analysis of Data".

2.3. Nox-Derived ROS Generation Contributes to CORM-2-Induced HO-1 Expression

CORM-2 mediates the Nox-dependent ROS generation, leading to HO-1 expression [8,9,24]. Thus, the roles of Noxs in the ROS-dependent HO-1 expression were investigated. The inhibitors of Nox (diphenyleneiodonium, DPI) and p47^{phox} (apocynin, APO) were used to investigate whether Noxs mediated ROS-dependent HO-1 expression in CORM-2-treated HTSMCs. As shown in Figure 3A,B, pretreatment with either DPI or APO significantly attenuated CORM-2-induced both HO-1 protein and mRNA expression. To further investigate whether CORM-2 stimulates Nox activity, as shown in Figure 3C, CORM-2 time-dependently stimulated Nox activity, significantly increased within 30 min and sustained up to 24 h, which was blocked by pretreatment with either DPI (10 µM) or APO (100 µM) (Figure 3D, grey bars), accompanied with inhibiting the ROS generation induced by CORM-2 (Figure 3D, open bars). These results were further supported by the data of DCF (dichlorofluorescein) (for H_2O_2) and DHE (for O_2^{-}) fluorescence images observed under a fluorescent microscope (Figure 3F), suggesting that CORM-2-stimulated Nox-derived ROS generation contributes to HO-1 expression. Moreover, we found that Nox/ROS generation stimulated by CORM-2 was mediated via Nox(1,2,4) or p47^{phox}, which was confirmed by using their own siRNAs in HTSMCs (Figure 3G).

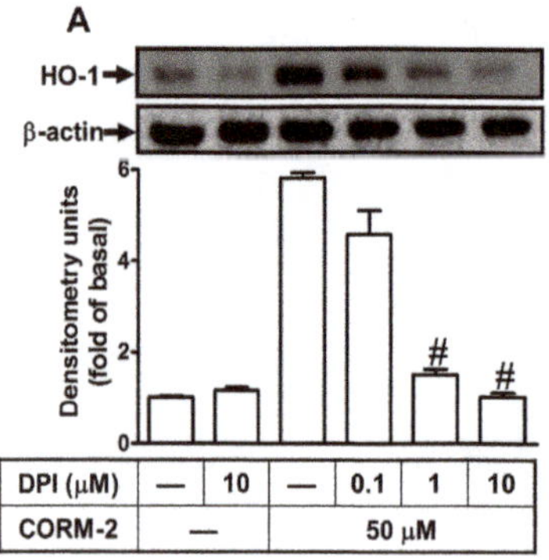
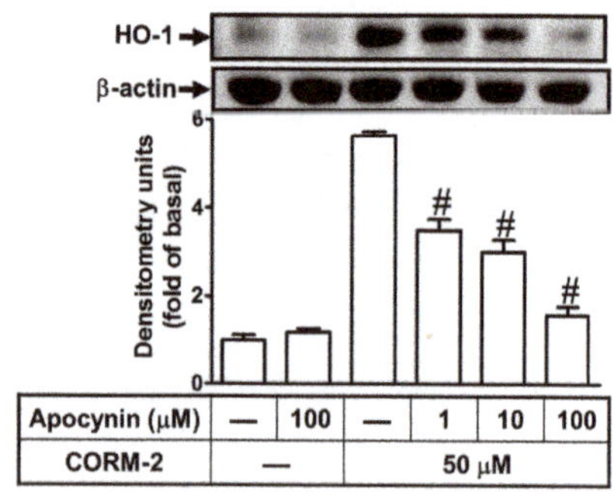

Figure 3. *Cont.*

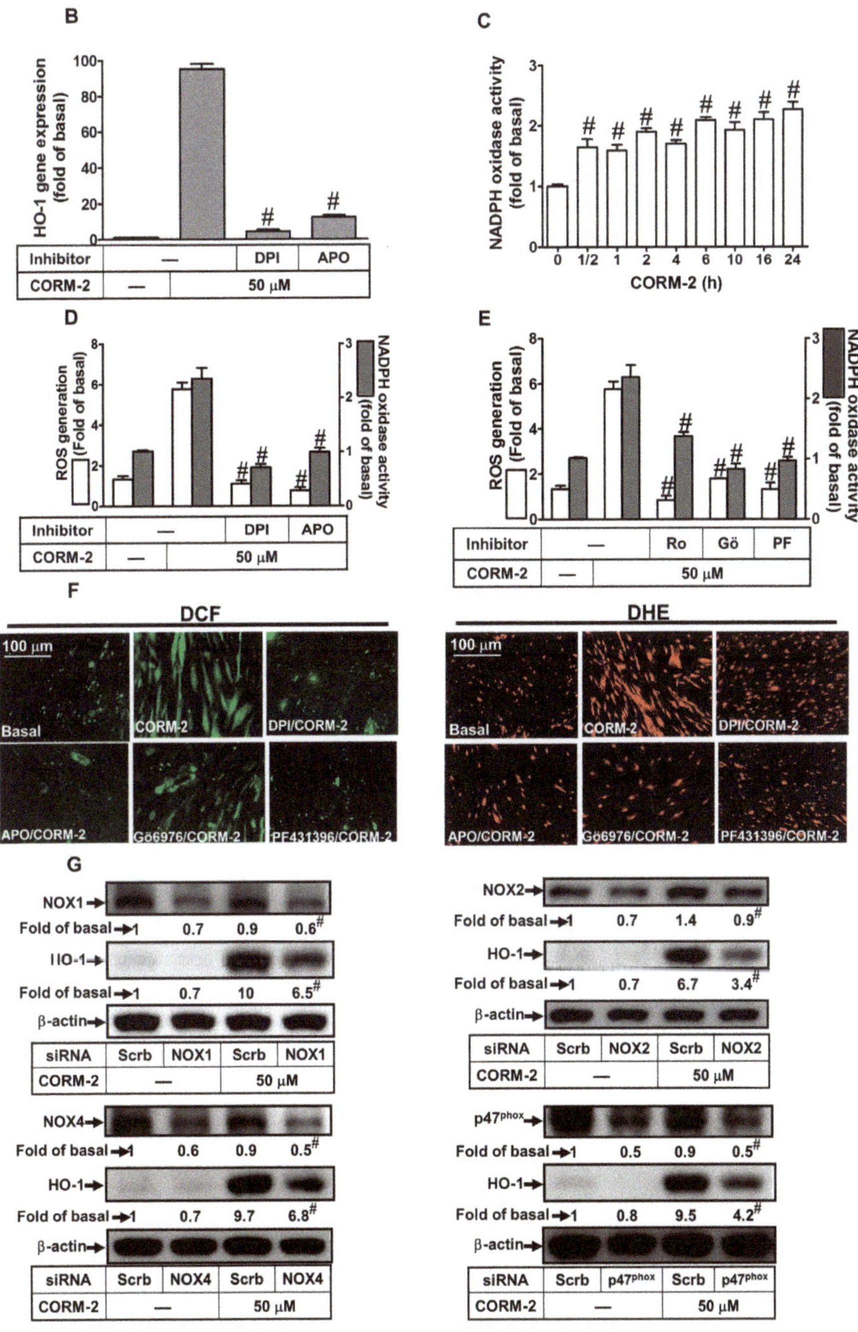

Figure 3. Nicotinamaide adenine dinucleotide phosphate (NADPH) oxidase-dependent ROS generation contributes to CORM-2-induced HO-1 expression in HTSMCs. (**A**) Cells were pretreated without or with diphenyleneiodonium (DPI) or apocynin (APO) for 1 h before exposure to CORM-2 for 24 h.

The protein levels of HO-1 and β-actin (served as an internal control) were determined by Western blot. (**B**) Cells were pretreated with DPI (10 μM) or APO (100 μM) for 1 h, and then incubated with CORM-2 for 6 h. The mRNA expression of HO-1 was determined by real-time PCR. (**C**) Cells were incubated with CORM-2 (50 μM) for the indicated time intervals. The Nox activity was analyzed. (**D,E**) Cells were pretreated without or with (**D**) DPI (10 μM) or APO (100 μM), and (**E**) Ro31-8220 (10 μM), Gö6976 (10 μM), or PF431396 (10 μM) for 1 h, and then incubated with CORM-2 for 6 h. The Nox activity and ROS generation were analyzed. (**F**) CMH2, DCF-DA, and DHE staining, cells were treated with CORM-2 for 6 h in the absence or presence of DPI (10 μM), APO (100 μM), Gö6976 (10 μM), or PF431396 (10 μM). The fluorescence images were detected by a fluorescence microscope. Image of fluorescence microscope, 400×. (**G**) Cells were transfected with either scrambled (Scrb), Nox-(1,2,4), or p47[phox] siRNA, and then incubated with CORM-2 for 24 h. The levels of Nox-(1,2,4), p47[phox], HO-1, and β-actin (served as an internal control) protein were determined by Western blot. Data are expressed as mean ± SEM of five independent experiments ($n = 5$). [#] $p < 0.05$, as compared with the cells exposed to vehicle (**C,G**) or CORM-2 alone (**A,B,D,E**). Data analysis and processing are described in the section "Statistical Analysis of Data".

2.4. CORM-2 Induces HO-1 Expression via PKCα

Our previous report and the others also indicate that PKCs are involved in HO-1 expression in brain astrocytes [25]. Thus, we investigated whether PKC members are involved in the CORM-2-induced HO-1 expression. We found that pretreatment of HTSMCs with Ro31-8220 (a pan-PKC inhibitor) concentration-dependently attenuated the HO-1 induction by CORM-2 (Figure 4A). Next, to determine which PKC isoforms, PKCα especially, mediate CORM-2-induced HO-1 expression, two selective PKCα inhibitors (Gö6976 and Gö6983) were used for these purposes. The usage of Gö6983, an ATP-competitive bisindolylmaleimide PKC inhibitor, blocks the PKC phosphorylation [34]. Our previous report also indicates that Gö6983 inhibits PKCα/βII phosphorylation stimulated by thrombin in SK–N–SH (human neuroblastoma) cells [35]. We found that pretreatment with either Gö6976 or Gö6983 concentration-dependently blocked CORM-2-induced HO-1 expression in HTSMCs (Figure 4A), indicating that PKCα was involved in CORM-2-induced HO-1 expression. Moreover, pretreatment with Ro31-8220 (10 μM), Gö6976 (10 μM), or Gö6983 (10 μM) significantly inhibited CORM-2-induced HO-1 mRNA expression (Figure 4B). The role of PKCα in CORM-2-induced HO-1 expression was further confirmed by transfection with PKCα siRNA, which significantly knocked down PKCα protein and blocked the CORM-2-induced HO-1 expression (Figure 4C). In addition, the activation of PKCα/βII in CORM-2-induced responses was confirmed by determining their phosphorylation. As shown in Figure 4D, CORM-2 time-dependently stimulated PKCα/βII phosphorylation with a maximal response within 2–4 h, which was attenuated by pretreatment with Gö6983 (10 μM), but not by pretreatment with APO, DPI, or NAC, indicating that PKCα/βII are the upstream components of Nox/ROS in HO-1 expression. This note was also supported by the results that pretreatment with either Ro31-8220 or Gö6976 inhibited the CORM-2-stimulated Nox activity and ROS generation (Figure 3E,F). These results suggested that HO-1 expression induced by CORM-2 is mediated via PKCα/βII-dependent Nox activation and ROS generation in HTSMCs.

2.5. Involvement of Pyk2 in CORM-2-Induced HO-1 Expression

Previous reports have indicated that PF431396 treatment attenuates proline-rich tyrosine kinase 2 (Pyk2) phosphorylation at Tyr (tyrosine)[402] in various types of cells [36,37]. To investigate the role of Pyk2 in HO-1 expression, PF431396 (a Pyk2 inhibitor) was used. In this study, we found that pretreatment with PF431396 concentration-dependently attenuated CORM-2-induced both of HO-1 protein (Figure 5A) and mRNA (Figure 5B) expression. To confirm the role of Pyk2 in CORM-2-induced HO-1 expression, Pyk2 siRNA transfection significantly knocked down the Pyk2 protein level and inhibited CORM-2-induced HO-1 expression (Figure 5C). We further determined whether CORM-2 stimulated activation of Pyk2; the phosphorylation of Pyk2 was detected by Western blot. As shown in Figure 5D, CORM-2 time-dependently stimulated Pyk2 phosphorylation, which was attenuated by pretreatment of PF431396 (10 µM), but not by APO, DPI, or NAC, indicating that Pyk2 is an upstream component of Nox/ROS in HO-1 expression. This note was also supported by the results that PF431396 attenuated the CORM-2-triggered Nox activity and ROS generation (Figure 3E,F). These findings indicated that HO-1 expression by CORM-2 is mediated via a Pyk2-dependent Nox/ROS activity in HTSMCs.

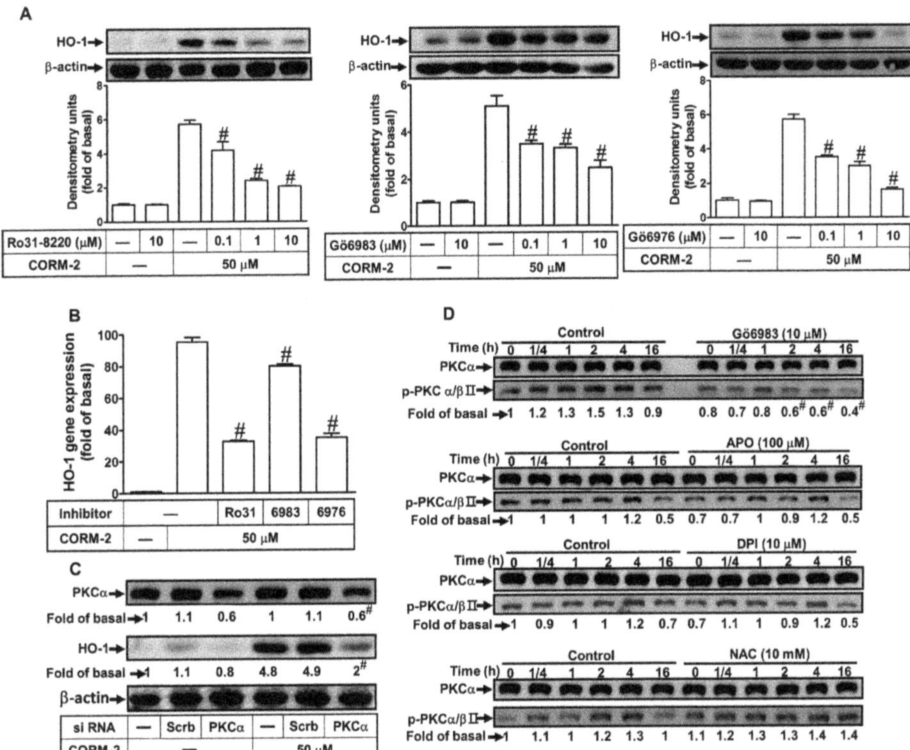

Figure 4. CORM-2 induces HO-1 expression via a PKCα-dependent pathway in HTSMCs. (**A**) Cells were pretreated with Ro31-8220, Gö6983, or Gö6976 for 1 h, and then incubated with CORM-2 (50 µM) for 24 h. The protein levels of HO-1 and β-actin (served as an internal control) were determined by Western blot. (**B**) Cells were pretreated with Ro31-8220 (10 µM), Gö6983 (10 µM), or Gö6976 (10 µM) for 1 h, and then incubated with CORM-2 for 6 h. The mRNA expression of HO-1 was determined by real-time PCR. (**C**) Cells were transfected with either scrambled (Scrb) or protein kinase C (PKCα) siRNA, and then incubated with CORM-2 for 24 h. The levels of PKCα, HO-1, and β-actin (served as an internal control) protein were determined by Western blot. (**D**) Cells were pretreated without or with Gö6983 (10 µM), APO (100 µM), DPI (10 µM), or NAC (10 mM) for 1 h, and then incubated with CORM-2 for the indicated time intervals. The levels of phospho-PKCα and PKCα were determined by Western blot. Data are expressed as mean ± SEM of five independent experiments ($n = 5$). $^{\#} p < 0.05$, as compared with the cells exposed to CORM-2 alone. Data analysis and processing are described in the section "Statistical Analysis of Data".

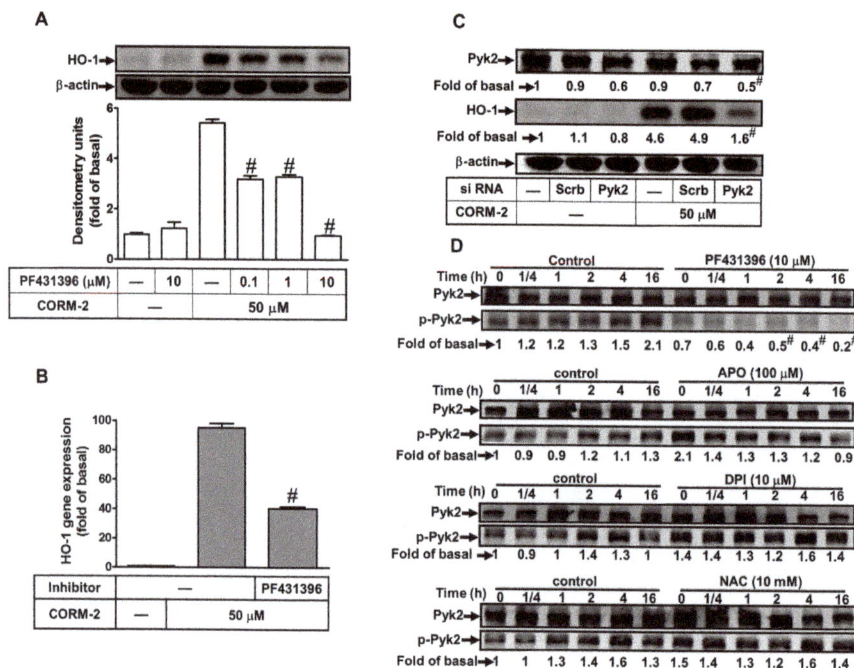

Figure 5. Proline-rich tyrosine kinase 2 (Pyk2) is involved in CORM-2-induced HO-1 expression in HTSMCs. (**A**) Cells were pretreated with PF431396 for 1 h, and then incubated with CORM-2 for 24 h. The protein levels of HO-1 and β-actin (served as an internal control) were determined by western blot. (**B**) Cells were pretreated with PF431396 (10 μM) for 1 h, and then incubated with CORM-2 for 6 h. The mRNA expression of HO-1 was determined by real-time PCR. (**C**) Cells were transfected with either scrambled (Scrb) or Pyk2 siRNA, and then incubated with CORM-2 for 24 h. The levels of Pyk2, HO-1, and β-actin (served as an internal control) protein were determined by Western blot. (**D**) Cells were pretreated without or with PF431396 (10 μM), APO (100 μM), DPI (10 μM), or NAC (10 mM) for 1 h, and then incubated with CORM-2 for the indicated time intervals. The levels of phospho-Pyk2 and Pyk2 (served as an internal control) were determined by Western blot. Data are expressed as mean $\pm$ SEM of five independent experiments (n = 5). $^{\#}$ $p < 0.05$, as compared with the cells exposed to CORM-2 alone. Data analysis and processing are described in the section "Statistical Analysis of Data".

2.6. Involvement of ERK1/2 in CORM-2-Induced HO-1 Expression

We have previously demonstrated that MAPKs participate in the HO-1 induction by LTA and CSPE in HTSMCs [20]. Therefore, we further approached the roles of p42/p44 MAPK in CORM-2-induced HO-1 expression in these cells. We found that pretreatment with a MEK1/2 inhibitor (U0126) significantly blocked HO-1 protein and mRNA expressions (Figure 6A,B), which were both CORM-2-induced. To further ensure the role of p42/p44 MAPK in CORM-2-induced HO-1 expression, transfection with p44 siRNA markedly knocked down the p44 protein level and blocked the CORM-2-induced HO-1 expression (Figure 6C). To confirm whether p42/p44 MAPK phosphorylation is necessary for CORM-2-induced HO-1 expression, activation of the kinases was assayed by Western blot using an antibody specific for the phosphorylated form of p42/p44 MAPK. As shown in Figure 6D, CORM-2 stimulated a time-dependent phosphorylation of p42/p44 MAPK, which was inhibited by pretreatment with U0126 (10 μM) during the period of observation. Further, pretreatment with Gö6983, PF431396, NAC, DPI, or APO significantly attenuated CORM-2-stimulated p42/p44 MAPK phosphorylation (Figure 5D), indicating that p42/p44 MAPK was a downstream component of PKCα, which is a

Pyk2-mediated Nox/ROS generation pathway. Our findings demonstrated that CORM-2-induced HO-1 expression is mediated through the activation of the PKCα, Pyk2/Nox/ROS/p42/p44 MAPK pathway in HTSMCs.

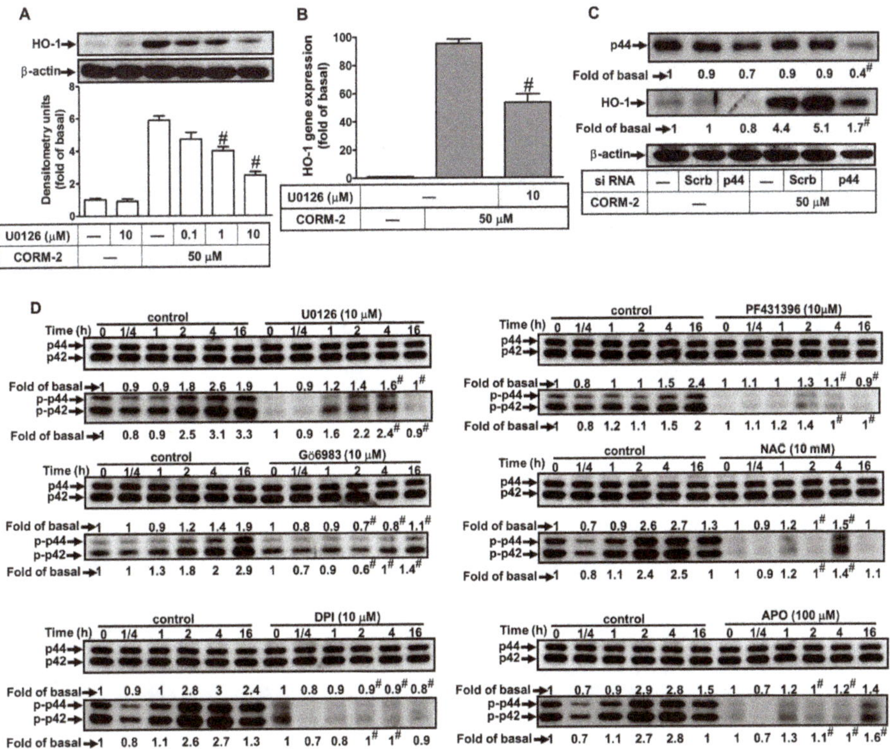

Figure 6. CORM-2-induced HO-1 expression is mediated through extracellular signal-regulated kinase 1/2 (ERK1/2) in HTSMCs. (**A**) Cells were pretreated with U0126 for 1 h, and then incubated with CORM-2 for 24 h. The protein levels of HO-1 and β-actin (served as an internal control) were determined by Western blot. (**B**) Cells were pretreated with U0126 (10 μM) for 1 h, and then incubated with CORM-2 for 6 h. The mRNA expression of HO-1 was determined by real-time PCR. (**C**) Cells were transfected with either scrambled (Scrb) or p44 siRNA, and then incubated with CORM-2 for 24 h. The levels of p44 and HO-1 protein were determined by Western blot. (**D**) Cells were pretreated with or without U0126 (10 μM), Gö6983 (10 μM), PF431396 (10 μM), NAC (10 mM), DPI (10 μM), or APO (100 μM), and then incubated with CORM-2 for the indicated time intervals. The levels of phospho-p42/p44 MAPK and p42/p44 MAPK (served as an internal control) were determined by Western blot. Data are expressed as mean ± SEM of five independent experiments ($n = 5$). $^{\#} p < 0.05$, as compared with the cells exposed to CORM-2 alone. Data analysis and processing are described in the section "Statistical Analysis of Data".

2.7. Induction of c-Fos and c-Jun/AP-1 is Required for CORM-2-Induced HO-1 Expression

Moreover, AP-1 has been shown to regulate HO-1 expression through binding to respective elements in the promoter region in response to oxidative stress [38,39]. Hence, we determined whether CORM-2-induced HO-1 expression was mediated via AP-1 using its inhibitor tanshinone IIA (TSIIA). As shown in Figure 7A,B, pretreatment with TSIIA attenuated CORM-2-induced HO-1 protein and mRNA expression, suggesting that in HTSMCs, activated AP-1 is an important event for CORM-2-induced HO-1 expression. To further approach the roles of c-Fos and c-Jun in CORM-2-induced HO-1 expression, transfection with either c-Fos or c-Jun siRNA significantly knocked down the c-Fos or c-Jun protein expression, respectively, and attenuated HO-1 induction in CORM-2-treated HTSMCs (Figure 7C). To study whether CORM-2 accelerated AP-1 transcription activity, a reporter plasmid construct containing the AP-1 response element was used to evaluate AP-1 activity in HTSMCs. We found that CORM-2 stimulated a time-dependent AP-1 promoter activity (Figure 7D), which was inhibited by pretreatment with Gö6983, PF431396, NAC, DPI, APO, U0126, or TSIIA (Figure 7E). These results suggested that HO-1 induction is mediated via PKCα, Pyk2/Nox/ROS/p42/p44 MAPK-dependent activation of AP-1(c-Fos/c-Jun) in CORM-2-treated HTSMCs.

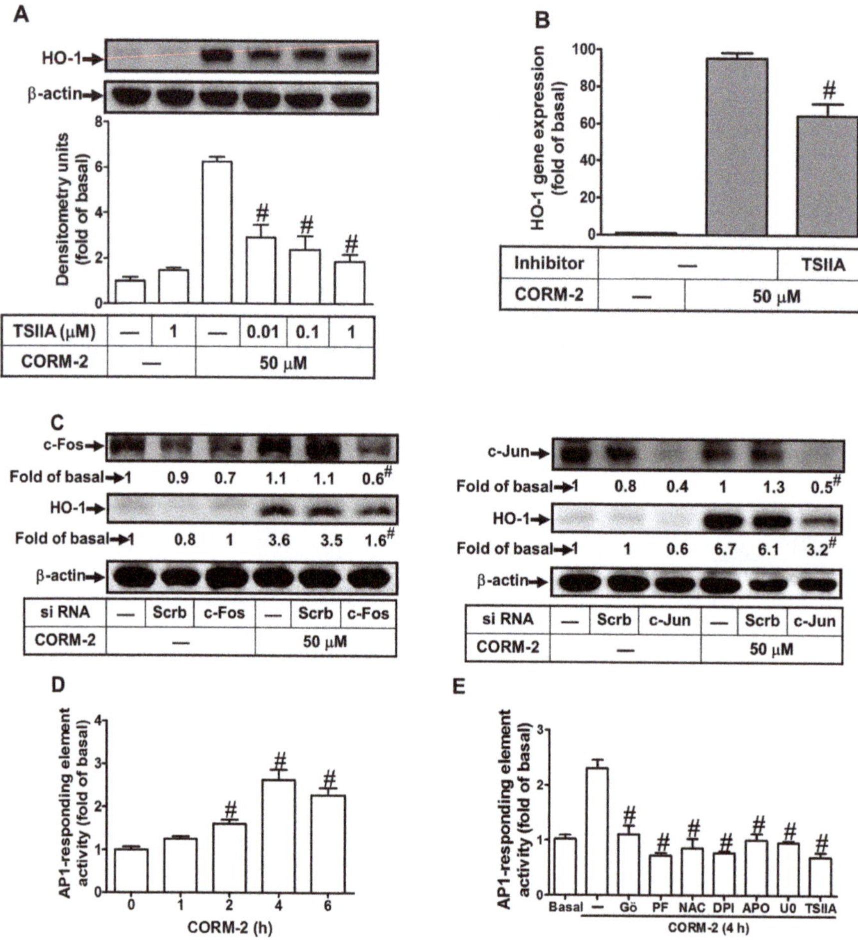

Figure 7. Involvement of c-Fos and c-Jun/activator protein 1 (AP-1) in CORM-2-mediated HO-1 expression. (**A**) Cells were pretreated with tanshinone IIA (TSIIA) for 1 h, and then incubated with CORM-2 for 24 h. The protein levels of HO-1 and β-actin (served as an internal control) were determined by Western blot. (**B**) Cells were pretreated with TSIIA (1 μM) for 1 h, and then incubated with CORM-2 for 6 h. The mRNA expression of HO-1 was determined by real-time PCR. (**C**) Cells were transfected with either scrambled (Scrb), c-Fos, or c-Jun siRNA, and then incubated with CORM-2 for 24 h. The levels of c-Fos, c-Jun, HO-1, and β-actin (served as an internal control) protein were determined by Western blot. (**D**) Cells were transiently cotransfected with pAP1-Luc (activator protein-1 luciferase reporter plasmid) and pGal (plasmid contains the reporter gene β-galactosidase) for 24 h, and then incubated with CORM-2 for the indicated time intervals. (**E**) Cells were pretreated with Gö6983 (10 μM),

PF431396 (10 μM), N-acetyl-cysteine (NAC, 10 mM), DPI (10 μM), APO (100 μM), U0126 (10 μM), or TSIIA (1 μM) for 1 h and then incubated with CORM-2 for 4 h. The AP-1 promoter activity in the cell lysates was determined as described in the Methods section. Data are expressed as mean ± SEM of five independent experiments ($n = 5$). [#] $p < 0.05$, as compared with the cells exposed to vehicle (**D**) or CORM-2 (**A–C,E**) alone. Data analysis and processing are described in the section "Statistical Analysis of Data".

2.8. CORM-2 Induces HO-1 Expression via Its Promoter Transcriptional Activity

CORM-2 has been shown to induce HO-1 gene regulation and activate the AP-1 activity. Thus, human HO-1 promoter was constructed in a luciferase reporter plasmid to evaluate the CORM-2 induced HO-1 transcription in HTSMCs. The HO-1 promoter contains several putative recognition elements for a variety of transcriptional factors, including the AP-1 site. As expected, CORM-2 stimulated the HO-1 promoter activity in a time-dependent manner with a maximal response within 6 h (Figure 8A), which was inhibited by pretreatment with Gö6983, PF431396, NAC, DPI, APO, U0126, or TSIIA (Figure 8B). These results confirmed that CORM-2 stimulates HO-1 promoter activity via AP-1 activation, which was mediated through PKCα or Pyk2-regulated Nox/ROS/p42/p44 MAPK-dependent c-Fos-c-Jun/AP-1 pathway in HTSMCs.

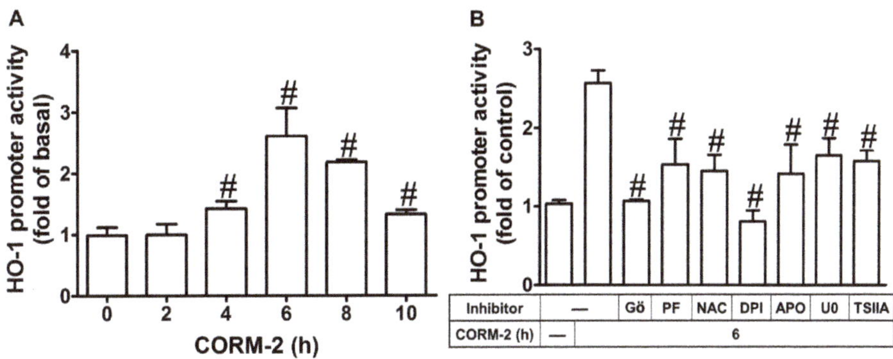

Figure 8. CORM-2 induces AP-1-dependent HO-1 expression via PKCα/Pyk2/Nox/ROS/ERK pathway in HTSMCs. (**A**) Cells were transiently cotransfected with pHO1-Luc and pGal for 24 h, and then incubated with CORM-2 (50 μM) for the indicated time intervals. (**B**) Cells were pretreated with Gö6983 (10 μM), PF431396 (10 μM), NAC (10 mM), DPI (10 μM), APO (100 μM), U0126 (10 μM), or TSIIA (10 μM) for 1 h and then incubated with CORM-2 for 6 h. The HO-1 promoter activity in the cell lysates was determined. Data are expressed as mean ± SEM of five independent experiments ($n = 5$). [#] $p < 0.05$, as compared with the cells exposed to vehicle (**A**) or CORM-2 alone. Data analysis and processing are described in the section "Statistical Analysis of Data".

3. Discussion

CORMs exert anti-inflammatory effects through the up-regulation of anti-oxidant enzymes such as HO-1 [26]. However, the roles of Nox/ROS involved in CORM-2-induced HO-1 expression were still unknown. Here, we observed that pretreatment with CORM-2 inhibited the LPS-induced airway inflammation via HO-1 induction in mice. Further, our results demonstrated that the levels of HO-1 protein, mRNA, and promoter activity were increased in HTSMCs challenged with CORM-2. CORM-2-induced HO-1 expression was mediated through PKCα and Pyk2/Nox/ROS/ERK1/2 linking to the AP-1 pathway in HTSMCs (Figure 9). These results suggested that HO-1 expression induced by CORM-2 is mediated via a PKCα and Pyk2/Nox/ROS/p42/p44 MAPK-dependent AP-1 pathway in HTSMCs.

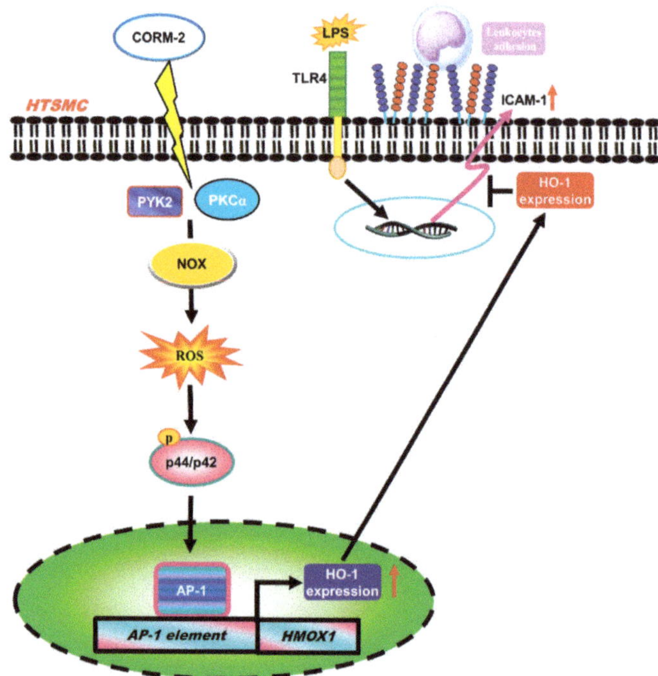

Figure 9. Schematic signaling pathways are involved in CORM-2-induced HO-1 expression in HTSMCs. CORM-2-induced HO-1 expression is mediated via a PKCα or Pyk2/Nox/ROS/ERK1/2 cascade linking to activation of c-Fos and c-Jun/AP-1. The up-regulation of HO-1 could protect against the LPS-induced airway inflammation.

ROS exert as a messenger in the normal physiological functions and the inflammatory responses dependent on their cellular concentrations [40]. The up-regulation of HO-1 due to Nox activity and ROS formation is induced by LPS and cytokines [8,19]. Others and our previous studies indicated that the CORMs mediate Nox-dependent ROS generation in astrocytes [21,22]. Therefore, Nox-dependent ROS generation is involved in HO-1 expression by CORM-2 in HTSMCs. We further clarified the role of ROS in HO-1 expression; a thiol-containing compound (NAC) was used to scavenge ROS. NAC has been shown to reduce the injurious effects of hydrogen peroxide in human alveolar and bronchial epithelial cells [41]. We also found that CORM-2-induced HO-1 expression was inhibited by NAC, and strongly supported the role of ROS in the CORM-2-induced HO-1 expression.

Moreover, two Nox-related inhibitors, DPI (a Nox inhibitor) and APO (a p47^{phox} inhibitor), have been shown to prevent p47^{phox} (a Nox subunit) translocation to the membrane and inhibit Nox activation [42]. Our results showed that pretreatment with either DPI or APO attenuated CORM-2-induced ROS generation and HO-1 expression. These data are consistent with previous reports showing that Nox-derived ROS generation is involved in HO-1 induction by LTA or CSPE in HTSMCs [5,20]. Indeed, low levels of ROS could regulate proliferation, gene expression, immunity, and wound healing [43]. Conversely, higher levels of ROS can exert antibacterial effect, and cause cell damage and death [23,44]. In addition, ROS generation could initiate HO-1 expression through the degradation of Keap1 and translocation of Nrf2 into the nucleus. In our previous study, CORM-2 has been shown to activate Nox and produce ROS in brain astrocytes [22]. Previous reports also indicate that CO release in mammal cells acts as a secondary messenger to mediate metabolism and gene expression, including HO-1 [45,46]. The members of Nox family such as Nox-(1,2,4) and p47^{phox} were shown to be involved in CORM-2-induced HO-1 expression, which were confirmed by using their own siRNAs. In this study, we also demonstrated that PKCα and Pyk2 are the upstream components of ROS generation by their inhibitors. Our results showed that PKCα/Pyk2-mediated Nox/ROS signal contributes to CORM-2-induced HO-1 expression in HTSMCs. However, how CORM-2 activated PKCα/Pyk2 and led to Nox/ROS generation is an important issue preserved for further investigation.

Abnormal MAPK activations are implicated in a variety of inflammatory responses and tissue injury, and the induction of several inflammatory mediators in different cell types [25,47]. Here, we demonstrated that ERK1/2 was required for the CORM-2-induced HO-1 expression, which was attenuated by a selective MEK1/2 inhibitor U0126 or transfection with p44 siRNA. These kinases involved in CORM-2-stimulated pathways were further confirmed by CORM-2-mediated ERK1/2 phosphorylation. These results are consistent with the HO-1 expression mediated by ERK1/2 to activate the antioxidant response element (ARE) region, which is the Nrf2 binding site in HepG2 (liver hepatocellular carcinoma) Cells [48] and heme-mediated neuronal injury [49]. Moreover, pretreatment with the inhibitor of PKCα, Pyk2, Nox, or ROS scavenger significantly attenuated ERK1/2 phosphorylation, suggesting that PKCα/Pyk2-dependent Nox/ROS are required for ERK1/2 phosphorylation. These results are consistent with reports that ROS-dependent MAPK pathways are involved in the regulation of cellular functions [38,50,51]. Indeed, our previous study found that the inhibition of JNK1/2 or p38 MAPK attenuates CORM-2-induced HO-1 expression via a c-Src/EGFR/PI3K/Akt pathway [27]. In contrast, CORM-2-stimulated JNK1/2 and p38 MAPK phosphorylation was not mediated via the PKCα/Pyk2-mediated Nox/ROS signal, although these two kinases also regulate CORM-2-induced HO-1 expression in HTSMCs.

The activated transcription factors interact with response elements on the HO-1 promoter to regulate gene transcription [44]. Here, we focused on the role of transcription factor AP-1, which is modulated during oxidative stress associated with inflammatory diseases [52]. The involvement of AP-1 in these responses was further supported by the results that CORM-2 induced c-Fos and c-Jun, AP-1 subunits, and activation via PKCα/Pyk2-mediated Nox/ROS linking to the ERK1/2 pathway. Moreover, the roles of c-Fos-c-Jun/AP-1 in CORM-2-induced HO-1 expression were confirmed by transfection with c-Fos or c-Jun siRNA to attenuate CORM-2-induced HO-1 expression. Several reports have shown that many regulatory elements of transcription factors, including AP-1, were analyzed on the 5′ region of the HO-1 promoter in several animal species [28,53]. Thus, we also demonstrated that CORM-2-stimulated HO-1 promoter activity was reduced by pretreatment with Gö6976, PF431396, NAC, DPI, APO, U0126, or TSIIA, indicating that CORM-2 induces HO-1 promoter activity via a PKCα/Pyk2-mediated Nox/ROS/ERK1/2/AP-1 pathway. These results are consistent with the reports that alpha-lipoic acid induced HO-1 expression in vascular smooth muscle cells [54], and BK induced HO-1 expression in brain astrocytes [38].

Previous reports indicated that the CORMs up-regulate the HO-1 activity and attenuate the LPS-induced inflammatory responses in macrophages [9] and animal study [8,31]. Moreover, the overexpression of HO-1 in ovalbumin (OVA)-sensitized guinea pigs effectively decreases inflammatory reaction, mucus secretion, and responsiveness to histamine in airways [55], suggesting that HO-1 exhibits protecting ability in the host during airway inflammation. In this study, the induction of HO-1 by CORM-2 protected against LPS-induced ICAM-1 expression and leukocytes infiltration in both in vitro and in vivo studies. Importantly, previous reports also indicated that CORM-2-derived CO release can attenuate the cell sequestration, NF-κB activity, and ICAM-1 expression of leukocyte after lung injury [36,56] and regulate the expressions of adhesion molecules on human umbilical vein endothelia cells to affect leukocyte attachment [54]. Our previous report also indicated that the induction of HO-1 by CoPPIX inhibits the TNF-α-induced ICAM-1 and VCAM-1 expression which is revered by zinc protoporphyrin IX (ZnPPIX, an inhibitor of HO-1 activity) in HTSMCs [14].

4. Materials and Methods

4.1. Reagents and Chemicals

DMEM/F-12 (Dbecco's Modified Eagle Medium/Nutrient Mixture F-12) medium, fetal bovine serum (FBS), TRIzol reagent, and PLUS-Lipofectamine were from Invitrogen (Carlsbad, CA, USA). Human siRNAs for PKCα (L-003523-00-0020) was from Dharmacon (Lafayette, CO, USA) and Pyk2 (SASI_Hs01_00032249), ERK1 (SASI_Hs01_00190617), HO-1 (SASI_Hs01_00035065), Nox-1 (SASI_Hs01_00342845), Nox-2 (SASI_Hs01_00086110), Nox-4 (SASI_Hs02_00349918), p47^{phox} (SASI_Hs02_00302212), and c-Fos (SASI_Hs01_00184572) were from Sigma (St. Louis, MO, USA). Hybond C membrane, enhanced chemiluminescence (ECL), and Western blotting detection system were from GE Healthcare Biosciences (Buckinghamshire, UK). PhosphoPlus PKCα (#9375), Pyk2 (#3291), and ERK1/2 (#9101) antibodies were from Cell Signaling (Danvers, MA, USA). The HO-1 (ADI-SPA-895) antibody was from Enzo (Farmingdale, NY, USA). PKCα (sc-208), Pyk2 (sc-9019), ERK1 (sc-94), c-Fos (sc-7202), and β-actin (sc-47778) antibodies were from Santa Cruz (Santa Cruz, CA, USA). Ro31-8220, Gö6983, Gö6976, PF431396, diphenyleneiodonium chloride (DPI), apocynin (APO), U0126, and tanshinone IIA (TSIIA) were from Biomol (Plymouth Meeting, PA, USA). The bicinchoninic acid (BCA) protein assay kit was from Pierce (Rockford, IL, USA). SDS-PAGE (sodium dodecyl sulfate polyacrylamide gel electrophoresis) reagents were from MDBio Inc (Taipei, Taiwan). Tricarbonyldichlororuthenium (II) dimer (CORM-2), ruthenium (III) chloride, RuCl$_3$ [inactive form of CORM-2, (iCORM-2)], N-acetyl-cysteine (NAC), lipopolysaccharide (LPS), enzymes, and other chemicals were from Sigma (St. Louis, MO, USA).

4.2. Animal Care and Experimental Procedures

Male ICR mice aged 6–8 weeks were purchased from the National Laboratory Animal Centre (Taipei, Taiwan) and handled according to the guidelines of Animal Care Committee of Chang Gung University (Approval Document No. Chang Gung University 16-046, 4 October 2016) and National Institute of Health (NIH) Guides for the Care and Use of Laboratory Animals. All the studies involving animals are reported in accordance with the ARRIVE guidelines [57,58]. Mice were assigned randomly into three groups: sham [0.1 mL of dimethyl sulfoxide (DMSO)-phosphate-buffered saline (PBS) (1:100) with 0.1% (*w/v*) bovine serum albumin (BSA) treated mice], LPS (LPS-treated mice), and CORM-2 + LPS; 5 mice in each group/cage and kept in standard individually ventilated cages in an animal facility under standardized conditions (12 h light/dark cycle, 21–24 °C, humidity of 50–60%) with food and water ad libitum. Mice were intraperitoneally (i.p.) injected with CORM-2 (8 mg/kg of body weight) for 24 h, and then anesthetized by i.p. injection of pentothal (50 mg/kg) placed individually on a board in a near-vertical position and the tongues withdrawn with a lined forceps. LPS (3 mg/kg) was placed posterior in the throat and aspirated into lungs for 16 h of development of a lung inflammation model [59]. At the end of the experimental period, mice were killed by a high dose of pentothal (100 mg/kg i.p.) for the collection of lung tissues extracted for protein (right superior lobe + post caval lobe) and mRNA (right middle lobe + right inferior lobe) expression of ICAM-1, HO-1, or β-actin analyses. BAL fluid was performed through a tracheal cannula using 1-mL aliquots of ice-cold PBS solution. BAL fluid was centrifuged at 500× *g* at 4 °C, and cell pellets were washed and re-suspended in PBS. Leukocyte count was determined by a hemocytometer, as previously described [7]. Data collection and evaluation of all the in vivo and in vitro experiments were performed blindly of the group identity.

4.3. Cell Culture and Treatment

HTSMCs were purchased from ScienCell Research Laboratories (San Diego, CA, USA). The cultured conditions and treatments were conducted as previously described [60]. Cells were plated onto 12-well culture plates and made quiescent at confluence by incubation in serum-free DMEM/F-12 for 24 h. Growth-arrested cells were incubated with or without CORM-2 at 37 °C for the indicated time intervals. Our previous report indicates that CORM-2 induces HO-1 expression in time-dependent and concentration-dependent manners. The HO-1 expression is up-regulated to a maximal response within 16–24 h treatment with 50 μM of CORM-2 [27]. Therefore, the concentration of CORM-2 at 50 μM was used throughout this study. When the inhibitors were used, cells were pretreated with the inhibitor for 1 h before exposure to CORM-2. Experiments were performed using cells from passages four to seven.

4.4. Preparation of Cell Extracts and Western Blot Analysis

After treatment, the cells were washed with ice-cold PBS, scraped, and collected by centrifugation at $16,000 \times g$ for 10 min at 4 °C to yield the whole cell extract, as previously described [7]. The supernatants were harvested and mixed with SDS-PAGE loading buffer (final concentration: 100 mM Tris-HCl pH 6.8, 1% SDS, 2.5% glycerol, 100 mM β-mercaptoethanol, 0.01% bromophenol blue). Samples were denatured, separated with 10% SDS-PAGE, and transferred to the nitrocellulose membrane. Membranes were probed overnight with an anti-phospho-PKCα, phospho-Pyk2, phospho-ERK1/2, HO-1, PKCα, Pyk2, ERK1, c-Fos, or β-actin antibody. Membranes were washed with Tween-Tris buffered solution (TTBS) four times for 5 min each, incubated with anti-rabbit or anti-mouse horseradish peroxidase antibody (1:2000) for 1 h. The immunoreactive bands were detected by ECL (enhanced chemiluminescence) reagents and captured by a UVP BioSpectrum 500 Imaging System (Upland, CA, USA). The image densitometry analysis was conducted using UN-SCAN-IT gel software (Orem, UT, USA).

4.5. Total RNA Extraction and Real Time-Quantitative PCR Analysis

Total RNA was isolated from HTSMCs treated with CORM-2 for the indicated time intervals in 10-cm culture dishes with TRIzol according to the protocol of the manufacturer. The mRNA was reverse-transcribed into cDNA and analyzed by real time-quantitative (q)PCR. Real time-qPCR was performed with the TaqMan gene expression assay system, using primers and probe mixes for HO-1, c-Fos, and endogenous GAPDH control genes. PCRs were performed using a 7500 Real Time-PCR System (Applied Biosystems, Foster City, CA, USA). Relative gene expression was determined by the ΔΔCt method, where Ct meant threshold cycle. All the experiments were performed in triplicate.

4.6. Plasmid Construction, Transfection, and Luciferase Reporter Gene Assays

For construction of the HO-1 luciferase (Luc) plasmid, human HO-1 promoter, a region spanning –3106 to +186 bp provided by Dr. Y. C. Liang (Graduate Institute of Biomedical Technology, Taipei Medical University, Taipei, Taiwan) was inserted into a pGL3-basic vector (Promega, Madison, WI, USA). Plasmid pAP1-Luc, the fragment of the AP-1-responding element was inserted into the pGL3 (plasmid contains the reporter gene β-galactosidase) promoter. The plasmid DNA was extracted by using QIAGEN plasmid DNA preparation kits and transfected into HTSMCs with Lipofectamine reagent according to the standard protocol of the manufacturer. The plasmid pCMV-β-gal was cotransfected to be the internal control. The HO-1 promoter-driven-Luc activity was analyzed by a luciferase assay system (Promega, Madison, WI, USA). Firefly luciferase activities were standardized with β-galactosidase activity.

4.7. Transient Transfection with siRNAs

HTSMCs (3×10^5 cells) were plated in 12-well culture plates for 24 h to about 80% confluence. Cells were washed once with PBS, and 0.4 mL of serum-free DMEM/F-12 medium was added to each well. The transient transfection of siRNAs (scrambled, PKCα, Pyk2, ERK1, and c-Fos, 100 nM) was performed by using Lipofectamine™ RNAiMAX reagent (from Sigma, St. Louis, MO, USA) according to the manufacturer's instructions.

4.8. Measurement of Intracellular ROS Generation

The peroxide-sensitive fluorescent probe CMH_2DCF-DA and DHE were used to assess the intracellular ROS generation [61] with minor modifications. Briefly, HTSMCs were incubated with 10 μM of CMH_2DCF-DA (in warm PBS) for 45 min at 37 °C. The medium was removed and replaced with fresh DMEM/F-12 media for CORM-2 treatments. CMH_2 DCF-DA interacted with cells, and then generated a non-fluorescent product: H_2DCF. CORM-2 induced the generation of a ROS oxidized product: DCF. Relative fluorescence intensity was recorded (0.5 to 24 h) by a fluorescent plate reader (Thermo, Appliskan; Waltham, MA, USA) at an excitation wavelength of 485 nm, and emission was measured at a wavelength of 530 nm. For DHE staining, cells were treated with CORM-2 for the indicated time intervals and then incubated with 10 μM of DHE (in DMEM/F12 medium) for 10 min. For immunofluorescence staining, the stained cells were washed three times with cold PBS, and then the fluorescence for DCF and DHE staining was detected at 495/529 and 518/605 nm, respectively, using a fluorescence microscope (Zeiss, Axiovert 200M; Oberkochen, Baden-Württemberg, Germany).

4.9. Determination of NADPH Oxidase Activity by Chemiluminescence Assay

The Nox activity in intact cells was assayed by lucigenin chemiluminescence [38]. After incubation, the cells were gently scraped and centrifuged at $400\times g$ for 10 min at 4 °C. The cell pellet was re-suspended in a known volume (35 μL/well) of ice-cold RPMI 1640 medium, and the cell suspension was kept on ice. To a final 200 μL of pre-warmed (37 °C) PBS containing either NADPH (1 μM) or lucigenin (20 μM), 5 μL of cell suspension (2×10^4 cells) was added to initiate the reaction followed by the immediate measurement of chemiluminescence using an Appliskan luminometer (Thermo®; Waltham, MA, USA) in an out-of-coincidence mode. Neither NADPH nor NADH enhanced the background chemiluminescence of lucigenin alone (30–40 counts/min). Chemiluminescence was continuously measured for 12 min, and the activity of Nox was expressed as counts per million cells. The calculated numbers of Nox activity were calibrated with protein concentration. The equal amount of warmed PBS medium (containing NADPH and lucigenin) was used as the blank, and the untreated cells were the basal group.

4.10. Statistical Analysis of Data

All the data were expressed as the mean ± SEM in at least five individual experiments ($n = 5$). Statistical analysis was performed by using GraphPad Prizm Program 6.0 software (GraphPad, San Diego, CA). We used one-way ANOVA followed by Dunnett's post hoc test when comparing more than two groups of data and a one-way ANOVA, non-parametric Kruskal–Wallis test, followed by Dunnett's post hoc test when comparing multiple independent groups. p values of 0.05 were considered to be statistically significant. Post tests were run only if F achieved $p < 0.05$ and there was no significant variance in homogeneity. Error bars were omitted when they fell within the dimensions of the symbols.

5. Conclusions

These results suggested that CORM-2-induced HO-1 expression is mediated through PKCα/Pyk2 and Nox/ROS-dependent activation of ERK1/2, linking to the up-regulation of AP-1 (c-Fos and c-Jun), which promotes HO-1 expression and enzymatic activity in HTSMCs. Based on the observations from literatures and our findings, we depict a model for the molecular mechanisms underlying CORM-2-induced HO-1 expression and activity in HTSMCs. The results obtained with cellular and animal experiments indicated that better understanding the mechanisms underlying CORM-2-induced HO-1 expression promotes the development of therapeutic strategies for airway inflammatory disorders.

Author Contributions: C.-C.L., L.-D.H., R.-L.C., and C.-M.Y. designed and conducted the study. C.-C.L., L.-D.H., and R.-L.C. performed and collected the data. C.-C.L., L.-D.H., R.-L.C., and C.-M.Y. analyzed and interpreted the data. C.-C.L. and C.-M.Y. prepared the manuscript. C.-C.L., L.-D.H., R.-L.C., and C.-M.Y. reviewed the manuscript. C.-C.L., L.-D.H., R.-L.C., and C.-M.Y. approved the final manuscript.

Funding: This work was supported by the Ministry of Education, Taiwan [grant number EMRPD1I0381]; the Ministry of Science and Technology, Taiwan [Grant numbers: MOST105-2320-B-182-005-MY3, MOST107-2320-B-182-020-MY2, and MOST107-2320-B-182A-011]; Chang Gung Medical Research Foundation, Taiwan [Grant numbers: CMRPD1F0023, CMRPD1F0552, CMRPD1F0553, CMRPD1I0051-3, CMRPG3F1533, CMRPG3H0062, CMRPG3H0063, and CMRPG5F0203, CMRPG5I0041].

Acknowledgments: We appreciated Chen-yu Wang for his suggestions and construction of plasmids applied in this study and YC Tai for her technical assistance.

Conflicts of Interest: The authors declare no conflict of interest.

Abbreviations

AP-1	activator protein 1
APO	apocynin
ARE	antioxidant response element
BAL	bronchoalveolar lavage
BCA	bicinchoninic acid
BSA	Bovine serum albumin
CO	carbon monoxide
CORM-2	carbon monoxide releasing molecule-2
CoPPIX	cobalt protoporphyrin
CSPE	cigarette smoke particle extract
DCF	dichlorofluorescein
DCF-DA	2′,7′-dichlorofluorescin diacetate
DHE	dihydroethidium
DMEM/F-12	Dubecco's Modified Eagle Medium/Nutrient Mixture F-12
DMSO	dimethyl sulfoxide
DPI	diphenyleneiodonium
ECL	enhanced chemiluminescence
EGFR	epidermal growth factor receptor
ERK1/2	extracellular signal-regulated kinase 1/2
FBS	fetal bovine serum
MCH$_2$DCF-DA	chloromethyl 2′,7′-dichloro fluorescein diacetate
H & E	hematoxylin & eosin
HO-1	heme oxygenase-1
HTSMCs	human tracheal smooth muscle cells
ICAM-1	intercellular adhesion molecule
ICR	Institute of Cancer Research
i.p.	intraperitoneally
LPS	lipopolysaccharide
LTA	lipotechoic acid
MAPKs	mitogen-activated protein kinases
mRNA	messenger ribonucleic acid
NAC	N-acetyl-cysteine
NADPH	nicotinamaide adenine dinucleotide phosphate
NF-κB	nuclear factor-κB
Nox	NADPH oxidase
Nrf2	nuclear factor erythroid 2-related factor 2
PBS	phosphate-buffered saline
PI3K	phosphoinositide 3-kinase
PKC	protein kinase C
Pyk2	proline-rich tyrosine kinase 2
ROS	reactive oxygen species

SDS-PAGE	sodium dodecyl sulfate polyacrylamide gel electrophoresis
siRNA	small interfering ribonucleic acid
TNF	tumor necrosis factor
TSIIA	tanshinone IIA
TTBS	Tween-Tris buffered solution
VCAM-1	vascular cell adhesion molecule
XTT	2,3-bis-(2-methoxy-4-nitro-5-sulfophenyl)-2H-tetrazolium-5-carboxanilide
ZnPP IX	zinc protoporphyrin IX

References

1. Maines, M.D. The heme oxygenase system: A regulator of second messenger gases. *Annu. Rev. Pharmacol. Toxicol.* **1997**, *37*, 517–554. [CrossRef]
2. Ryter, S.W.; Alam, J.; Choi, A.M. Heme oxygenase-1/carbon monoxide: From basic science to therapeutic applications. *Physiol. Rev.* **2006**, *86*, 583–650. [CrossRef] [PubMed]
3. Tenhunen, R.; Marver, H.S.; Schmid, R. The enzymatic conversion of heme to bilirubin by microsomal heme oxygenase. *Proc. Natl. Acad. Sci. USA* **1968**, *61*, 748–755. [CrossRef] [PubMed]
4. Otterbein, L.E.; Soares, M.P.; Yamashita, K.; Bach, F.H. Heme oxygenase-1: Unleashing the protective properties of heme. *Trends Immunol.* **2003**, *24*, 449–455. [CrossRef]
5. Lee, I.T.; Wang, S.W.; Lee, C.W.; Chang, C.C.; Lin, C.C.; Luo, S.F.; Yang, C.M. Lipoteichoic acid induces HO-1 expression via the TLR2/MyD88/c-Src/NADPH oxidase pathway and Nrf2 in human tracheal smooth muscle cells. *J. Immunol.* **2008**, *181*, 5098–5110. [CrossRef] [PubMed]
6. Balla, G.; Jacob, H.S.; Balla, J.; Rosenberg, M.; Nath, K.; Apple, F.; Eaton, J.W.; Vercellotti, G.M. Ferritin: A cytoprotective antioxidant strategem of endothelium. *J. Biol. Chem.* **1992**, *267*, 18148–18153. [PubMed]
7. Matsumoto, H.; Ishikawa, K.; Itabe, H.; Maruyama, Y. Carbon monoxide and bilirubin from heme oxygenase-1 suppresses reactive oxygen species generation and plasminogen activator inhibitor-1 induction. *Mol. Cell. Biochem.* **2006**, *291*, 21–28. [CrossRef]
8. Jamal Uddin, M.; Joe, Y.; Kim, S.K.; Oh Jeong, S.; Ryter, S.W.; Pae, H.O.; Chung, H.T. IRG1 induced by heme oxygenase-1/carbon monoxide inhibits LPS-mediated sepsis and pro-inflammatory cytokine production. *Cell. Mol. Immunol.* **2016**, *13*, 170–179. [CrossRef]
9. Sawle, P.; Foresti, R.; Mann, B.E.; Johnson, T.R.; Green, C.J.; Motterlini, R. Carbon monoxide-releasing molecules (CO-RMs) attenuate the inflammatory response elicited by lipopolysaccharide in RAW264.7 murine macrophages. *Br. J. Pharmacol.* **2005**, *145*, 8008–8010. [CrossRef]
10. Rushworth, S.A.; Chen, X.L.; Mackman, N.; Ogborne, R.M.; O'Connell, M.A. Lipopolysaccharide-induced heme oxygenase-1 expression in human monocytic cells is mediated via Nrf2 and protein kinase C. *J. Immunol.* **2005**, *175*, 4408–4415. [CrossRef]
11. Aggeli, I.K.; Gaitanaki, C.; Beis, I. Involvement of JNKs and p38-MAPK/MSK1 pathways in H_2O_2-induced upregulation of heme oxygenase-1 mRNA in H9c2 cells. *Cell Signal.* **2006**, *18*, 1801–1812. [CrossRef] [PubMed]
12. Fredenburgh, L.E.; Perrella, M.A.; Mitsialis, S.A. The role of heme oxygenase-1 in pulmonary disease. *Am. J. Respir. Cell. Mol. Biol.* **2007**, *36*, 158–165. [CrossRef]
13. Ferrandiz, M.L.; Devesa, I. Inducers of heme oxygenase-1. *Curr. Pharm. Des.* **2008**, *14*, 473–486. [CrossRef] [PubMed]
14. Lee, I.T.; Luo, S.F.; Lee, C.W.; Wang, S.W.; Lin, C.C.; Chang, C.C.; Chen, Y.L.; Chau, L.Y.; Yang, C.M. Overexpression of HO-1 protects against TNF-alpha-mediated airway inflammation by down-regulation of TNFR1-dependent oxidative stress. *Am. J. Pathol.* **2009**, *175*, 519–532. [CrossRef] [PubMed]
15. Lee, I.T.; Yang, C.M. Role of NADPH oxidase/ROS in pro-inflammatory mediators-induced airway and pulmonary diseases. *Biochem. Pharmacol.* **2012**, *84*, 581–590. [CrossRef] [PubMed]
16. Zuo, L.; Otenbaker, N.P.; Rose, B.A.; Salisbury, K.S. Molecular mechanisms of reactive oxygen species-related pulmonary inflammation and asthma. *Mol. Immunol.* **2013**, *56*, 57–63. [CrossRef] [PubMed]
17. Otterbein, L.E.; Kolls, J.K.; Mantell, L.L.; Cook, J.L.; Alam, J.; Choi, A.M. Exogenous administration of heme oxygenase-1 by gene transfer provides protection against hyperoxia-induced lung injury. *J. Clin. Investig.* **1999**, *103*, 1047–1054. [CrossRef] [PubMed]

18. Li, F.J.; Duggal, R.N.; Oliva, O.M.; Karki, S.; Surolia, R.; Wang, Z.; Watson, R.D.; Thannickal, V.J.; Powell, M.; Watts, S.; et al. Heme oxygenase-1 protects corexit 9500A-induced respiratory epithelial injury across species. *PLoS ONE* **2015**, *10*, e0122275. [CrossRef] [PubMed]

19. Srisook, K.; Han, S.S.; Cho, H.S.; Li, M.H.; Ueda, H.; Kim, C.; Cha, Y.N. CO from enhanced HO activity or from CORM-2 inhibits both O2- and NO production and downregulates HO-1 expression in LPS-stimulated macrophages. *Biochem. Pharmacol.* **2006**, *71*, 307–318. [CrossRef]

20. Cheng, S.E.; Lee, I.T.; Lin, C.C.; Kou, Y.R.; Yang, C.M. Cigarette smoke particle-phase extract induces HO-1 expression in human tracheal smooth muscle cells: Role of the c-Src/NADPH oxidase/MAPK/Nrf2 signaling pathway. *Free Radic. Biol. Med.* **2010**, *48*, 1410–1422. [CrossRef]

21. Choi, Y.K.; Por, E.D.; Kwon, Y.G.; Kim, Y.M. Regulation of ROS production and vascular function by carbon monoxide. *Oxid. Med. Cell. Longev.* **2012**, *2012*, 794237. [CrossRef] [PubMed]

22. Chi, P.L.; Lin, C.C.; Chen, Y.W.; Hsiao, L.D.; Yang, C.M. CO Induces Nrf2-Dependent Heme Oxygenase-1 Transcription by Cooperating with Sp1 and c-Jun in Rat Brain Astrocytes. *Mol. Neurobiol.* **2015**, *52*, 277–292. [CrossRef] [PubMed]

23. Taille, C.; El-Benna, J.; Lanone, S.; Boczkowski, J.; Motterlini, R. Mitochondrial respiratory chain and NAD(P)H oxidase are targets for the antiproliferative effect of carbon monoxide in human airway smooth muscle. *J. Biol. Chem.* **2005**, *280*, 25350–25360. [CrossRef] [PubMed]

24. Motterlini, R.; Green, C.J.; Foresti, R. Regulation of heme oxygenase-1 by redox signals involving nitric oxide. *Antioxid. Redox Signal.* **2002**, *4*, 615–624. [CrossRef] [PubMed]

25. Lee, I.T.; Yang, C.M. Inflammatory signalings involved in airway and pulmonary diseases. *Mediat. Inflamm.* **2013**, *2013*, 791231. [CrossRef] [PubMed]

26. Motterlini, R.; Otterbein, L.E. The therapeutic potential of carbon monoxide. *Nat. Rev. Drug Discov.* **2010**, *9*, 728–743. [CrossRef] [PubMed]

27. Yang, C.M.; Lin, C.C.; Lee, I.T.; Hsu, C.K.; Tai, Y.C.; Hsieh, H.L.; Chi, P.L.; Hsiao, L.D. c-Src-dependent transactivation of EGFR mediates CORM-2-induced HO-1 expression in human tracheal smooth muscle cells. *J. Cell. Physiol.* **2015**, *230*, 2351–2361. [CrossRef]

28. Alam, J.; Cook, J.L. How many transcription factors does it take to turn on the heme oxygenase-1 gene? *Am. J. Respir. Cell Mol. Biol.* **2007**, *36*, 166–174. [CrossRef]

29. Alam, J.; Stewart, D.; Touchard, C.; Boinapally, S.; Choi, A.M.; Cook, J.L. Nrf2, a Cap'n'Collar transcription factor, regulates induction of the heme oxygenase-1 gene. *J. Biol. Chem.* **1999**, *274*, 26071–26078. [CrossRef]

30. Constantin, M.; Choi, A.J.; Cloonan, S.M.; Ryter, S.W. Therapeutic potential of heme oxygenase-1/carbon monoxide in lung disease. *Int. J. Hypertens.* **2012**, *2012*, 859235. [CrossRef]

31. Xue, J.; Habtezion, A. Carbon monoxide-based therapy ameliorates acute pancreatitis via TLR4 inhibition. *J. Clin. Investig.* **2014**, *124*, 437–447. [CrossRef] [PubMed]

32. Qin, W.; Zhang, J.; Lv, W.; Wang, X.; Sun, B. Effect of carbon monoxide-releasing molecules II-liberated CO on suppressing inflammatory response in sepsis by interfering with nuclear factor kappa B activation. *PLoS ONE* **2013**, *8*, e75840. [CrossRef] [PubMed]

33. Takasuka, H.; Hayashi, S.; Koyama, M.; Yasuda, M.; Aihara, E.; Amagase, K.; Takeuchi, K. Carbon monoxide involved in modulating HCO_3^- secretion in rat duodenum. *J. Pharmacol. Exp. Ther.* **2011**, *337*, 293–300. [CrossRef] [PubMed]

34. Wu-Zhang, A.X.; Newton, A.C. Protein kinase C pharmacology: Refining the toolbox. *Biochem. J.* **2013**, *452*, 195–209. [CrossRef] [PubMed]

35. Yang, C.C.; Lin, C.C.; Chien, P.T.; Hsiao, L.D.; Yan, C.M. Thrombin/Matrix Metalloproteinase-9-Dependent SK-N-SH Cell Migration is Mediated Through a PLC/PKC/MAPKs/NF-kappaB Cascade. *Mol. Neurobiol.* **2016**, *53*, 5833–5846. [CrossRef] [PubMed]

36. Mills, R.D.; Mita, M.; Nakagawa, J.; Shoji, M.; Sutherland, C.; Walsh, M.P. A role for the tyrosine kinase Pyk2 in depolarization-induced contraction of vascular smooth muscle. *J. Biol. Chem.* **2015**, *290*, 8677–8692. [CrossRef] [PubMed]

37. Rhee, I.; Davidson, D.; Souza, C.M.; Vacher, J.; Veillette, A. Macrophage fusion is controlled by the cytoplasmic protein tyrosine phosphatase PTP-PEST/PTPN12. *Mol. Cell. Biol.* **2013**, *33*, 2458–2469. [CrossRef]

38. Hsieh, H.L.; Wang, H.H.; Wu, C.Y.; Yang, C.M. Reactive Oxygen Species-Dependent c-Fos/Activator Protein 1 Induction Upregulates Heme Oxygenase-1 Expression by Bradykinin in Brain Astrocytes. *Antioxid. Redox Signal.* **2010**, *13*, 1829–1844. [CrossRef]

39. Rochette, L.; Cottin, Y.; Zeller, M.; Vergely, C. Carbon monoxide: Mechanisms of action and potential clinical implications. *Pharmacol. Ther.* **2013**, *137*, 133–152. [CrossRef]

40. Kamata, H.; Hirata, H. Redox regulation of cellular signalling. *Cell Signal.* **1999**, *11*, 1–14. [CrossRef]

41. Mulier, B.; Rahman, I.; Watchorn, T.; Donaldson, K.; MacNe, W.; Jeffery, P.K. Hydrogen peroxide-induced epithelial injury: The protective role of intracellular nonprotein thiols (NPSH). *Eur. Respir. J.* **1998**, *11*, 384–391. [CrossRef] [PubMed]

42. Barbieri, S.S.; Cavalca, V.; Eligini, S.; Brambilla, M.; Caian, A.; Tremoli, E.; Colli, S. Apocynin prevents cyclooxygenase 2 expression in human monocytes through NADPH oxidase and glutathione redox-dependent mechanisms. *Free Radic. Biol. Med.* **2004**, *37*, 156–165. [CrossRef] [PubMed]

43. Stanley, A.; Hynes, A.; Brakebusch, C.; Quondamatteo, F. Rho GTPases and Nox dependent ROS production in skin. Is there a connection? *Histol. Histopathol.* **2012**, *27*, 1395–1406. [PubMed]

44. Tavares, A.F.; Teixeira, M.; Romao, C.C.; Seixas, J.D.; Nobre, L.S.; Saraiva, L.M. Reactive oxygen species mediate bactericidal killing elicited by carbon monoxide-releasing molecules. *J. Biol. Chem.* **2011**, *286*, 26708–26717. [CrossRef] [PubMed]

45. Prabhakar, N.R. NO and CO as second messengers in oxygen sensing in the carotid body. *Respir. Physiol.* **1999**, *115*, 161–168. [CrossRef]

46. Choi, Y.K.; Maki, T.; Mandeville, E.T.; Koh, S.H.; Hayakawa, K.; Arai, K.; Kim, Y.M.; Whalen, M.J.; Xing, C.; Wang, X.; et al. Dual effects of carbon monoxide on pericytes and neurogenesis in traumatic brain injury. *Nat. Med.* **2016**, *22*, 1335–1341. [CrossRef]

47. Alam, R.; Gorska, M.M. Mitogen-activated protein kinase signalling and ERK1/2 bistability in asthma. *Clin. Exp. Allergy* **2011**, *41*, 149–159. [CrossRef]

48. Yuan, X.; Xu, C.; Pan, Z.; Keum, Y.S.; Kim, J.H.; Shen, G.; Yu, S.; Oo, K.T.; Ma, J.; Kong, A.N. Butylated hydroxyanisole regulates ARE-mediated gene expression via Nrf2 coupled with ERK and JNK signaling pathway in HepG2 cells. *Mol. Carcinog.* **2006**, *45*, 841–850. [CrossRef]

49. Foresti, R.; Bani-Hani, M.G.; Motterlini, R. Use of carbon monoxide as a therapeutic agent: Promises and challenges. *Intensive Care Med.* **2008**, *34*, 649–658. [CrossRef]

50. Pawate, S.; Shen, Q.; Fan, F.; Bhat, N.R. Redox regulation of glial inflammatory response to lipopolysaccharide and interferongamma. *J. Neurosci. Res.* **2004**, *77*, 540–551. [CrossRef]

51. Cheng, P.Y.; Lee, Y.M.; Shih, N.L.; Chen, Y.C.; Yen, M.H. Heme oxygenase-1 contributes to the cytoprotection of alpha-lipoic acid via activation of p44/42 mitogen-activated protein kinase in vascular smooth muscle cells. *Free Radic. Biol. Med.* **2006**, *40*, 1313–1322. [CrossRef] [PubMed]

52. Sen, C.K.; Packer, L. Antioxidant and redox regulation of gene transcription. *FASEB J.* **1996**, *10*, 709–720. [CrossRef] [PubMed]

53. Alam, J.; Igarashi, K.; Immenschuh, S.; Shibahara, S.; Tyrrell, R.M. Regulation of heme oxygenase-1 gene transcription: Recent advances and highlights from the International Conference (Uppsala, 2003) on Heme Oxygenase. *Antioxid. Redox Signal.* **2004**, *6*, 924–933. [PubMed]

54. Urquhart, P.; Rosignoli, G.; Cooper, D.; Motterlini, R.; Perretti, M. Carbon monoxide-releasing molecules modulate leukocyte-endothelial interactions under flow. *J. Pharmacol. Exp. Ther.* **2007**, *321*, 656–662. [CrossRef] [PubMed]

55. Almolki, A.; Taillé, C.; Martin, G.F.; Jose, P.J.; Zedda, C.; Conti, M.; Megret, J.; Henin, D.; Aubier, M.; Boczkowski, J. Heme oxygenase attenuates allergen-induced airway inflammation and hyperreactivity in guinea pigs. *Am. J. Physiol. Lung Cell. Mol. Physiol.* **2004**, *287*, L26–L34. [CrossRef] [PubMed]

56. Sun, B.; Sun, H.; Liu, C.; Shen, J.; Chen, Z.; Chen, X. Role of CO-releasing molecules liberated CO in attenuating leukocytes sequestration and inflammatory responses in the lung of thermally injured mice. *J. Surg. Res.* **2007**, *139*, 128–135. [CrossRef]

57. Kilkenny, C.; Browne, W.; Cuthill, I.C.; Emerson, M.; Altman, D.G.; Group NCRRGW. Animal research: Reporting in vivo experiments: The ARRIVE guidelines. *Br. J. Pharmacol.* **2010**, *160*, 1577–1579. [CrossRef] [PubMed]

58. McGrath, K.C.; Li, X.H.; McRobb, L.S.; Heather, A.K. Inhibitory Effect of a French Maritime Pine Bark Extract-Based Nutritional Supplement on TNF-alpha-Induced Inflammation and Oxidative Stress in Human Coronary Artery Endothelial Cells. *Evid. Based Complement. Altern. Med.* **2015**, *2015*, 260530. [CrossRef] [PubMed]

59. Hsu, C.K.; Lee, I.T.; Lin, C.C.; Hsiao, L.D.; Yang, C.M. Nox2/ROS-dependent human antigen R translocation contributes to TNF-alpha-induced SOCS-3 expression in human tracheal smooth muscle cells. *Am. J. Physiol. Lung Cell. Mol. Physiol.* **2014**, *306*, L521–L533. [CrossRef]

60. Lee, C.W.; Chien, C.S.; Yang, C.M. Lipoteichoic acid-stimulated p42/p44 MAPK activation via Toll-like receptor 2 in tracheal smooth muscle cells. *Am. J. Physiol. Lung Cell. Mol. Physiol.* **2004**, *286*, L921–L930. [CrossRef]

61. Hsieh, H.L.; Lin, C.C.; Hsiao, L.D.; Yang, C.M. High glucose induces reactive oxygen species-dependent matrix metalloproteinase-9 expression and cell migration in brain astrocytes. *Mol. Neurobiol.* **2013**, *48*, 601–614. [CrossRef] [PubMed]

International Journal of
Molecular Sciences

Article

Key Role of Reactive Oxygen Species (ROS) in Indirubin Derivative-Induced Cell Death in Cutaneous T-Cell Lymphoma Cells

Marwa Y. Soltan [1,2], Uly Sumarni [1], Chalid Assaf [1,3], Peter Langer [4,5], Ulrich Reidel [1] and Jürgen Eberle [1,*]

[1] Skin Cancer Centre Charité, Department of Dermatology and Allergy, Charité—Universitätsmedizin Berlin, Charitéplatz 1, 10117 Berlin, Germany; Marwayassin@med.asu.edu.eg (M.Y.S.); uly.sumarni@googlemail.com (U.S.); chalid.assaf@helios-gesundheit.de (C.A.); ulrich.reidel@charite.de (U.R.)
[2] Department of Dermatology and Venereology, Faculty of Medicine, Ain Shams University, Cairo 11591, Egypt
[3] Clinic for Dermatology and Venereology, Helios Klinikum Krefeld, Lutherplatz 40, 47805 Krefeld, Germany
[4] Institute of Chemistry, University of Rostock, Albert-Einstein-Str. 3a, 18059 Rostock, Germany; peter.langer@uni-rostock.de
[5] Leibniz Institute of Catalysis at the University of Rostock e.V., Albert-Einstein-Str. 29a, 18059 Rostock, Germany
* Correspondence: juergen.eberle@charite.de; Tel.: +49-30-450-518-383

Received: 8 February 2019; Accepted: 2 March 2019; Published: 7 March 2019

Abstract: Cutaneous T-cell lymphoma (CTCL) may develop a highly malignant phenotype in its late phase, and patients may profit from innovative therapies. The plant extract indirubin and its chemical derivatives represent new and promising antitumor strategies. This first report on the effects of an indirubin derivative in CTCL cells shows a strong decrease of cell proliferation and cell viability as well as an induction of apoptosis, suggesting indirubin derivatives for therapy of CTCL. As concerning the mode of activity, the indirubin derivative DKP-071 activated the extrinsic apoptosis cascade via caspase-8 and caspase-3 through downregulation of the caspase antagonistic proteins c-FLIP and XIAP. Importantly, a strong increase of reactive oxygen species (ROS) was observed as an immediate early effect in response to DKP-071 treatment. The use of antioxidative pre-treatment proved the decisive role of ROS, which turned out upstream of all other proapoptotic effects monitored. Thus, reactive oxygen species appear as a highly active proapoptotic pathway in CTCL, which may be promising for therapeutic intervention. This pathway can be efficiently activated by an indirubin derivative.

Keywords: CTCL; apoptosis; cell viability; c-FLIP; XIAP

---

1. Introduction

Reactive oxygen species (ROS) play important roles in tissue damage and aging, as also addressed in this special issue. On the other hand, an increasing number of scientific studies in recent years indicated a particular role of ROS in apoptosis regulation in cancer cells. The mechanism(s) are still under discussion. Here, we give an example of cutaneous T-cell lymphoma (CTCL), where ROS is induced by the drug candidate indirubin.

Non-Hodgkin's lymphomas (NHL) have shown increasing incidence in the last decades. About 5% of NHL are characterized by primary cutaneous manifestation through clonal proliferation of skin-homing memory T cells. This group of cutaneous T-cell lymphomas (CTCL) encloses Mycosis fungoides, Sézary syndrome and CD30(+) lymphoproliferative disorders. Cutaneous T-cell lymphomas represent a clinically and biologically distinct group of NHL without evidence for systemic disease at the time of first diagnosis. In clinical appearance and prognosis, they are clearly different from the histotypically cognate systemic lymphomas and their possible secondary cutaneous manifestations. Typically, they have the immunophenotype of CD3+ CD4+ CD45RO+ memory T-lymphocytes. While in its early stage CTCL may show an also indolent clinical course, it frequently transforms to a rapidly growing, malignant phenotype in later phases [1,2]. New treatments are needed particularly for these patients.

The elimination of tumor cells through the induction of apoptosis represents a principle goal in cancer treatment, and therapy resistance can thus be frequently explained by apoptosis deficiency [3]. Also, established therapies for CTCL as UV radiation or extracorporal photopheresis aim at an induction of apoptosis in tumor cells [4]. Extrinsic proapoptotic pathways are initiated by death ligands as CD95L/FasL or TRAIL (TNF-related apoptosis-inducing ligand), which also contribute to self-control of lymphocytes. Their binding to death receptors results in the formation of a death-inducing signaling complex, where initiator caspase-8 is activated [5]. Caspase-8 activation can be prevented by the competitive inhibitor protein c-FLIP (cellular FLICE-inhibitory protein) [6]. Initiator caspase-8 may cleave and activate the main effector caspase-3, which itself cleaves a large number of death substrates with the final result of DNA fragmentation [7]. Caspase-3 is negatively regulated through the binding of XIAP (chromosome X-linked inhibitor of apoptosis protein) [8].

In CTCL cells, the activation of the extrinsic caspase cascade plays a decisive role in controlling apoptosis. Thus, apoptosis resistance is correlated with reduced expression of the death receptor CD95/FAS [9] as well as with high and constitutive expression of c-FLIP [10]. Also, activation of the pro-survival transcription factors NF-κB [11] and STAT3 [12] were reported. In particular, different therapeutic strategies as NSAIDs, SAHA (suberoylanilide hydroxamic) and pentoxifylline resulted in downregulation of c-FLIP and XIAP in CTCL cells [13–15]. Finally, reactive oxygen species (ROS) may contribute to the regulation of apoptosis [16,17]. This is also suggested by enhanced levels of singlet oxygen ($^1O_2$) in the course of photodynamic therapy (PDT), used for the treatment of actinic keratosis [18,19] and also considered for CTCL [20]. However, the relation of ROS with described apoptosis pathways is still largely elusive.

The natural compound indirubin and a number of reported chemical derivatives are considered as candidates for cancer therapy. Indirubin was identified as an active component in a traditional Chinese medicine remedy (Danggui Longhui Wan), also applied for chronic myeloid leukemia. In clinical trials, indirubin has shown significant antitumor activity in chronic myeloid and chronic granulocytic leukemia [21,22]. Explaining the mode of action, a large number of intracellular targets have been described for indirubin derivatives, including cyclin-dependent kinases (CDK1, CDK2, CDK4 and CDK5), pRb, glycogen synthase kinase 3 (GSK-3), STAT3 (Signal transducer and activator of transcription), EGFR (Epidermal growth factor receptor), c-Jun, and JNK2 [23–26]. The activation of extrinsic apoptosis pathways by indirubin derivatives was found in melanoma cells [27,28]. To improve the anti-cancer activity of indirubin, we have previously introduced a series of chemical modifications [29–31]. Here, we investigated the direct effects of the indirubin derivative DKP-071 in CTCL cells. We furthermore unraveled its mode of action, which is based on caspase activation, downregulation of the caspase antagonists c-FLIP and XIAP and, in particular, on early production of ROS.

2. Results

2.1. Decreased Cell Proliferation and Viability Along with Induced Apoptosis by DKP-071

Synthesis and structural aspects of the indirubin derivative DKP-071/substance 9d (Figure 1a) have been reported previously [31]. Here, its effects in three CTCL cell lines MyLa, HuT-78 and HH were investigated. These cell lines are characterized by the formation of cell clusters, a typical lymphocyte differentiation step [32,33]. In a first approach, we observed reduced cell cluster size in response to DKP-071, likely indicating reduced T-cell activity (Figure 1b). In line with this, cell proliferation was significantly reduced at 24 h, as determined by WST-1 assay (Figure 1c). Reduced cell numbers were however not due to direct cytotoxicity, as lactate dehydrogenase (LDH) release assays at 24 h did not show a significant increase in MyLa or in HH cells (Figure 1d).

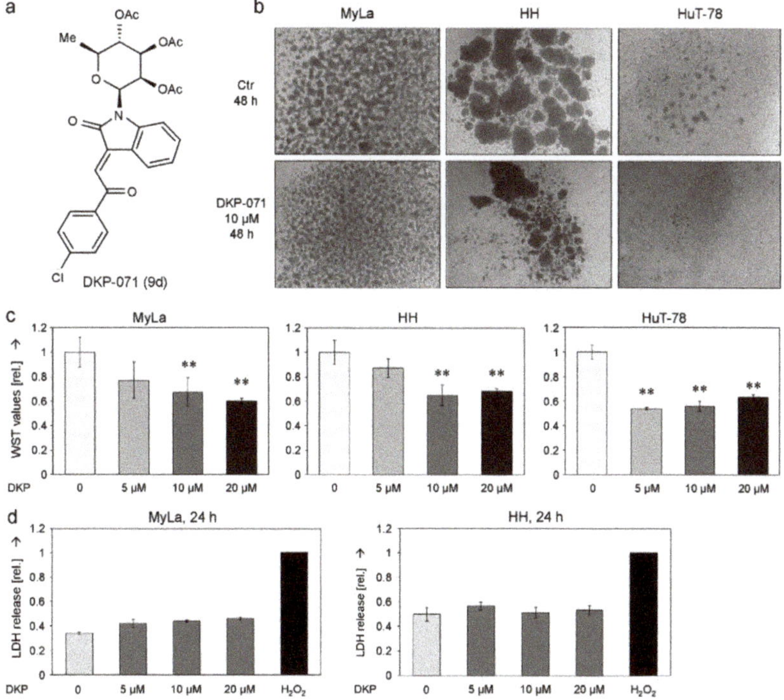

Figure 1. Decreased cell proliferation of CTCL cells by DKP-071. (**a**) Chemical structure of the indirubin derivative DKP-071 (termed substance 9d in [31]). (**b**) Cell cluster formation in CTCL cell lines MyLa, HuT-78 and HH. Control cells (Ctr) are shown vs. cells treated for 48 h with 10 µM DKP-071 (magnification: 1:40). Many independent experiments showed the same result. (**c**) Cell proliferation of MyLa, HuT-78 and HH, at 24 h in response to treatment with 5, 10 and 20 µM DKP-071 (DKP). Cell proliferation data were determined by WST-1 assay, and values are shown in relation (rel) to negative controls (0), which were set to "1". Statistical significance is indicated (** $p < 0.01$). (**d**) Cytotoxicity was determined at 24 h in MyLa and in HH cells by LDH release assay. Values are shown in relation (rel) to H_2O_2-treated positive controls, which were set to "1".

Cell viability, as determined by calcein staining, was strongly decreased. A dose dependency (5–20 µM) was shown for MyLa and HH cells. At 48 h of treatment, 10 µM DKP-071 reduced the numbers of viable cells to 23% (MyLa), 9% (HuT-78) and 38% (HH), respectively (Figure 2a). Based on cell viability data, we calculated IC50 values of 7 µM DKP-071 for Myla and 11 µM for HH. For HuT-78, we used the WST data of Figure 1c, which resulted in an IC50 value of 8 µM for HuT-78. Loss of cell viability went along with an induction of apoptosis, which was determined by counting sub-G1 cells in cell cycle analyses. Induction of apoptosis showed a comparable dose dependency. At 48 h of treatment, 10 µM DKP-071 induced apoptosis in 17% (MyLa), 24% (HuT-78) and 22% of HH cells, respectively (Figure 2b). The concentration of 10 µM was selected for subsequent experiments.

2.2. Changes of Mitochondrial Membrane Potential and ROS Production

Questioning the mechanisms that mediate the antineoplastic effects of DKP-071 in CTCL cells, we determined the relative changes in the mitochondrial membrane potential (MMP) as well as relative levels of reactive oxygen species (ROS) in response to treatment. Loss of MMP, indicative for an activation of mitochondrial apoptosis pathways, already started in the three cell lines at 5 h (31–49%) but was much more evident at later time (24 h, 90% cells with low MMP; Figure 3a).

Reactive oxygen species (ROS) may mediate independent cell death pathways in cancer cells which are not yet completely understood [16]. Earlier than the loss of MMP, ROS levels were already strongly enhanced after 2 h. Thus, 87%, 83% and 57% of MyLa, HuT-78 and HH cells, respectively, showed high ROS levels at 2 h of DKP-071 treatment (Figure 3b).

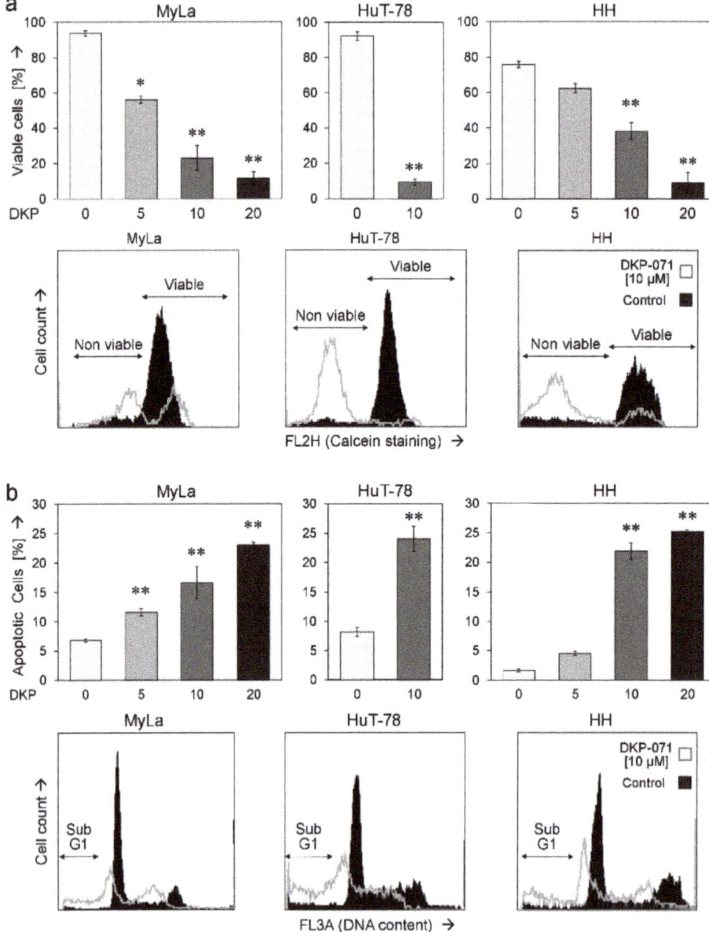

Figure 2. Reduced cell viability and induction of apoptosis. (**a**) Cell viability and (**b**) apoptosis were determined in three cell lines, in response to 48 h treatment with DKP-071 (5, 10 and 20 µM for MyLa and HH as well as 10 µM for HuT-78). Values were determined by calcein staining (**a**) and propidiumiodide staining (**b**), respectively. Characteristic histograms are shown for each cell line (10 µM treatment, overlays with controls); fractions of non-viable and viable as well as of apoptotic cells (sub-G1) are indicated. Mean values of triplicates +/− SDs of a representative experiment are shown. Statistical significance is indicated (treated cells vs. controls; * $p < 0.05$; ** $p < 0.01$).

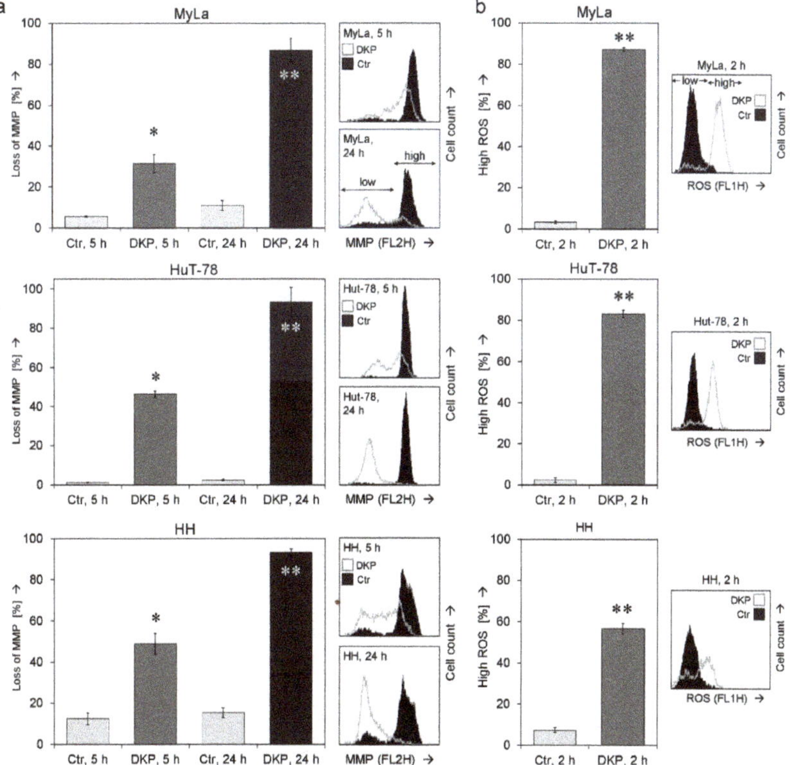

Figure 3. Effects on mitochondrial membrane potential and on ROS levels. (**a**) Relative changes in mitochondrial membrane potential (MMP) were determined at 5 h and 24 h in three CTCL cell lines in response to treatment with DKP-071 (10 µM). Mean values of triplicates +/− SD are shown; a second independent experiment series of MyLa revealed highly comparable results. Representative histograms (overlays of treated cells vs. controls) are given on the right side. (**b**) ROS levels were determined at 2 h of treatment. Mean values of triplicates +/− SD are shown; for MyLa, three independent experiments, each one with triplicates, revealed highly comparable results. Representative histograms (overlays of treated cells vs. controls) are given on the right side. Statistical significance is indicated (treated cells vs. controls; * $p < 0.05$; ** $p < 0.01$).

2.3. Critical role of ROS for Proapoptotic Effects of DKP-071

To prove the significance of ROS as well as of caspase activation for the antineoplastic effects, the antioxidants tocopherol (vitamin E, VE) and N-acetyl cysteine (NAC) as well as the pan-caspase inhibitor QVD-Oph were applied. ROS production in response to DKP-071 was slightly reduced by VE and was strongly reduced by NAC, as shown in MyLa at 2 h. Most effective was a combination of VE and NAC (both at 2 mM, VE/NAC), which completely abolished ROS production after DKP-071 treatment in the three cell lines. QVD-Oph remained without effect on ROS, indicating that ROS was independent of caspase activity (Figure 4).

ROS scavenging by VE/NAC proved the significant and upstream role of ROS for DKP-071-mediated effects. Thus, cell proliferation, which was decreased by DKP-071 at 24 h, was restored in three cell lines by VE/NAC (Figure 5a). Similarly, the effects of DKP-071 on cell viability were strongly diminished by VE/NAC as shown in MyLa at 48 h (from 4% to 78% viable cells, Figure 5b). Finally, the induction of apoptosis, induced by DKP-071 in MyLa at 48 h (35%), was completely prevented by VE/NAC (3%, Figure 5c). Caspase inhibition through QVD-Oph also diminished apoptosis and loss of cell viability, which was, however, less effective than the antioxydative treatment (Figure 4b,c). These findings support the explanation that ROS production was an upstream step.

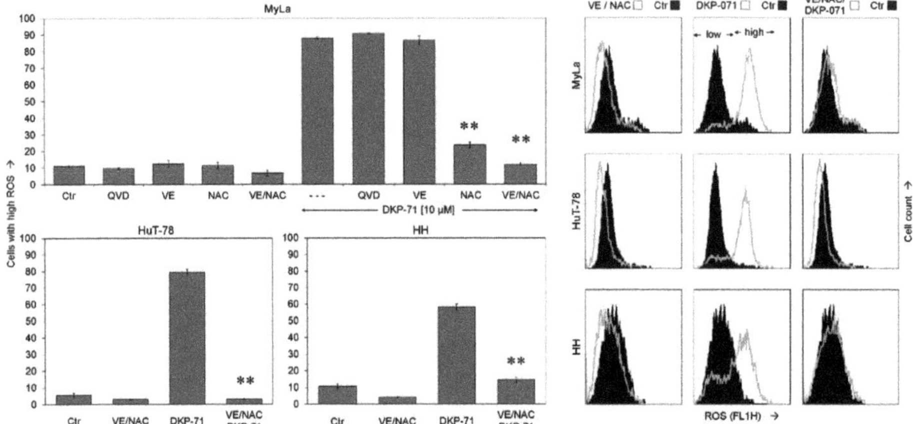

Figure 4. ROS suppression by antioxidative treatment. ROS levels are shown in MyLa in response to DKP-071 (10 µM). In addition, antagonists as vitamin E (VE, 1 mM), N-acetyl cysteine (NAC, 1 mM), the pancaspase inhibitor QVD-Oph (QVD, 10 µM), as well as combined NAC and VE (each 2 mM) were applied 1 h before DKP-071 treatment was started. Cells which received only DKP-071 but no antagonist are indicated by (- - -). The antioxidative effect was also shown in HuT-78 and in HH by the use of VE/NAC. Mean values of triplicates +/− SD of a representative experiment are shown; for MyLa, three independent experiments, each one with triplicates, revealed highly comparable results. Examples of flow cytometry measurement are shown on the right side as overlays versus control. Statistical significance of the differences of DKP-071/NAC-treated cells as well as DKP-071/VE/NAC-treated cells is indicated, each compared to the cells that received only DKP-071 (** $p < 0.01$).

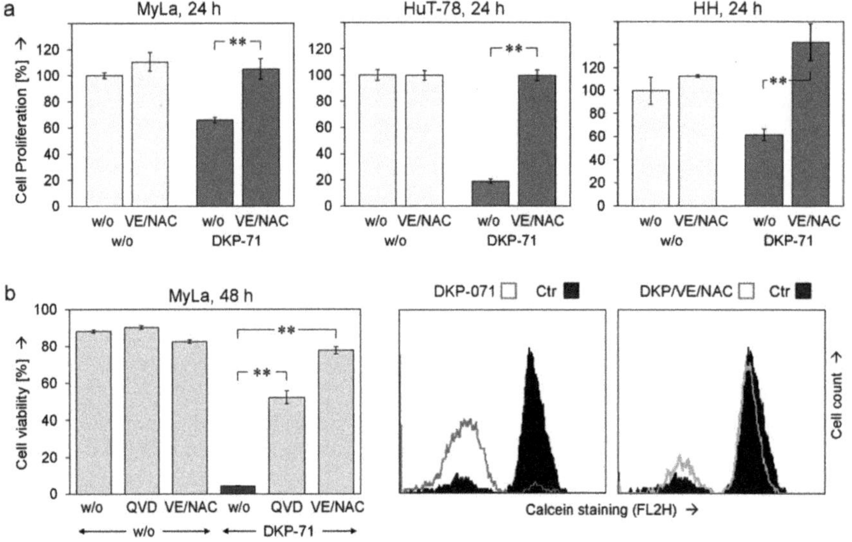

Figure 5. *Cont.*

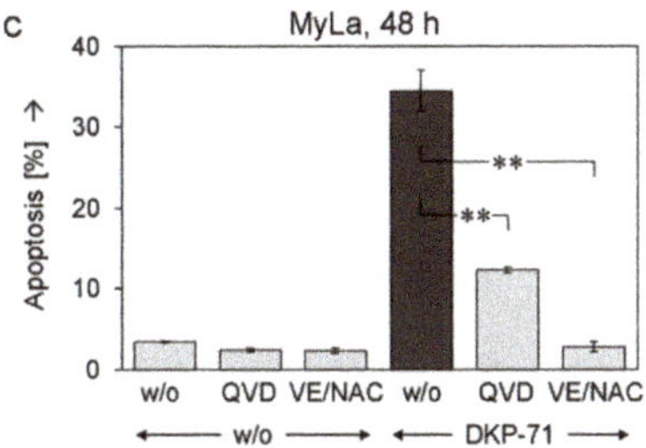
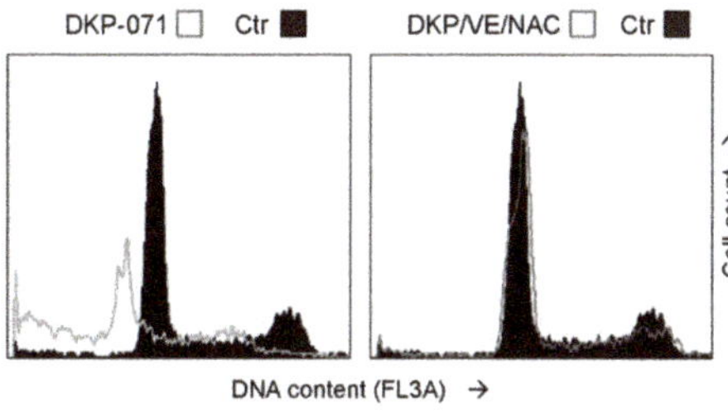

Figure 5. ROS suppression prevents apoptosis and restores cell viability. (**a**) Cell proliferation was determined at 24 h in response to DKP-071 as well as in response to the combination of NAC and VitE (2 mM; VE/NAC). Values were normalized to non-treated controls (100%). (**b**,**c**) Effects of agonists and antagonists on apoptosis induction (**b**) and cell viability (**c**), both at 48 h, are shown for MyLa cells. Examples of flow cytometry measurement are shown on the right side as overlays of treated cells versus controls. Mean values of triplicates +/− SD of representative experiments are shown; at least two independent experiments, each one with triplicates, revealed highly comparable results. Statistical significance of the differences between DKP-071/VE/NAC-treated cells to cells that received only DKP-071 is indicated (** $p < 0.01$).

2.4. Role of Caspases and Caspase Antagonistic Proteins

ROS also appeared upstream of MMP loss. Thus, VE/NAC almost completely prevented the loss of MMP in MyLa at 5 h (24% to 7%) whereas caspase inhibition was less effective here (18%, Figure 6a). Extrinsic caspase pathways are of major importance for apoptosis regulation in lymphoma cells, also including CTCL [10]. Thus, we investigated by Western blotting the activation/processing of initiator caspase-8 and the main effector caspase-3 as well as the expression of the caspase-3 antagonist XIAP and the caspase-8 antagonist c-FLIP.

In response to 40 h treatment with DKP-071, the proform of caspase-8 (53/55 kD) disappeared, indicating its complete processing. In parallel, Caspase-3 was processed to its mature, active cleavage product of 15 kDa. Importantly, VE/NAC strongly reduced the processing both of caspase-8 and caspase-3. In particular, no active cleavage product of caspase-3 (15 kDa) was detected, but the processing was stopped at an intermediate product of 19 kDa. In clear contrast, QVD-Oph remained without effect on caspase-8. It, however, prevented caspase-3 autoprocessing and halted the cascade at the 19 and 17 kDa intermediate cleavage products (Figure 6b, top). These findings clearly showed that ROS was upstream of any caspase regulation, while QVD-Oph acts downstream as a caspase-3 antagonist.

Explaining the activation of caspases, DKP-071 strongly reduced the expression of two most characteristic caspase antagonists, namely XIAP (51 kDa) and c-FLIP (long isoform, 52 kDa and short isoform, 25 kDa). Of particular note, this downregulation was completely prevented by antioxidants (VE/NAC), while caspase-3 inhibition through QVD-Oph remained without effect on c-FLIP and XIAP (Figure 6b, bottom). Thus, ROS was also upstream of the downregulation of c-FLIP and XIAP. These findings suggest a cascade in CTCL cells, leading from ROS production in response to DKP-071 treatment to c-FLIP and XIAP downregulation and further to caspase activation and apoptosis (Figure 6c).

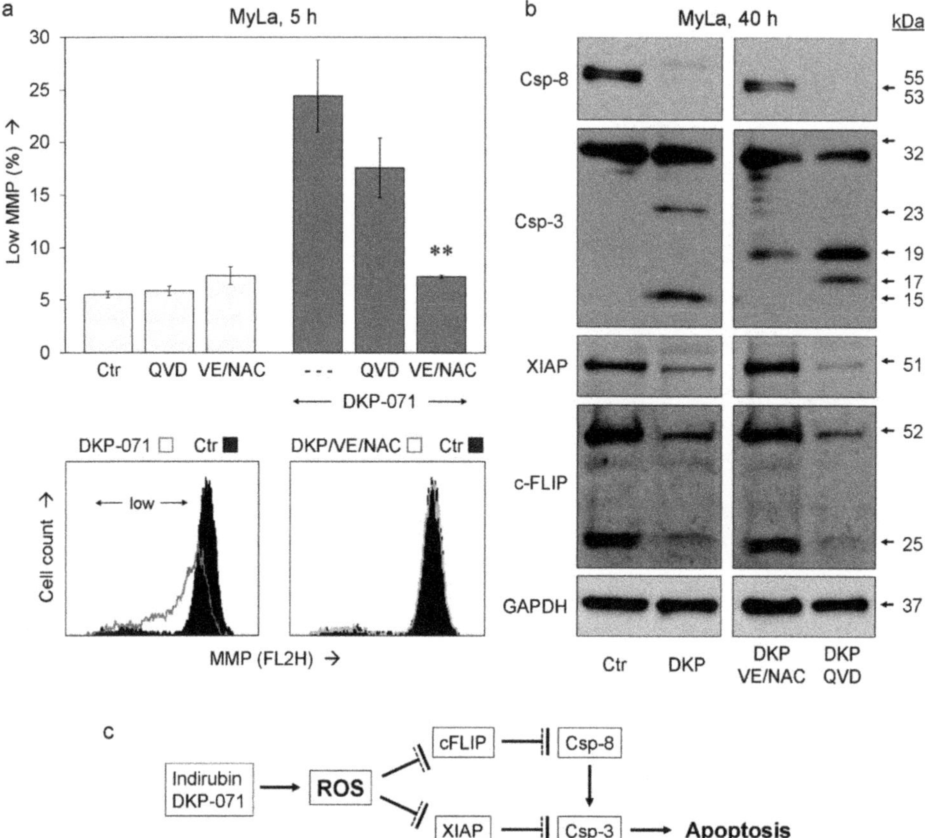

Figure 6. Effects on MMP and caspase cascade. (**a**) Effects of antioxidative treatment (VE/NAC, 2 mM) as well as of QVD-Oph (QVD, 10 μM) on mitochondrial membrane potential (MMP) are shown for MyLa cells at 5 h of DKP-071 treatment. Cells, which received only DKP-071 but no antagonist are indicated by (- - -). Mean values of triplicates +/− SD of a representative experiment are shown; two independent experiments, each one with triplicates, revealed highly comparable results. Statistical significance of the differences between DKP-071/VE/NAC-treated cells to cells that received only DKP-071 is indicated (** $p < 0.01$). Examples of flow cytometry measurement are shown below as overlays versus control. The cell population with low MMP is indicated. (**b**) Effects of DKP-071 and antioxidative pre-treatment on the expression of characteristic regulatory proteins of the extrinsic apoptosis caspase cascade were investigated by Western blotting. Each 30 μg of protein was loaded per lane, and blots were probed with antibodies for extrinsic initiator caspase-8 (proform, 53/55 kDa), main effector caspase-3 (proform, 32 kDa; cleavage products, 23, 19, 17, 15 kDa), caspase-3 antagonist XIAP (51 kDa) and caspase-8 antagonist c-FLIP (FLIP$_L$, long, 52 kDa; FLIP$_S$, short, 25 kDa). The housekeeping protein glyceraldehyde 3-phosphate dehydrogenase (GAPDH, 37 kDa) was used as loading control. Three independent series of protein extracts and Western blotting experiments revealed highly comparable results. (**c**) The pathway suggested for indirubin DKP-071-mediated apoptosis. Arrows indicate activation; blunt end lines indicate inhibition.

3. Discussion

Many new therapies established in recent decades for most tumors have sometimes dramatically enhanced patients' survival. Also for CTCL, present therapeutic options as topical steroids, bexarotene, phototherapy, interferon or some forms of targeted therapy have strongly improved the clinical outcome and often allow long-term survival. Nevertheless, in its late phase, CTCL may transform to a rapidly growing and ulcerating phenotype, characterized also by pronounced therapy resistance [1,2]. As apoptosis deficiency represents an in principal decisive issue in therapy resistance, the specific targeting of apoptosis pathways appears as a promising strategy [3].

Indirubin has been identified as the major component of a traditional Chinese medicine remedy, also applied for leukemia. In clinical trials for chronic myeloid and chronic granulocytic leukemia, it revealed significant antitumor activity, resulting in partial or complete remissions [21,22]. Chemical modifications of indirubin may further enhance its activity. Thus, we have recently reported the synthesis of new indirubin derivatives characterized by N-glycosylated 3-alkylideneoxindol structures [31]. This first report on the effects of an indirubin derivative in CTCL cells shows a particular high activity resulting in decreased cell proliferation and cell viability as well as induction of apoptosis. There is no cutaneous non-tumorigenic T-cell line, which could be used for investigations in vitro and prove the tumor-specificity of the effects. However, due to the only moderate side effects reported for indirubin in clinical trials [21,22], the here investigated indirubin derivative may be suggested as a potential new therapeutic option for CTCL. A proapoptotic strategy, as by DKP-071, may also apply for early CTCL, characterized by an indolent clinical course and apoptosis susceptibility.

As concerning the pathways, by which indirubins may exert their effects, multiple targets have been suggested including CDKs, GSK-3β, pRb, c-Src, FGF-R1, VEGFR, STAT3, c-Jun and JNK2 [23–26]. In melanoma cells, we have previously shown that indurubin derivatives enhance extrinsic apoptosis pathways as induced by TRAIL (TNF-related apoptosis-inducing ligand [27,28]. Extrinsic apoptosis pathways via caspase-8/caspase-3 are of particular importance for apoptosis regulation in CTCL cells [10,34]. Also in CTCL, DKP-071 mediated its proapoptotic effects via activation of the extrinsic caspase cascade, as shown by caspase-8 and caspase-3 processing as well as by the inhibition of the effects of indirubin through the caspase-3 antagonist QVD-Oph. Caspase activity is controlled by several antiapoptotic proteins such as the competitive caspase-8 antagonist c-FLIP and the protein family of cIAPs (cellular inhibitor of apoptosis proteins), e.g., the caspase-3 antagonist XIAP (chromosome X-linked) [6,8]. These two antagonists appeared as essentially involved in the present setting, as strongly downregulated by DKP-071 in CTCL cells. Downregulation of these two factors in CTCL has also been reported by other treatments. Thus, c-FLIP was downregulated in response to SAHA (suberoylanilide hydroxamic) and NSAIDs, while XIAP was downregulated in response to SAHA and pentoxifylline [13–15].

Reactive oxygen species (ROS) play important roles in tissue damage and aging, and antioxidative strategies were established to prevent these negative effects. Besides this, ROS may also be involved in proapoptotic signaling in cancer cells [16,28,35,36]. As an example from the clinic, photodynamic therapy (PDT), used for the treatment of actinic keratosis, results in the production of high amounts of singlet oxygen. However, the role of ROS in PDT is still controversially discussed [18–20]. ROS may act as a signaling molecule e.g., by affecting the mitochondrial membrane and thus activating intrinsic, mitochondrial apoptosis pathways, or it may provoke cellular damage e.g., by oxidizing membrane lipids resulting in necrosis or activation of a damage response.

Here, we show even another alternative. In CTCL, ROS acts in cellular signaling clearly upstream of the extrinsic apoptosis pathway. It is induced already at 2 h in response to indirubin treatment and resides upstream of the downregulation of XIAP and c-FLIP as well as upstream of the loss of MMP, suggesting a signaling cascade as shown in Figure 6c. Also for other treatments in CTCL cells, relations of ROS and apoptosis induction have been reported, such as for an inhibitor of the antioxidative protein mucin 1 [37], for silencing of the enhancer of zeste homolog 2 (EZH2) [38] and for doxycycline [39]. Thus, ROS production appears to be an important cellular signaling step related to the induction of apoptosis. The new proapoptotic pathways behind this may be useful for targeted cancer therapy. In CTCL, our data suggest a strong relation between ROS, caspase antagonists and the activation of the extrinsic apoptosis pathway. This signaling cascade is efficiently triggered by the indirubin derivative DKP-071, suggesting it as a therapeutic strategy for CTCL.

4. Materials and Methods

4.1. Cell Culture and Treatments

Three CTCL cell lines have been used here, which derive from patients with Mycosis fungoides and Sézary syndrome, respectively: Cell line MyLa was kindly supplied by Prof. K. Kaltoft, Århus University, Århus, DK. It derived from a plaque biopsy of a patient with MF [40]; HuT-78 was kindly supplied by Prof. R.C. Gallo, University of Maryland, Baltimore, MD, USA. [41] It derived from PBMCs of a patient with Sézary syndrome; and HH (CRL-2105, ATCC, Manassas, VA, USA) derived from peripheral blood of a patient with aggressive CTCL [42]. Cells were grown at 37 °C, 5% CO_2 in RPMI 1640 medium supplemented with L-glutamine, 10% FCS and antibiotics (Biochrom, Berlin, Germany). Under these applied conditions, the cell lines revealed a similar growth behavior and proliferation rate; only the formation of cell clusters, typical for T-cells, varied considerably (Figure 1a).

For assays, cells were seeded in 24-well culture plates (100,000 cells and 500 µL per well) or in 96-well pates (40,000 cells and 200 µL per well). For protein extraction, cells were seeded in 6-well plates (400,000 per well, 2 mL). Treatments started at 24 h after seeding.

The indirubin derivative DKP-071 (termed substance 9d in [31]; Figure 1a) was used in concentrations of 5–20 µM. For caspase inhibition, cells were pre-incubated for 1 h with the pan-caspase inhibitor QVD-Oph (Sigma-Aldrich, Taufkirchen, Germany, 10 µM). For ROS scavenging, cells were pre-treated for 1 h with 1 mM α-tocopherol (vitamin E, Fluka, Steinheim, Germany), with N-acetylcysteine (NAC; Sigma-Aldrich, Taufkirchen, Germany; 1 mM) or with a combination of vitamin E and NAC (VE/NAC; each 2 mM).

4.2. Assays for Apoptosis, Cytotoxicity, Cell Viability and Cell Proliferation

Apoptotic cells were quantified as sub-G1 cells (less DNA than G1 cells) in cell cycle analyses. The assay reveals less DNA in apoptotic cells, because small DNA fragments are washed out from isolated nuclei [43]. Cells were harvested by centrifugation, lysed and stained for at least 1 h in hypotonic PI solution (40 µg/mL propidium iodide, Sigma-Aldrich, in 0.1% sodium citrate, 0.1% Triton X-100). Stained isolated nuclei were analyzed by flow cytometry at FL3A with a FACS Calibur (BD Bioscience, Bedford, MA, USA). Cells in G1, G2 and S-phase as well as sub-G1 cells were determined.

Cell cytotoxicity was determined by the analysis of cell supernatants for the activity of lactate dehydrogenase (LDH), which is released from cytotoxic cells in culture. Aliquots of cell supernatants were collected at 24 h, and LDH activity was quantified with a LDH activity assay (Cytotoxicity detection assay; Roche Diagnostics, Penzberg, Germany), following the protocol of the supplier. As positive controls, Triton X-100 (0.7%)-treated cells were used. Relative values were determined in an ELISA reader. The increased LDH activity in treated wells traces back to damaged, cytotoxic cells.

Cell viability was determined by staining cells with calcein-AM (PromoCell, Heidelberg, Germany), which is a cell-permeant non-fluorescent substance that is converted to green-fluorescent calcein in live cells by the activity of intracellular esterases. Cells, grown and treated in 24-well plates, were harvested after 24 h or 48 h by centrifugation. After washing 1× with PBS, they were resuspended in 200 µL of 2.5 µg/mL calcein-AM in PBS. Staining was done at 37 °C for 1 h. Labeled cells were washed again with 1 mL of PBS and were resuspended in 200 µL PBS. The percentage of viable cells was determined by flow cytometry (FL2H). In the calcein assay, the total cell number does not contribute to the results, only the percentage of active cells in the remaining cell population is determined.

Cell proliferation was determined by WST-1 assay (Roche Diagnostics). It depends on the cleavage of the water-soluble tetrazolium salt by mitochondrial dehydrogenases in metabolically active cells. CTCL cells were seeded in 96-well plates, and treatments were started after 24 h. At the time of analysis (24–48 h), WST-1 reagent was added for 1–2 h, and the absorbance at 450 nm was determined in an ELISA reader. Data were reported as the percentage of non-treated controls. The WST-1 assay gives an overview of all the antiproliferative effects of a given drug, which includes a lack of cells due to decreased cell proliferation as well as a lack of cells due to the induction of apoptosis or induction of cytotoxicity. Furthermore, the cell activity of the remaining cells contributes to the results.

4.3. Mitochondrial Membrane Potential (MMP) and Reactive Oxygen Species (ROS)

Changes to MMP over time or in response to drug treatment can be assessed by staining cells with the fluorescent dye TMRM$^+$ (Tetramethylrhodamin-methylester, Sigma-Aldrich). Due to its positive charge, TMRM$^+$ accumulates in negatively charged mitochondria. In the course of depolarisation of the mitochondria, (e.g., at the beginning of apoptosis), anions are released and TMRM$^+$ concentrations decrease. This staining may also be used to determine the effectiveness of therapeutic compounds [44]. Cells grown and treated in 24-well plates were harvested and stained for 20 min at 37 °C with 1 μM TMRM$^+$. After two-times washing with PBS, cells were analyzed by flow cytometry (FL2H).

Intracellular ROS levels were determined by the cell-permeable substance H$_2$DCFDA (2′,7′-dichlorodihydrofluorescein diacetate; Thermo Fisher Scientific, Hennigsdorf, Germany). In cells with high ROS, the non-fluorescent H$_2$DCFDA is oxidized and thus converted in the strongly fluorescent DCF (2′,7′-Dichlorfluorescein). Cells grown in 24-well plates were pre-treated for 1 h with H$_2$DCFDA (10 μM), before starting treatment with effectors. After treatment, cells were harvested by centrifugation, washed with 1 mL PBS, resuspended in PBS, and analyzed by flow cytometry (FL1H). As positive controls, cells were treated with H$_2$O$_2$ (1 mM, 1 h).

4.4. Western Blotting

For Western blotting, total protein extracts were obtained by cell lysis in 150 mM NaCl, 1 mM EDTA, 2 mM PMSF, 1 mM leupeptin, 1 mM pepstatin, 0.5% SDS, 0.5% NP-40 and 10 mM Tris-HCl, pH 7.5. Western blotting on nitrocellulose membranes was performed as described previously [45]. Primary antibodies: Cleaved caspase-3 (9664, rabbit, 1:1000, Cell Signaling, Danvers, MA, USA); XIAP (#2042, rabbit, 1:1000, Cell Signaling); caspase-8 (#9746; mouse, 1:1000, Cell Signaling); c-FLIP (G-11, sc-5276; Santa Cruz, Heidelberg, Germany; 1:50); and GAPDH (sc-32233, mouse, 1:1000, Santa Cruz Biotech). Secondary antibodies: peroxidase-labelled goat anti-rabbit and goat anti-mouse (Dako, Hamburg, Germany; 1:5000).

4.5. Statistics

Assays described above were performed in triplicate determinations, and usually at least two completely independent series were performed for each experiment. For the determination of statistical significance, the values of all individual experiments were given together, after normalizing according to the controls. A Student's *t*-test (two-tailed, heteroscedastic) was applied, and p-values of <0.05 were considered as statistically significant, while p-values of <0.01 were considered as highly significant. When applicable, significance is indicated in the bar charts (* for $p < 0.05$; ** for $p < 0.01$). The results of Western blots were reproduced in three independent series of experiments.

5. Conclusions

In summary, cutaneous T-cell lymphoma cells can be targeted by the induction of ROS. This results in an activation of the extrinsic apoptosis pathway via downregulation of c-FLIP and XIAP. This pathway is efficiently activated by an indirubin derivative, which thus represents an interesting candidate for therapy of cutaneous T-cell lymphoma.

Author Contributions: All authors have contributed to this work. M.Y.S. performed experiments, evaluations, literature screening and helped writing the manuscript; U.S. performed several experiments; C.A., P.L. and U.R. discussed the project and helped with writing the manuscript; P.L. also contributed with essential reagents; J.E. suggested the project, helped with the experiments, controlled all evaluations and wrote the manuscript.

Acknowledgments: Marwa Soltan received a scholarship from the Berlin Foundation for Dermatology (Berliner Stiftung für Dermatologie, BSD).

Conflicts of Interest: The authors declare no conflict of interest.

References

1. Willemze, R.; Jaffe, E.S.; Burg, G.; Cerroni, L.; Berti, E.; Swerdlow, S.H.; Ralfkiaer, E.; Chimenti, S.; Diaz-Perez, J.L.; Duncan, L.M.; et al. WHO-EORTC classification for cutaneous lymphomas. *Blood* **2005**, *105*, 3768–3785. [CrossRef] [PubMed]

2. Jawed, S.I.; Myskowski, P.L.; Horwitz, S.; Moskowitz, A.; Querfeld, C. Primary cutaneous T-cell lymphoma (mycosis fungoides and Sezary syndrome): Part II. Prognosis, management, and future directions. *J. Am. Acad. Dermatol.* **2014**, *70*, 223.e1–223.e17. [CrossRef] [PubMed]

3. Eberle, J.; Kurbanov, B.M.; Hossini, A.M.; Trefzer, U.; Fecker, L.F. Overcoming apoptosis deficiency of melanoma-hope for new therapeutic approaches. *Drug Resist. Updates* **2007**, *10*, 218–234. [CrossRef] [PubMed]

4. Bladon, J.; Taylor, P.C. Extracorporeal photopheresis induces apoptosis in the lymphocytes of cutaneous T-cell lymphoma and graft-versus-host disease patients. *Br. J. Haematol.* **1999**, *107*, 707–711. [CrossRef] [PubMed]

5. Krammer, P.H.; Arnold, R.; Lavrik, I.N. Life and death in peripheral T cells. *Nat. Rev. Immunol.* **2007**, *7*, 532–542. [CrossRef] [PubMed]

6. Irmler, M.; Thome, M.; Hahne, M.; Schneider, P.; Hofmann, K.; Steiner, V.; Bodmer, J.L.; Schroter, M.; Burns, K.; Mattmann, C.; et al. Inhibition of death receptor signals by cellular FLIP. *Nature* **1997**, *388*, 190–195. [CrossRef] [PubMed]

7. Fischer, U.; Janicke, R.U.; Schulze-Osthoff, K. Many cuts to ruin: A comprehensive update of caspase substrates. *Cell Death Differ.* **2003**, *10*, 76–100. [CrossRef] [PubMed]

8. Deveraux, Q.L.; Takahashi, R.; Salvesen, G.S.; Reed, J.C. X-linked IAP is a direct inhibitor of cell-death proteases. *Nature* **1997**, *388*, 300–304. [CrossRef] [PubMed]

9. Wu, J.; Nihal, M.; Siddiqui, J.; Vonderheid, E.C.; Wood, G.S. Low FAS/CD95 expression by CTCL correlates with reduced sensitivity to apoptosis that can be restored by FAS upregulation. *J. Investig. Dermatol.* **2009**, *129*, 1165–1173. [CrossRef] [PubMed]

10. Braun, F.K.; Fecker, L.F.; Schwarz, C.; Walden, P.; Assaf, C.; Durkop, H.; Sterry, W.; Eberle, J. Blockade of death receptor-mediated pathways early in the signaling cascade coincides with distinct apoptosis resistance in cutaneous T-cell lymphoma cells. *J. Investig. Dermatol.* **2007**, *127*, 2425–2437. [CrossRef] [PubMed]

11. Sors, A.; Jean-Louis, F.; Pellet, C.; Laroche, L.; Dubertret, L.; Courtois, G.; Bachelez, H.; Michel, L. Down-regulating constitutive activation of the NF-kappaB canonical pathway overcomes the resistance of cutaneous T-cell lymphoma to apoptosis. *Blood* **2006**, *107*, 2354–2363. [CrossRef] [PubMed]

12. Eriksen, K.W.; Kaltoft, K.; Mikkelsen, G.; Nielsen, M.; Zhang, Q.; Geisler, C.; Nissen, M.H.; Ropke, C.; Wasik, M.A.; Odum, N. Constitutive STAT3-activation in Sezary syndrome: Tyrphostin AG490 inhibits STAT3-activation, interleukin-2 receptor expression and growth of leukemic Sezary cells. *Leukemia* **2001**, *15*, 787–793. [CrossRef] [PubMed]

13. Braun, F.K.; Al-Yacoub, N.; Plotz, M.; Mobs, M.; Sterry, W.; Eberle, J. Nonsteroidal anti-inflammatory drugs induce apoptosis in cutaneous T-cell lymphoma cells and enhance their sensitivity for TNF-related apoptosis-inducing ligand. *J. Investig. Dermatol.* **2012**, *132*, 429–439. [CrossRef] [PubMed]

14. Al-Yacoub, N.; Fecker, L.F.; Mobs, M.; Plotz, M.; Braun, F.K.; Sterry, W.; Eberle, J. Apoptosis induction by SAHA in cutaneous T-cell lymphoma cells is related to downregulation of c-FLIP and enhanced TRAIL signaling. *J. Investig. Dermatol.* **2012**, *132*, 2263–2274. [CrossRef] [PubMed]

15. Gahlot, S.; Khan, M.A.; Rishi, L.; Majumdar, S. Pentoxifylline augments TRAIL/Apo2L mediated apoptosis in cutaneous T cell lymphoma (HuT-78 and MyLa) by modulating the expression of antiapoptotic proteins and death receptors. *Biochem. Pharmacol.* **2010**, *80*, 1650–1661. [CrossRef] [PubMed]

16. Quast, S.A.; Berger, A.; Eberle, J. ROS-dependent phosphorylation of Bax by wortmannin sensitizes melanoma cells for TRAIL-induced apoptosis. *Cell Death Dis.* **2013**, *4*, e839. [CrossRef] [PubMed]

17. Bauer, D.; Werth, F.; Nguyen, H.A.; Kiecker, F.; Eberle, J. Critical role of reactive oxygen species (ROS) for synergistic enhancement of apoptosis by vemurafenib and the potassium channel inhibitor TRAM-34 in melanoma cells. *Cell Death Dis.* **2017**, *8*, e2594. [CrossRef] [PubMed]

18. Zou, Z.; Chang, H.; Li, H.; Wang, S. Induction of reactive oxygen species: An emerging approach for cancer therapy. *Apoptosis* **2017**, *22*, 1321–1335. [CrossRef] [PubMed]

19. Salmeron, M.L.; Quintana-Aguiar, J.; De La Rosa, J.V.; Lopez-Blanco, F.; Castrillo, A.; Gallardo, G.; Tabraue, C. Phenalenone-photodynamic therapy induces apoptosis on human tumor cells mediated by caspase-8 and p38-MAPK activation. *Mol. Carcinog.* **2018**, *57*, 1525–1539. [CrossRef] [PubMed]

20. Salva, K.A.; Kim, Y.H.; Rahbar, Z.; Wood, G.S. Epigenetically Enhanced PDT Induces Significantly Higher Levels of Multiple Extrinsic Pathway Apoptotic Factors than Standard PDT, Resulting in Greater Extrinsic and Overall Apoptosis of Cutaneous T-cell Lymphoma. *Photochem. Photobiol.* **2018**, *94*, 1058–1065. [CrossRef] [PubMed]

21. Xiao, Z.; Hao, Y.; Liu, B.; Qian, L. Indirubin and meisoindigo in the treatment of chronic myelogenous leukemia in China. *Leuk. Lymphoma* **2002**, *43*, 1763–1768. [CrossRef] [PubMed]

22. Blazevic, T.; Heiss, E.H.; Atanasov, A.G.; Breuss, J.M.; Dirsch, V.M.; Uhrin, P. Indirubin and Indirubin Derivatives for Counteracting Proliferative Diseases. *Evid. Based Complement. Altern. Med.* **2015**, *2015*, 654098. [CrossRef] [PubMed]

23. Perabo, F.G.; Frossler, C.; Landwehrs, G.; Schmidt, D.H.; von Rucker, A.; Wirger, A.; Muller, S.C. Indirubin-3′-monoxime, a CDK inhibitor induces growth inhibition and apoptosis-independent up-regulation of survivin in transitional cell cancer. *Anticancer Res.* **2006**, *26*, 2129–2135. [PubMed]

24. Meijer, L.; Skaltsounis, A.L.; Magiatis, P.; Polychronopoulos, P.; Knockaert, M.; Leost, M.; Ryan, X.P.; Vonica, C.A.; Brivanlou, A.; Dajani, R.; et al. GSK-3-selective inhibitors derived from Tyrian purple indirubins. *Chem. Biol.* **2003**, *10*, 1255–1266. [CrossRef] [PubMed]

25. Zhen, Y.; Sorensen, V.; Jin, Y.; Suo, Z.; Wiedlocha, A. Indirubin-3′-monoxime inhibits autophosphorylation of FGFR1 and stimulates ERK1/2 activity via p38 MAPK. *Oncogene* **2007**, *26*, 6372–6385. [CrossRef] [PubMed]

26. Zhang, Y.; Du, Z.; Zhuang, Z.; Wang, Y.; Wang, F.; Liu, S.; Wang, H.; Feng, H.; Li, H.; Wang, L.; et al. E804 induces growth arrest, differentiation and apoptosis of glioblastoma cells by blocking Stat3 signaling. *J. Neurooncol.* **2015**, *125*, 265–275. [CrossRef] [PubMed]

27. Berger, A.; Quast, S.A.; Plotz, M.; Hein, M.; Kunz, M.; Langer, P.; Eberle, J. Sensitization of melanoma cells for death ligand-induced apoptosis by an indirubin derivative—Enhancement of both extrinsic and intrinsic apoptosis pathways. *Biochem. Pharmacol.* **2011**, *81*, 71–81. [CrossRef] [PubMed]

28. Zhivkova, V.; Kiecker, F.; Langer, P.; Eberle, J. Crucial role of reactive oxygen species (ROS) for the proapoptotic effects of indirubin derivative DKP-073 in melanoma cells. *Mol. Carcinog.* **2019**, *58*, 258–269. [CrossRef] [PubMed]

29. Libnow, S.; Methling, K.; Hein, M.; Michalik, D.; Harms, M.; Wende, K.; Flemming, A.; Kockerling, M.; Reinke, H.; Bednarski, P.J.; et al. Synthesis of indirubin-N′-glycosides and their anti-proliferative activity against human cancer cell lines. *Bioorg. Med. Chem.* **2008**, *16*, 5570–5583. [CrossRef] [PubMed]

30. Kunz, M.; Driller, K.M.; Hein, M.; Libnow, S.; Hohensee, I.; Ramer, R.; Hinz, B.; Berger, A.; Eberle, J.; Langer, P. Synthesis of thia-analogous indirubin N-Glycosides and their influence on melanoma cell growth and apoptosis. *ChemMedChem* **2010**, *5*, 534–539. [CrossRef] [PubMed]

31. Kleeblatt, D.; Becker, M.; Plötz, M.; Schonherr, M.; Villinger, A.; Hein, M.; Eberle, J.; Kunz, M.; Rahman, Q.; Langer, P. Synthesis and Bioactivity of N-glycosylated 3-(2-Oxo-2- arylethylidene)-indolin-2-ones. *RSC Adv.* **2015**, *5*, 20623–20633. [CrossRef]

32. Bang, K.; Lund, M.; Mogensen, S.C.; Thestrup-Pedersen, K. In vitro culture of skin-homing T lymphocytes from inflammatory skin diseases. *Exp. Dermatol.* **2005**, *14*, 391–397. [CrossRef] [PubMed]

33. Drexler, H.G.; Pommerenke, C.; Eberth, S.; Nagel, S. Hodgkin lymphoma cell lines: To separate the wheat from the chaff. *Biol. Chem.* **2018**, *399*, 511–523. [CrossRef] [PubMed]

34. Braun, F.K.; Hirsch, B.; Al-Yacoub, N.; Durkop, H.; Assaf, C.; Kadin, M.E.; Sterry, W.; Eberle, J. Resistance of cutaneous anaplastic large-cell lymphoma cells to apoptosis by death ligands is enhanced by CD30-mediated overexpression of c-FLIP. *J. Investig. Dermatol.* **2010**, *130*, 826–840. [CrossRef] [PubMed]

35. Chen, J.C.; Zhang, Y.; Jie, X.M.; She, J.; Dongye, G.Z.; Zhong, Y.; Deng, Y.Y.; Wang, J.; Guo, B.Y.; Chen, L.M. Ruthenium(II) salicylate complexes inducing ROS-mediated apoptosis by targeting thioredoxin reductase. *J. Inorg. Biochem.* **2019**, *193*, 112–123. [CrossRef] [PubMed]

36. Takasaki, T.; Hagihara, K.; Satoh, R.; Sugiura, R. More than Just an Immunosuppressant: The Emerging Role of FTY720 as a Novel Inducer of ROS and Apoptosis. *Oxid. Med. Cell. Longev.* **2018**, *2018*, 4397159. [CrossRef] [PubMed]

37. Jain, S.; Washington, A.; Leaf, R.K.; Bhargava, P.; Clark, R.A.; Kupper, T.S.; Stroopinsky, D.; Pyzer, A.; Cole, L.; Nahas, M.; et al. Decitabine Priming Enhances Mucin 1 Inhibition Mediated Disruption of Redox Homeostasis in Cutaneous T-Cell Lymphoma. *Mol. Cancer Ther.* **2017**, *16*, 2304–2314. [CrossRef] [PubMed]
38. Yi, S.; Sun, J.; Qiu, L.; Fu, W.; Wang, A.; Liu, X.; Yang, Y.; Kadin, M.E.; Tu, P.; Wang, Y. Dual Role of EZH2 in Cutaneous Anaplastic Large Cell Lymphoma: Promoting Tumor Cell Survival and Regulating Tumor Microenvironment. *J. Investig. Dermatol.* **2017**, *138*, 1126–1136. [CrossRef] [PubMed]
39. Alexander-Savino, C.V.; Hayden, M.S.; Richardson, C.; Zhao, J.; Poligone, B. Doxycycline is an NF-kappaB inhibitor that induces apoptotic cell death in malignant T-cells. *Oncotarget* **2016**, *7*, 75954–75967. [CrossRef] [PubMed]
40. Kaltoft, K.; Bisballe, S.; Dyrberg, T.; Boel, E.; Rasmussen, P.B.; Thestrup-Pedersen, K. Establishment of two continuous T-cell strains from a single plaque of a patient with mycosis fungoides. *In Vitro Cell. Dev. Biol.* **1992**, *28A*, 161–167. [CrossRef] [PubMed]
41. Gootenberg, J.E.; Ruscetti, F.W.; Mier, J.W.; Gazdar, A.; Gallo, R.C. Human cutaneous T cell lymphoma and leukemia cell lines produce and respond to T cell growth factor. *J. Exp. Med.* **1981**, *154*, 1403–1418. [CrossRef] [PubMed]
42. Starkebaum, G.; Loughran, T.P., Jr.; Waters, C.A.; Ruscetti, F.W. Establishment of an IL-2 independent, human T-cell line possessing only the p70 IL-2 receptor. *Int. J. Cancer* **1991**, *49*, 246–253. [CrossRef] [PubMed]
43. Riccardi, C.; Nicoletti, I. Analysis of apoptosis by propidium iodide staining and flow cytometry. *Nat. Protoc.* **2006**, *1*, 1458–1461. [CrossRef] [PubMed]
44. Creed, S.; McKenzie, M. Measurement of Mitochondrial Membrane Potential with the Fluorescent Dye Tetramethylrhodamine Methyl Ester (TMRM). *Methods Mol. Biol.* **2019**, *1928*, 69–76. [PubMed]
45. Eberle, J.; Fecker, L.F.; Hossini, A.M.; Wieder, T.; Daniel, P.T.; Orfanos, C.E.; Geilen, C.C. CD95/Fas signaling in human melanoma cells: Conditional expression of CD95L/FasL overcomes the intrinsic apoptosis resistance of malignant melanoma and inhibits growth and progression of human melanoma xenotransplants. *Oncogene* **2003**, *22*, 9131–9141. [CrossRef] [PubMed]

International Journal of
Molecular Sciences

Article

Artemisinin Attenuated Hydrogen Peroxide (H$_2$O$_2$)-Induced Oxidative Injury in SH-SY5Y and Hippocampal Neurons via the Activation of AMPK Pathway

Xia Zhao [†], Jiankang Fang [†], Shuai Li [†], Uma Gaur, Xingan Xing, Huan Wang and Wenhua Zheng [*]

Center of Reproduction, Development & Aging and Institute of Translation Medicine, Faculty of Health Sciences, University of Macau, Taipa, Macau 999078, China; yb77625@um.edu.mo (X.Z.); yb57646@um.edu.mo (J.F.); yb67619@um.edu.mo (S.L.); gaur.uma2906@gmail.com (U.G.); yb77638@um.edu.mo (X.X.); yb77624@um.edu.mo (H.W.)
* Correspondence: wenhuazheng@umac.mo; Tel.: +853-88224919
† These authors contribute equally to this work.

Received: 29 April 2019; Accepted: 22 May 2019; Published: 31 May 2019

Abstract: Oxidative stress is believed to be one of the main causes of neurodegenerative diseases such as Alzheimer's disease (AD). The pathogenesis of AD is still not elucidated clearly but oxidative stress is one of the key hypotheses. Here, we found that artemisinin, an anti-malarial Chinese medicine, possesses neuroprotective effects. However, the antioxidative effects of artemisinin remain to be explored. In this study, we found that artemisinin rescued SH-SY5Y and hippocampal neuronal cells from hydrogen peroxide (H$_2$O$_2$)-induced cell death at clinically relevant doses in a concentration-dependent manner. Further studies showed that artemisinin significantly restored the nuclear morphology, improved the abnormal changes in intracellular reactive oxygen species (ROS), reduced the mitochondrial membrane potential, and caspase-3 activation, thereby attenuating apoptosis. Artemisinin also stimulated the phosphorylation of the adenosine monophosphate -activated protein kinase (AMPK) pathway in SH-SY5Y cells in a time- and concentration-dependent manner. Inhibition of the AMPK pathway attenuated the protective effect of artemisinin. These data put together suggested that artemisinin has the potential to protect neuronal cells. Similar results were obtained in primary cultured hippocampal neurons. Cumulatively, these results indicated that artemisinin protected neuronal cells from oxidative damage, at least in part through the activation of AMPK. Our findings support the role of artemisinin as a potential therapeutic agent for neurodegenerative diseases.

Keywords: artemisinin; SH-SY5Y cells; hippocampal neurons; H$_2$O$_2$; AMPK pathway

---

1. Introduction

Neuronal damage caused by oxidative stress (mainly reactive oxygen species) is one of the major causes of neurodegenerative diseases such as Alzheimer's disease (AD) [1–3]. Excessive levels of reactive oxygen species (ROS) enhances cellular oxidative stress, leading to lipid peroxidation, protein denaturation, and DNA damage, disrupting cell function and integrity, leading to cell apoptosis [4]. Oxidative stress is an imbalance between ROS production and antioxidant defense and has been found to be associated with aging and aging-related neurodegenerative diseases [5–7]. H$_2$O$_2$ is an important source of ROS and is widely used as an inducer of oxidative stress in cell models for the study of various neurodegenerative diseases caused by oxidative stress [8,9].

Artemisinin is extracted from the plant *Artemisia annua*. It is one of the most effective anti-malarial drug which has saved the lives of millions of malaria patients worldwide [10,11]. In addition to anti-malarial benefits, artemisinin has many other biological and pharmacological properties, including antioxidant, anti-inflammatory, antiviral and anti-tumor effects [12–14]. In addition, we have recently shown that artemisinin have neuroprotective potential [15–17]. However, the effect and underlying action mechanism of artemisinin against oxidative stress in SH-SY5Y neuronal cells and primary cultured neurons are still unknown. Artemisinin can cross the blood–brain barrier (BBB) and has no observable toxicity to the central nervous system itself. This support the hypothesis that artemisinin can have a favorable effect in the treatment of neurological diseases [18].

Decreased mitochondrial membrane potential is a marker of early cell apoptosis and is closely related to cellular oxidative damage and apoptosis.The production of ROS occurs mainly in mitochondria and it accumulates during aging ultimately becoming a major cause of cellular damage. AMP-activated protein kinase (AMPK) is an energy sensor which plays a key role in regulating complex signaling networks of mitochondrial biogenesis [19]. Emerging evidence have shown that AMPK not only maintains energy metabolism balance, but also regulates oxidative stress, endoplasmic reticulum stress, autophagy and inflammation, thereby increasing stress resistance [20]. Moreover, AMPK activation seems to decline during aging. In addition, activation of the AMPK cascade have been shown to be associated with improved glucose and lipid metabolism, as well as neuroprotective and anticancer effects [21,22]. The effects of age associated decline in AMPK activity on mitochondrial dysfunction and age related oxidative damage have also been verified previously [23].

In the present study, we evaluated whether artemisinin can protect against H_2O_2-induced cytotoxicity and the potential mechanisms behind the protective effects of artemisinin in SH-SY5Y cells and primary hippocampal neurons. Our results showed that artemisinin protected SH-SY5Y cells from H_2O_2-induced injury as indicated by3-(4,5-dimethylthiazol-2-yl)-2,5-diphenyl tetrazolium bromide (MTT) assay and mitochondrial membrane potential assay. Furthermore the nuclear morphology and abnormal changes in intracellular ROS were also restored to normal by artemisinin treatment. In addition, caspase-3 which is a key performer of apoptosis and was reported to be activated by H_2O_2 [24], was shown to be downregulated upon artemisinin treatment in present study. The AMPK signaling pathway plays an important role in systemic energy balance and metabolism and regulation of age-related diseases [11,25,26]. Finally, we also delineated the role of the AMPK signaling pathway in the protective effect of artemisinin. All of these may provide interesting insights into the potential applications of artemisinin in future AD research.

2. Results

2.1. Artemisinin Attenuated the Cell Viability Induced by H_2O_2 in SH-SY5Y Cells

SH-SY5Y cells were incubated with different concentrations of artemisinin or H_2O_2 for 24 h and the cell viability was determined by MTT assay. No cytotoxic effect was observed upon artemisinin treatment in SH-SY5Y cells up to 200 μM concentration (Figure 1B). Compared with the control group, H_2O_2 caused significant cytotoxicity in SH-SY5Y cells 600 μM onwards (Figure 1C), and therefore 600 μM H_2O_2 concentration was used in further experiments. To test the protective effects of artemisinin, SH-SY5Y cells were pretreated with artemisinin for 2 h before being exposed to H_2O_2 for another 24 h. The result showed that pre-treatment with artemisinin significantly reduced H_2O_2-induced cell death (Figure 1D) and 12.5 μM artemisinin was used in further experiments.

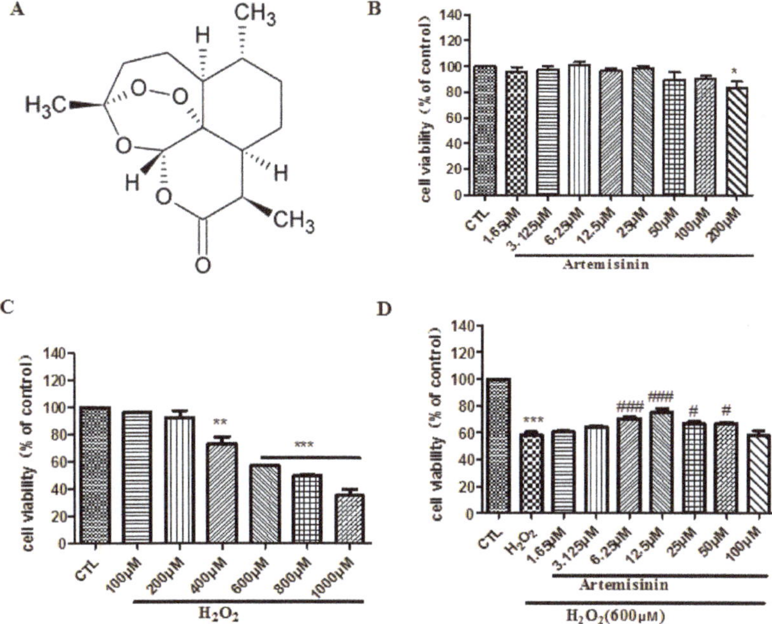

Figure 1. Artemisinin attenuated the decrease in cell viability caused by H_2O_2 in SH-SY5Y cells. (**A**) The structure of artemisinin. (**B**) The cytotoxicity of artemisinin, cells were treated with artemisinin (1.65–200 µM) for 24 h and cell viability was measured using the MTT(3-(4,5-dimethylthiazol-2-yl)-2,5-diphenyl tetrazolium bromide) assay. (**C**) The cytotoxicity of H_2O_2. (**D**) Cells were pretreated with artemisinin at indicated concentrations and then induced with or without 600 µM H_2O_2 for a further 24 h and cell viability was measured using the MTT assay. Data represent means ± SD, * $p < 0.05$, ** $p < 0.01$, *** $p < 0.001$, # $p < 0.05$, ### $p < 0.001$.

2.2. Artemisinin Pretreatment Attenuated H_2O_2-Induced Apoptosis in SH-SY5Y Cells

Both apoptosis and necrosis contribute to the cell viability loss during cell injuries. We tested if the protective effect of artemisinin against H_2O_2 insult was mediated by its anti-apoptosis effects. Nuclei condensation was observed in SH-SY5Y cells after exposure to 600 µM H_2O_2 in Hoechst 33342 staining assay. However, pre-treatment with 12.5 µM artemisinin significantly improved these changes (Figure 2A,B). The result was further confirmed using flow cytometry for Annexin V-FITC/PI-positive cells and data from these experiments indicated that H_2O_2 exposure markedly increased apoptosis in SH-SY5Y cells, while 12.5 µM artemisinin pretreatment significantly reduced the apoptosis caused by H_2O_2 (Figure 2C,D). Caspase-3 plays an important role in apoptosis and in order to further verify the anti-apoptosis effect of artemisinin we checked the caspase-3 activity. We found that artemisinin reversed H_2O_2-induced increase in the activity of caspase—(Figure 2E).

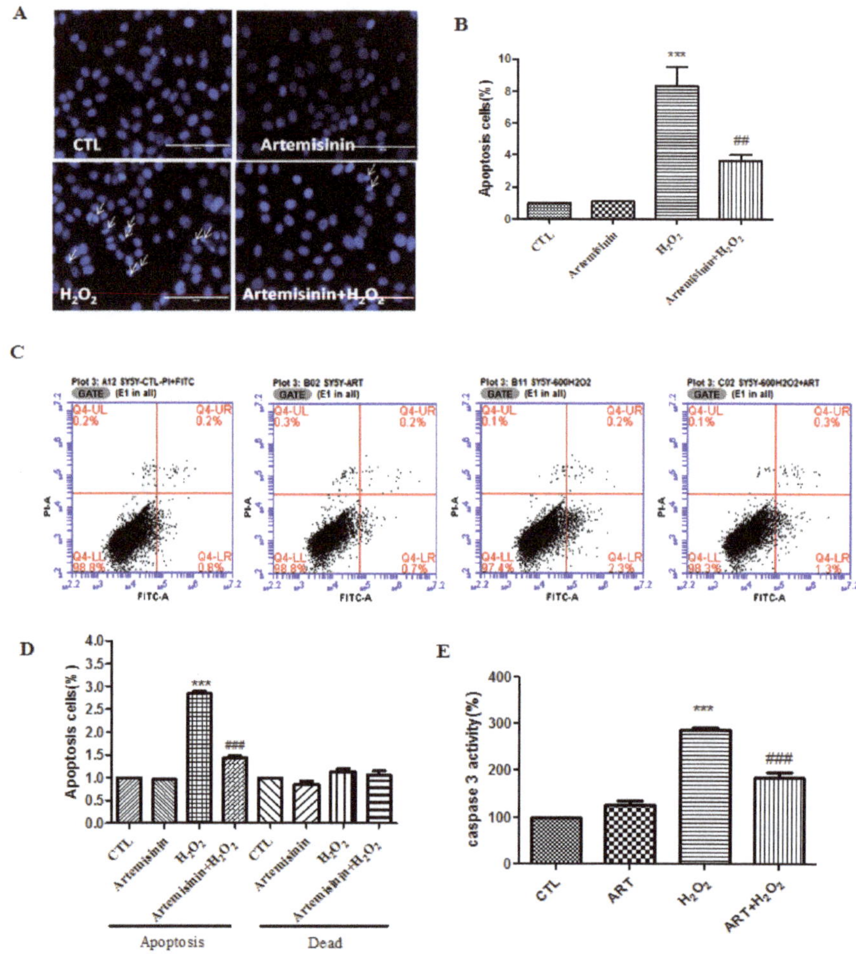

Figure 2. Artemisinin suppressed H_2O_2-induced apoptosis in SH-SY5Y cells. Cells were pre-treated with 12.5 µM artemisinin for 2 h and then induced with or without 600 µM H_2O_2 for another 24 h. The pictures have been taken at a magnification of 40× (100 µm). (**A**) Photographs of representative cultures measured by Hoechst staining. Apoptotic cells are marked with white arrows (**B**) Quantitative analysis of (**A**). (**C**) Photographs of representative cultures measured by flow cytometry. (**D**) Quantitative analysis of (**C**). (**E**) The activity of caspase-3 was monitored by caspase assay. Data represent means ± SD, *** $p < 0.001$, ## $p < 0.01$, ### $p < 0.001$.

2.3. Artemisinin Inhibited H_2O_2-Induced Increase in ROS Level and Restored the Mitochondrial Membrane Potential in SH-SY5Y Cells

Loss of mitochondrial membrane potential ($\Delta\psi$m) due to mitochondrial inhibition was involved in the cell apoptosis caused by H_2O_2. In further study, we elucidated whether artemisinin could reduce H_2O_2-induced $\Delta\psi$m loss. The $\Delta\psi$m in SH-SY5Y cells was assessed by analyzing the red/green fluorescent intensity ratio of JC-1 staining. The results revealed that artemisinin pretreatment significantly prevented the decline of $\Delta\psi$m induced by H_2O_2 (Figure 3A,C). The generation of excess ROS is considered to be among the main causes of cell apoptosis induced by H_2O_2. Therefore, we investigated whether artemisinin blocked H_2O_2-induced oxidative stress in SH-SY5Y cells. Cellular oxidative stress was determined by the CellROXs Deep Red Reagent. SH-SY5Y cells pretreated with or without 12.5 μM artemisinin for 2 h were treated with 600 μM H_2O_2 for 24 h. As expected (Figure 3B,D), artemisinin significantly decreased the intracellular ROS production induced by H_2O_2.

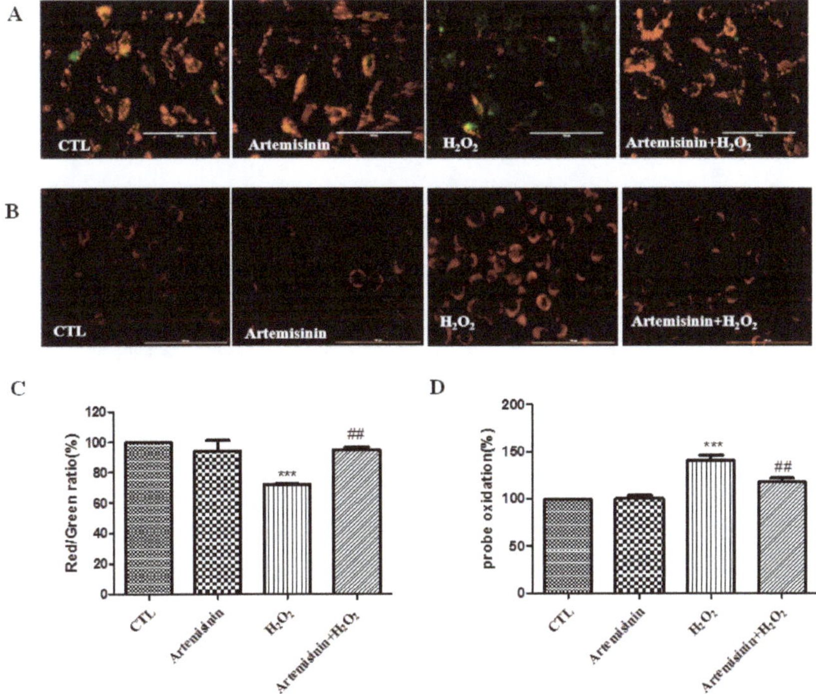

Figure 3. Artemisinin inhibited H_2O_2-induced increase of reactive oxygen species (ROS) level and restored the mitochondrial membrane potential in SH-SY5Y cells. (**A**). After pre-treatment with 12.5 μM artemisinin or 0.1% DMSO (vehicle control) for 2 h, SH-SY5Y cells were incubated with or without 600 μM H_2O_2 for another 24 h. The decline in the membrane potential was reflected by the shift of fluorescence from red to green indicated by JC-1. The pictures have been taken at a magnification of 40× (100 μm). (**B**). Intracellular ROS level was measured by the CellROXs Deep Red Reagent. The pictures has been taken on 40× (100 μm). (**C**). Quantitative analysis of (**A**). (**D**). Quantitative analysis of (**B**). The data were represented as the mean ± SD of three independent experiments. *** $p < 0.001$, ## $p < 0.01$.

2.4. Artemisinin Stimulated the Phosphorylation of AMPK in SH-SY5Y Cells

AMPK is a highly conserved regulator of cellular energy metabolism that plays an important role in regulating cell growth, proliferation, survival, and regulation of energy metabolism in the body. We therefore tested whether AMPK is involved in protective effect of artemisinin in SH-SY5Y cells. As shown in Figure 4, after treatment with different doses of artemisinin for different time points, the phosphorylation of AMPK was gradually increased (Figure 4A–D).

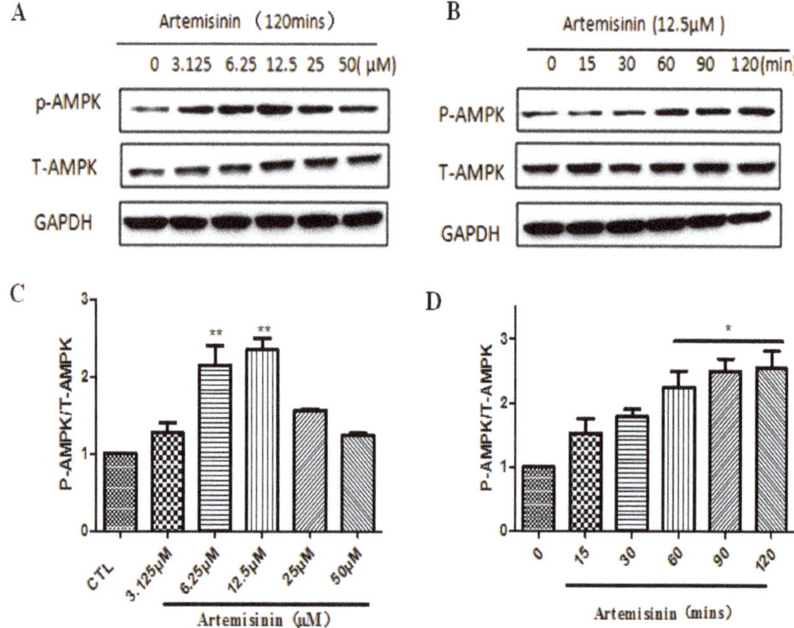

Figure 4. Artemisinin stimulated the phosphorylation of AMP-activated protein kinase (AMPK) in SH-SY5Y cells. (**A,C**) The SH-SY5Y cells were collected with artemisinin treatment for different times (0, 15, 30, 60, 90 and 120 min) at 12.5 µM, and at different concentrations (3.15, 6.25, 12.5, 25 and 50 µM) for 120 min. The expression of phosphocreatine AMPK, total AMPK and glyceraldehyde-3-phosphate dehydrogenase (GAPDH) were detected by western blot. (**B,D**) Quantification of representative protein band from western blotting. The data were represented as the mean ± SD. * $p < 0.05$, ** $p < 0.01$, versus the control group was considered significantly different.

2.5. AMPK Mediated the Protective Effects of Artemisinin in SH-SY5Y Cells

In order to examine whether AMPK is involved in the survival promoting effect of artemisinin on cell apoptosis induced by H_2O_2, we knocked down AMPKα by using shRNA plasmid specific for AMPK in SH-SY5Y cells (Figure 5A). After 24 h of transfection, cells were treated with or without artemisinin, then incubated with 600 µM H_2O_2 for 24 h, and the cell viability was determined by MTT assay. As shown in Figure 5B, knockdown of AMPKα by shRNA significantly attenuated the protective effect of artemisinin. To further confirm the effect of AMPK, we used the compound C (a specific inhibitor for the AMPK) to pretreat cells for 30 min. MTT assay showed that pretreatment with Compound C blocked the protective effects of artemisinin (Figure 5C). We got the similar result from TUNEL staining assay (Figure 5D,E). Western blot also showed that AMPK pathway played an important role in the protective effect of artemisinin (Figure 5G,H). At the same time, blocking the AMPK signaling pathway also restored the activity of caspase-3 (Figure 5G,I).

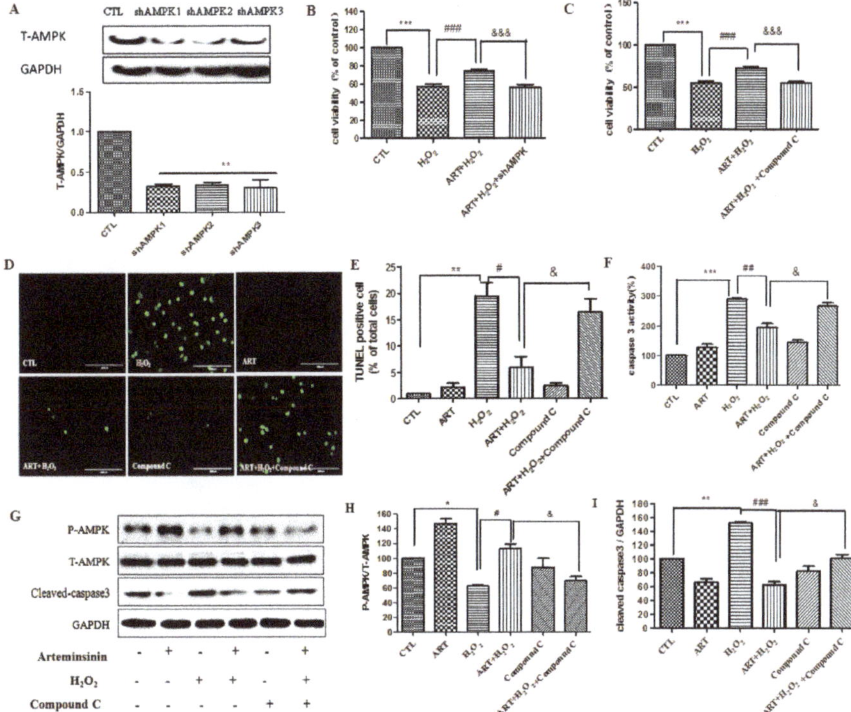

Figure 5. AMPK pathway mediated the protective effects of artemisinin in SH-SY5Y cells. (**A**). Cells were pretreated with shAMPKα 1–3 plasmid for 24 h, The knockout efficiency was determined by western blot. (**B**). Cells were pretreated with shAMPKα1 plasmid for 24 h before 12.5 μM artemisinin treatment for 2 h, and then incubated with or without 600 μM H_2O_2 for a further 24 h. Cell viability was evaluated using the MTT assay. (**C**). Cells were pre-treated with 5 μM Compound C (AMPK inhibitor) for 30 min, and treated with 12.5 μM artemisinin for 2 h, then incubated with or without 600 μM H_2O_2 for another 24 h, and the cell viability was determined by MTT assay. (**D,E**). TUNEL staining manifested that artemisinin attenuated H_2O_2-induced cell apoptosis significantly. The pictures have been taken at a magnification of 40× (100 μm) (**F**) The activity of caspase 3 was measured by caspase assay. (**G**). The expression of phosphorylated AMPK, total AMPK, cleaved caspase-3 and GAPDH were detected by western blot. (**H,I**). Quantitative analysis of (**G**). The data was represented as the mean ± SD. * $p < 0.05$, ** $p < 0.01$, *** $p < 0.001$, ## $p < 0.01$, ### $p < 0.001$, & $p < 0.05$, &&& $p < 0.001$.

2.6. Artemisinin Protected Primary Cultured Hippocampal Neurons Against H_2O_2 Induced Injury via AMPK Kinase

In order to verify whether the neuroprotective effects of artemisinin against H_2O_2 are limited to SH-SY5Y, we also checked its protective effects on primary cultured hippocampal neurons. In concordance with SH-SY5Y, artemisinin successfully protected hippocampal neurons against H_2O_2 insult in a concentration-dependent manner as shown in Figure 6A–C and Figure S1. Similar results was found in cortex neuron (Figure S2).The primary cultured neurons were more sensitive to both H_2O_2 insult and artemisinin treatment and therefore artemisinin achieved its significant protective effect at a lower concentration. Moreover, the protective effect of artemisinin is also inhibited by AMPK inhibitor compound C in these primary neuron cells (Figure 6E–H). The protein NeuN is localized in the nucleus and perinuclear cytoplasm of most neurons in the central nervous system. It is widely used to assess the functional status of neurons and their pathological states. Our results show that artemisinin can improve neuronal damage caused by H_2O_2 (Figure 6D). These results are consistent with our results from SH-SY5Y cells and further confirmed that artemisinin was able to protect neuronal cells from oxidative stress via the activation of AMPK kinase.

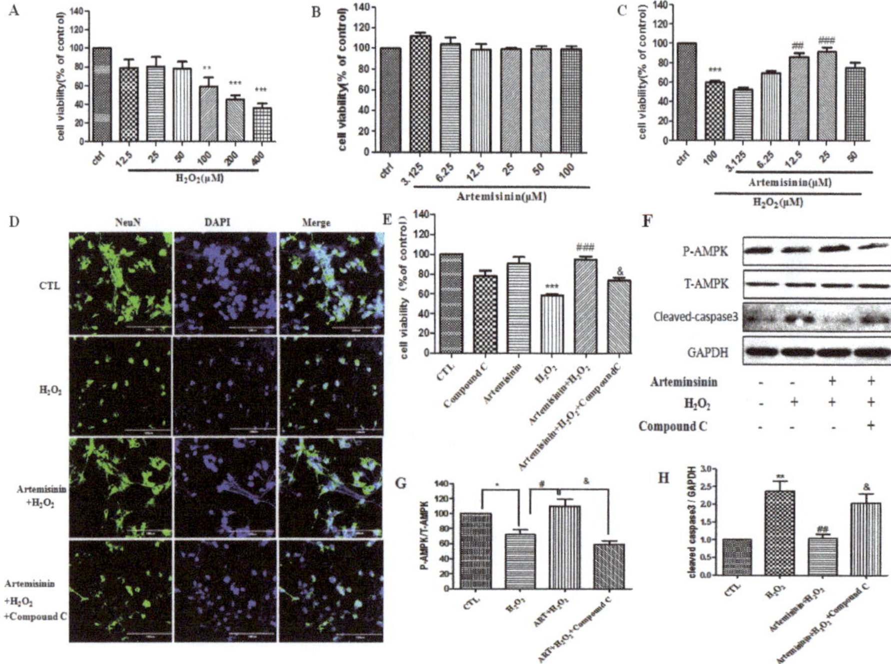

Figure 6. Artemisinin attenuated the decrease in cell viability caused by H_2O_2 in neuronal cells. (**A**) The cytotoxicity of H_2O_2 on neuronal cells. (**B**) Dose of artemisinin. (**C**) Primary cultured hippocampal neurons were pretreated with artemisinin at indicated concentrations and then induced with or without 100 μM H_2O_2 for another 24 h, and cell viability was measured using the MTT assay. (**D**) Primary cultured hippocampal neurons pretreated with 2.5 μM compound C for 30 min were treated with 25 μM artemisinin for 2 h, and then incubated with or without 100 μM H_2O_2 for another 24 h. Immunocytochemistry of NeuN in each group was detected. The pictures were taken at a magnification of 40× (100 μm). (**E**) Primary cultured hippocampal neurons pretreated with 2.5 μM compound C for 30 min were treated with 25 μM artemisinin for 2 h, and then incubated with or without 100 μM H_2O_2 for another 24 h. Cell viability was measured by the MTT assay. (**F**) The expression of phosphorylated AMPK, total AMPK, cleaved caspase-3 and GAPDH were detected by western blot. (**G,H**) Quantitative analysis of (**F**). The data were represented as the mean ± SD. ** $p < 0.01$, *** $p < 0.001$, ## $p < 0.01$, ### $p < 0.001$.

3. Discussion

In this study, we demonstrated that artemisinin blocked H_2O_2-induced neuronal damage by activating the AMPK pathway. This study discovered a novel neuroprotective effect of artemisinin, suggesting that artemisinin may have potential for the treatment of neurodegenerative diseases. In addition to antimalarial effect, recently, artemisinin has also demonstrated to possess anti-tumor and anti-inflammatory properties [27]. The hippocampus is one of the most vulnerable parts of the brain and is susceptible to numerous pathological conditions [28]. In the current study, we used H_2O_2 injury cellular model SH-SY5Y and hippocampal neurons to study the effects of artemisinin on oxidative stress. We found that pretreatment with artemisinin provided protection to SH-SY5Y and hippocampal neurons from H_2O_2-induced damage. Consistent with others and our previous reports, we found that H_2O_2 is cytotoxic to SH-SY5Y cells in a dose-dependent pattern [29]. In addition, data from Hoechst

33342 staining assay and flow cytometry showed that H_2O_2 also induced apoptosis in SH-SY5Y cells. The cell viability of SH-SY5Y cells incubated with H_2O_2 was significantly increased when pretreated with artemisinin. In addition, artemisinin can also attenuate H_2O_2-induced apoptosis of SH-SY5Y cells, which further suggests that artemisinin has a protective effect on H_2O_2-induced SH-SY5Y cells.

Mitochondria are the major site of ROS production [30], and H_2O_2 increases oxidative stress damage by increasing ROS production [29]. We concluded that pretreatment with artemisinin reduced H_2O_2-induced ROS accumulation. Mitochondrial dysfunction caused by oxidative stress plays a key role in the pathogenesis of aging-related neurodegenerative diseases [31]. Strategies to block mitochondrial dysfunction have been declared to be a potential therapy for preventing cell death [32]. The results of the current study have shown that artemisinin pretreatment can inhibit H_2O_2-induced $\Delta\psi m$ loss.

The AMPK signaling pathway plays a major role in cell survival, apoptosis and senescence prevention [33]. This pathway has been shown to be a key signaling pathway that induces cellular antioxidant mechanisms. Our data suggest that artemisinin pretreatment stimulates phosphorylation of AMPK and activates the AMPK pathway and plays a key role in the neuroprotective effects of H_2O_2 induced injury, consistent with activation of the AMPK cascade. The protective effect of artemisinin was attenuated after blocking the AMPK signaling pathway with the specific knockout plasmids of AMPK and specific pharmacological inhibitor Compound C. Consistently, pretreatment with artemisinin reversed [34] the increase in caspase-3 activity caused by H_2O_2. After blocking AMPK, the effect of artemisinin on caspase-3 activity disappeared. A major mechanism in the cellular defense against oxidative is activation of the Nrf2-antioxidant response element signaling pathway, but when blocked AMPK pathway Nrf2 expression was decreased (Figure S3).Taken together, these results provided mechanistic evidence to support the view that artemisinin-mediated protection against H_2O_2 induced injury occurs through AMPK activation. We got consistent results in hippocampal neurons as well (Figure 6). Our results suggest that the possible regulatory mechanism of artemisinin operates by protecting neuronal SH-SY5Y cells and hippocampal neurons from H_2O_2-induced oxidative damage via activating the AMPK signaling pathway (Figure 7), including inhibition of intracellular ROS production; restoration of mitochondrial membrane potential ($\Delta\psi m$); activation of AMPK signaling pathway; and reduction of caspase-3 activity. In addition, we also checked some of the other derivatives of artemisinin, like dihydroartemisinin (DHA), performed cellular viability assays, and found that DHA was less effective. The results are attached in the supplementary data (Figure S4). When we compared both the compound cellular viability assay results we found that the artemisinin was better than DHA. Owing to its lipid-soluble nature, artemisinin can pass the blood brain barrier and maintain a higher concentration in the central nervous system [35], giving it advantage over other neurological drugs. Also as an FDA-approved anti-malarial drug, artemisinin has been clinically used for decades with no significant adverse effects [36].

Our results found that artemisinin can eliminate ROS and protect the SH-SY5Y and hippocampal neurons from H_2O_2-induced oxidative damage. Therefore, artemisinin has the potential to be a novel drug to prevent and treat neurodegenerative diseases.

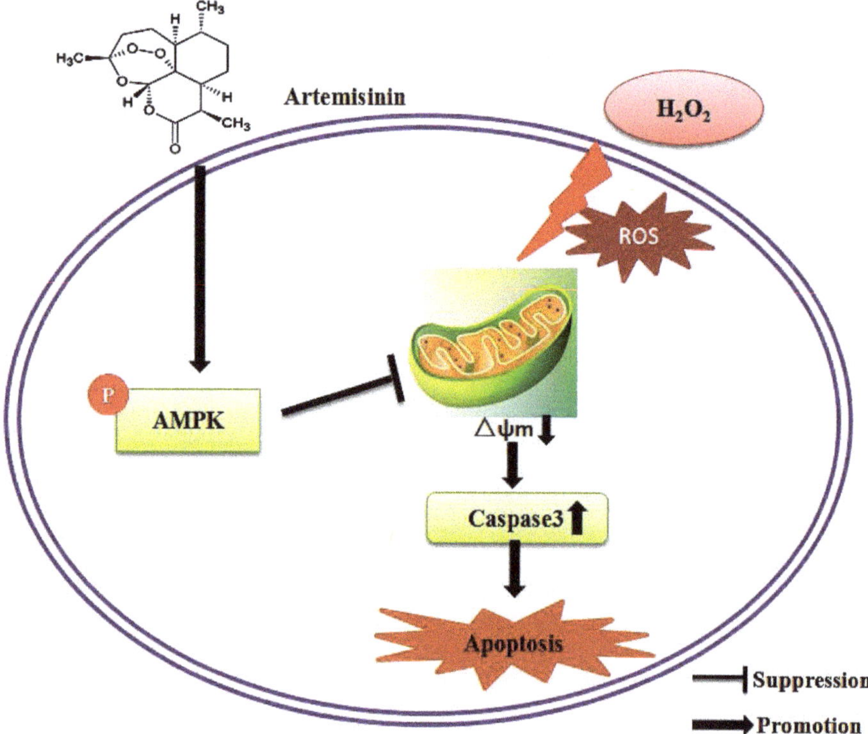

Figure 7. The possible mechanism of artemisinin. Excessive H_2O_2 leads to excessive ROS production and abnormal mitochondrial membrane potential, which in turn produces oxidative stress. Oxidative stress further leads to activation of caspase-3, which ultimately initiates apoptosis. Administration of artemisinin significantly inhibited H_2O_2-induced apoptosis and artemisinin ameliorated abnormal changes in these markers by activating the AMPK pathway. By blocking AMPK, the protective effect of artemisinin disappears. Therefore, artemisinin can be considered as a promising candidate for the treatment of neurodegenerative diseases caused by oxidative stress.

4. Materials and Methods

4.1. Materials

Analytical grade artemisinin was purchased from Meilunbio (Dalian China). Dimethyl sulfoxide (DMSO), Dulbecco's modified Eagle's medium (DMEM) and hydrogen peroxide (H_2O_2) were procured from Sigma (St. Louis, MO, USA). Poly-D-lysine, 3-(4,5-Dimethylthiazol-2-yl)-2,5-diphenyltetrazolium bromide (MTT), 5,5′,6,6′-tetrachloro-1,1′,3,3′-tetraethyl-benzimidazolyl-carbocyanineiodide (JC-1) and Hoechst 33258 were purchased from Molecular Probes (Eugene, OR, USA). anti-phospho-AMPK (2525, 1:1000 Rabbit IgG), anti-AMPK (5831, 1:1000 Rabbit IgG), NeuN (24307, 1:100 Rabbit IgG) and GAPDH antibodies (2683s, 1:1000 Rabbit IgG) were purchased from Cell Signaling Technology (Woburn, MA, USA). CellROX Deep Red Reagent were ordered from Thermo Fisher Scientific (Rockford, IL, USA). Annexin V-FITC/PI Apoptosis Detection Kit was obtained from BD Biosciences (San Diego, CA, USA). Fetal bovine serum (FBS) and 0.25% Trypsin were purchased from Life Technologies (Grand Island, NY, USA). AMPK-ShRNA was purchased from Shanghai Genechem Co; Ltd. (Shanghai, China). AMPK inhibitor (Compound C) was ordered from Calbiochem (San Diego, CA, USA).

4.2. Methods

4.2.1. SH-SY5Y Cell Culture

Human neuroblastoma SH-SY5Y cells were cultured in 75-cm^2 flasks in DMEM supplemented with 10% heat-inactivated FBS and 0.1% penicillin/streptomycin at 37 °C with 5% CO_2 humidified atmosphere. The medium was replaced every 2–3 days, and cells were sub-cultured once 80–90% confluency was reached. After digestion with 0.25% trypsin, cells were collected by centrifugation at 1000 rpm for 5 min and resuspended in fresh medium. Cells were seeded into 96-well, 12-well or six-well plates and grown overnight. Adherent cells were used for further experiments.

4.2.2. Hippocampal Neurons Culture

Newborn C57BL/6 mice were procured from the animal facility of University of Macau. The whole body was disinfected with 75% alcohol and the brain was surgically removed and stored into cold HBSS (Ca^{2+}, Mg^{2+} free) balance solution. The whole hippocampus region was dissected using a glass rod which was bent on both sides. The hippocampus was cleared of the blood and the mixed blood vessels by washing thrice with HBSS. Then the hippocampus was chopped into 1 mm^3 pieces using scissors and after washing thrice with HBSS the tissue was digested with 0.125% of trypsin at 37 °C for 15 min. The enzymatic digestion was stopped with 10% FBS and 5 mL of Neurobasal A (Gibco, USA) was added to the digested hippocampus tissue in a 15 mL centrifuge tube. The turbid tissue supernatant was collected in another 15 mL centrifuge tube and centrifuged at 1000 rpm for 10 min. The resulting cell pellet was resuspended in Neurobasal A/B27 (Gibco, Carlsbad, CA, USA) and seeded in poly-D-lysine treated plates at a density of about 1–2 × 10^5 cells / mL and incubated for growth at 37 °C in 5% CO_2 humidified atmosphere.

4.2.3. MTT Assay

For the MTT assay, SH-SY5Y cells were seeded at a density of 6–8 × 10^3 cells/well in 96-well plates with 1% FBS medium for 24 h. After serum starvation, the cells were incubated with drugs or inhibitors for appropriate time, and treated with H_2O_2 for another 24 h. The cells were then incubated with MTT (0.5 mg/mL) for additional 3–4 h. The medium was aspirated from each well and DMSO (100 µL) was added to dissolve the dark-blue formazan crystals that were formed in intact cells and absorbance of each well solution was measured with a microplate reader (SpectraMax 250, Molecular Device, Sunnyvale, CA, USA). The data were presented as Optical Density (OD) at a wavelength of 570 nm.

4.2.4. Hoechst 33258 Staining

Apoptosis of cells was examined by staining with the DNA binding dye. SH-SY5Y cells were pretreated with 12.5 µM artemisinin for 2h before being exposed to 600 µM H_2O_2 for another 24 h, followed by fixing the cells in 4% formaldehyde in PBS for 10 min at 4 °C. The fixed cells were further incubated for 10 min with 10 µg/mL of Hoechst 33258 in order to stain the nuclei. After rinsing twice with PBS, the apoptotic cells were visualized under a fluorescent microscope (Olympus, Japan). Cells exhibiting apoptosis hallmarks such as condensed chromatin or fragmented nuclei were scored as apoptotic cells. A minimum of 200 cells in five random fields were collected and quantified for each Hoechst staining experiment. The data was statistically presented as percentage of apoptotic cells.

4.2.5. Annexin V-FITC/PI Staining for Apoptosis Evaluation

Flow cytometry using Annexin V-FITC/PI staining was carried out for apoptosis evaluation. Flow cytometry was performed as described in the guidelines of assay kit. Briefly, SH-SY5Y cells were pretreated with 12.5 µM artemisinin for 2 h before being exposed to 600 µM H_2O_2 for another 24 h, then the cells were trypsinized and washed twice with ice-cold PBS then centrifuged at 1000 rpm for 5 min and re-suspended in Annexin V-FITC/PI binding buffer (195 µL). Annexin V-FITC (5 µL) was supplemented and the cells were incubated in the dark at room temperature for 10 min. The cells were then centrifuged at 1000 rpm for 5 min and re-suspended in Annexin V-FITC/PI binding buffer (190 µL); propidium iodide (PI) (10 µL) was further added, followed by incubation in the dark for 5 min. Apoptosis was quantified using Flow cytometry. Cell Quest Pro software (BD AccuriC6, BD, USA) was used for analyzing apoptosis condition.

4.2.6. Measurement of Intracellular ROS Levels

Intracellular ROS generation was assessed by CellROXs Deep Red Reagent. After pretreatment with 12.5 µM artemisinin for 2 h before being exposed to 600µM H_2O_2 for another 24 h, the SH-SY5Y cells were exposed to CellROXs Deep Red Reagent (5 mM) in DMEM for 1 h in the dark. After rinsing twice with 1× PBS solution, the Fluorescence was observed and recorded on a fluorescent microscope at excitation and emission wavelengths of 640 nm and 665 nm respectively. The semi quantification of ROS level was analyzed using Image-J software (version 1.48; NIH, Bethesda, MD, USA) and all the values of ROS levels were normalized to the control group.

4.2.7. Measurement of Mitochondrial Membrane Potential ($\Delta\psi$m)

JC-1 dye was utilized to monitor mitochondrial integrity. In brief, SH-SY5Y cells were seeded into 96-well plates (1×10^4 cells/well) in dark. After pretreatment with 12.5 µM artemisinin for 2 h before being exposed to 600 µM H_2O_2 for another 24 h, the cells were treated with JC-1 dye (10 µg/mL in medium) for 15 min at 37 °C and rinsed twice with PBS. For quantification of the signal, the intensities of red (excitation 560 nm, emission 595 nm) and green fluorescence (excitation 485 nm, emission 535 nm) were assessed using an Infinite M200 PRO Multimode Microplate reader. $\Delta\psi$m was calculated as the ratio of red/green fluorescence intensity and the values were normalized with respect to the control group. The fluorescent signal in the cells was also recorded with a fluorescent microscope.

4.2.8. Immunocytochemistry (ICC)

ICC is a method of detecting specific antigens in cells using an appropriate antibody labeling strategy. After drug treatment, cells were washed twice in PBS and then fixed in 4% paraformaldehyde for 15 min at room temperature to maintain cell morphology. The cells were then rinsed three times with PBS, and incubated in PBST (0.1% Triton X-100 in 1× PBS) for 20 min at room temperature. Following this, the cells were blocked with 1% BSA for 1 h at room temperature which can help reduce non-specific hydrophobic interactions. And then the primary antibody/antibodies were added and kept at 4 °C overnight. The next day after washing cells with PBS three times, the fluorophore-conjugated secondary antibodies were added for 2 h at room temperature, away from light. A drop of mounting medium was added to each slide. Then, samples were observed by confocal laser scanning microscopy.

4.2.9. TUNEL Assay

SH-SY5Y cells were pre-treated with 5 µM Compound C (AMPK inhibitor) for 30 min, and treated with 12.5 µM artemisinin for 2 h, then incubated with or without 600 µM H_2O_2 for a further 24 h. After treatment, the cells were fixed with 4% PFA for 30 min and washed with 0.1% Triton-X PBS for 3 times. Then cells were incubated with TUNEL test solution (C1086, Beyotime, China) at 37 °C for 60 min in the dark following the manufacturer's instruction. After washing with 1× PBS, images were taken with a fluorescence microscope.

4.2.10. Caspase-3 Activity Assay

After drug treatment, SH-SY5Y cells were digested with trypsin and harvested by centrifugation at 600 g for 5 min at 4 °C. The supernatant was carefully aspirated and washed once with PBS. According to the manufacturer's protocol, 100 µL of lysate was added per two million cells, and the pellet was resuspended. The supernatant was transferred to an ice-cold centrifuge tube and lysed on ice for

15 min. Then centrifuged at 16,000–20,000× g for 15 min at 4 °C. The reaction system was set up according to the manufacturer's instructions. The detection buffer was added first followed by the sample to be tested, and finally 10 μL of Ac-DEVD-pNA (2 mM) was added, mixed well avoiding the bubbles. The reaction system was then incubated in a working solution for 60–120 min at 37 °C. A405 was then determined by Infinite M200 PRO Multimode Microplate. All values of caspase-3 activities were normalized to the control group.

4.2.11. Western Blotting

Western blotting was performed following the standard procedure [15]. The experimental cells under different treatments were lysed in ice-cold RIPA lysis buffer and the protein concentration was assessed using a BCA protein assay kit according to the manufacturer's instructions. Samples with the same amount of proteins were separated on 10% polyacrylamide gels, and later transferred to PVDF membrane. After blocking with 3% BSA for 1 h, the membranes were incubated with selective primary antibodies overnight. Following day, the primary antibody was washed with 1× TBST thrice, and incubated with secondary antibody for another 2 h. After exposure with BCL, the intensity of the bands was quantified using Image J software.

4.2.12. Transfection of ShRNA Plasmid

The shRNA of AMPKα were designed and synthesized by GenPharma Co., Ltd. (Shanghai, China). One day before the transfection, SH-SY5Y cells were seeded into a 12-well plate at a density of 1×10^5 cells/well, grown overnight, and transfected when the cell density reached 80%. SH-SY5Y cells were transfected with Lipofectamine 2000 reagent according to the manufacturer's protocol. Protein samples were collected after 48 h of transfection. The AMPKα knockdown efficiency was verified by western blot.

4.2.13. Data Analysis and Statistics

Statistical analysis was performed using GraphPad Prism 5.0 statistical software (GraphPad software, Inc., San Diego, CA, USA). All experiments were performed in triplicates. Data are expressed as mean ± standard deviation (SD). Statistical analysis was carried out using one-way ANOVA followed by Tukey's multiple comparison, with $p < 0.05$ considered statistically significant.

Supplementary Materials: Supplementary materials can be found at http://www.mdpi.com/1422-0067/20/11/2680/s1.

Author Contributions: X.Z. and J.F. performed the experiments and drafted the manuscript. S.L. and X.X. performed part of experiments, H.W. and U.G. revised the manuscript. W.Z. conceived the hypothesis, designed the experiments and revised the manuscript. All authors read and approved the final manuscript.

Funding: The project was supported by National Natural Science Foundation of China (No. 31771128), The FHS Internal Collaboration Proposals 2017, MYRG2016-00052-FHS and MYRG2018-00134-FHS from the University of Macau, and the Science and Technology Development Fund (FDCT) of Macao (016/2016/A1). We are grateful to UM-FHS and core facilities for equipment and administrative support for this study.

Conflicts of Interest: Authors declare that there are no conflicts of interest.

Abbreviations

H_2O_2	Hydrogen peroxide
AMPK	Adenosine Monophosphate -activated protein kinase
DMSO	Dimethyl sulfoxide
FBS	Fatalbovineserum
AD	Alzheimer's disease
BCA	Bicinchoninic acid
RIPA	Radioimmunoprecipitation assay
shAMPK	AMP-activated protein kinase knockdown plasmid

Compound C   Adenosine Monophosphate -activated protein kinase inhibitor
MTT   3-(4,5-Dimethylthiazol-2-yl)-2,5-diphenyltetrazolium bromide
JC-1   5,5′,6,6′-tetrachloro-1,1′,3,3′-tetraethyl-benzimidazolyl-carbocyanine iodide
ICC   Immunocytochemistry analysis
NeuN   Neuronal nuclear protein

References

1.   Ferrini, M.G.; Gonzalez-Cadavid, N.F.; Rajfer, J. Aging related erectile dysfunction—potential mechanism to halt or delay its onset. *Transl. Androl. Urol.* **2017**, *6*, 20. [CrossRef] [PubMed]

2.   Tezil, T.; Basaga, H. Modulation of cell death in age-related diseases. *Curr. Pharm. Des.* **2014**, *20*, 3052–3067. [CrossRef] [PubMed]

3.   Pistollato, F.; Iglesias, R.C.; Ruiz, R.; Aparicio, S.; Crespo, J.; Lopez, L.D.; Manna, P.P.; Giampieri, F.; Battino, M. Nutritional patterns associated with the maintenance of neurocognitive functions and the risk of dementia and Alzheimer's disease: A focus on human studies. *Pharmacol. Res.* **2018**, *131*, 32–43. [CrossRef] [PubMed]

4.   Deng, G.; Su, J.H.; Ivins, K.J.; Van Houten, B.; Cotman, C.W. Bcl-2 facilitates recovery from DNA damage after oxidative stress. *Exp. Neurol.* **1999**, *159*, 309–318. [CrossRef] [PubMed]

5.   Gorman, A.; McGowan, A.; O'Neill, C.; Cotter, T. Oxidative stress and apoptosis in neurodegeneration. *J. Neurol. Sci.* **1996**, *139*, 45–52. [CrossRef]

6.   Hsuuw, Y.D.; Chang, C.K.; Chan, W.H.; Yu, J.S. Curcumin prevents methylglyoxal-induced oxidative stress and apoptosis in mouse embryonic stem cells and blastocysts. *J. Cell. Physiol.* **2005**, *205*, 379–386. [CrossRef] [PubMed]

7.   Gasparrini, M.; Afrin, S.; Forbes-Hernandez, T.Y.; Cianciosi, D.; Reboredo-Rodriguez, P.; Amici, A.; Battino, M.; Giampieri, F. Protective effects of Manuka honey on LPS-treated RAW 264.7 macrophages. Part 2: Control of oxidative stress induced damage, increase of antioxidant enzyme activities and attenuation of inflammation. *Food Chem. Toxicol.* **2018**, *120*, 578–587. [CrossRef] [PubMed]

8.   Whittemore, E.; Loo, D.; Watt, J.; Cotmans, C. A detailed analysis of hydrogen peroxide-induced cell death in primary neuronal culture. *Neuroscience* **1995**, *67*, 921–932. [CrossRef]

9.   Lindenboim, L.; Haviv, R.; Stein, R. Bcl-xL inhibits different apoptotic pathways in rat PC12 cells. *Neurosci. Lett.* **1998**, *253*, 37–40. [CrossRef]

10. Tu, Y. The discovery of artemisinin (qinghaosu) and gifts from Chinese medicine. *Nature Med.* **2011**, *17*, 1217. [CrossRef]

11. Burkewitz, K.; Zhang, Y.; Mair, W.B. AMPK at the nexus of energetics and aging. *Cell Metab.* **2014**, *20*, 10–25. [CrossRef]

12. Steely, A.M.; Willoughby, J.A.; Sundar, S.N.; Aivaliotis, V.I.; Firestone, G.L. Artemisinin disrupts androgen responsiveness of human prostate cancer cells by stimulating the 26S proteasome-mediated degradation of the androgen receptor protein. *Anti-cancer Drugs* **2017**, *28*, 1018–1031. [CrossRef] [PubMed]

13. Schmuck, G.; Roehrdanz, E.; Haynes, R.K.; Kahl, R. Neurotoxic mode of action of artemisinin. *Antimicrob. Agents Chemother.* **2002**, *46*, 821–827. [CrossRef] [PubMed]

14. Gardner, A.M.; Xu, F.-h.; Fady, C.; Jacoby, F.J.; Duffey, D.C.; Tu, Y.; Lichtenstein, A. Apoptotic vs. nonapoptotic cytotoxicity induced by hydrogen peroxide. *Free Radic. Biol. Med.* **1997**, *22*, 73–83. [CrossRef]

15. Zeng, Z.; Xu, J.; Zheng, W. Artemisinin protects PC12 cells against β-amyloid-induced apoptosis through activation of the ERK1/2 signaling pathway. *Redox Biol.* **2017**, *12*, 625–633. [CrossRef] [PubMed]

16. Yagi, Y.; Nakano, O.; Hashimoto, T.; Kimura, K.; Asakawa, Y.; Zhong, M.; Narimatsu, S.; Gohda, E. Induction of neurite outgrowth in PC12 cells by artemisinin through activation of ERK and p38 MAPK signaling pathways. *Brain Res.* **2013**, *1490*, 61–71.

17. Zheng, W.; Chong, C.-M.; Wang, H.; Zhou, X.; Zhang, L.; Wang, R.; Meng, Q.; Lazarovici, P.; Fang, J. Artemisinin conferred ERK mediated neuroprotection to PC12 cells and cortical neurons exposed to sodium nitroprusside-induced oxidative insult. *Free Radic. Biol. Med.* **2016**, *97*, 158–167. [CrossRef]

18. Zuo, S.; Li, Q.; Liu, X.; Feng, H.; Chen, Y. The potential therapeutic effects of artesunate on stroke and other central nervous system diseases. *BioMed. Res. Int.* **2016**, *2016*. [CrossRef]

19. Hardie, D.G.; Ross, F.A.; Hawley, S.A. AMPK: A nutrient and energy sensor that maintains energy homeostasis. *Nature Rev. Mol. Cell Biol.* **2012**, *13*, 251. [CrossRef] [PubMed]

20. Giampieri, F.; Alvarez-Suarez, J.M.; Cordero, M.D.; Gasparrini, M.; Forbes-Hernandez, T.Y.; Afrin, S.; Santos-Buelga, C.; González-Paramás, A.M.; Astolfi, P.; Rubini, C. Strawberry consumption improves aging-associated impairments, mitochondrial biogenesis and functionality through the AMP-Activated Protein Kinase signaling cascade. *Food Chem.* **2017**, *234*, 464–471. [CrossRef] [PubMed]

21. Shukitt-Hale, B.; Bielinski, D.F.; Lau, F.C.; Willis, L.M.; Carey, A.N.; Joseph, J.A. The beneficial effects of berries on cognition, motor behaviour and neuronal function in ageing. *Br. J. Nutr.* **2015**, *114*, 1542–1549. [CrossRef]

22. Zhang, Y.; Wang, X.; Wang, Y.; Liu, Y.; Xia, M. Supplementation of Cyanidin-3-O-β-Glucoside Promotes Endothelial Repair and Prevents Enhanced Atherogenesis in Diabetic Apolipoprotein E–Deficient Mice–3. *J. Nutr.* **2013**, *143*, 1248–1253. [CrossRef] [PubMed]

23. Reznick, R.M.; Zong, H.; Li, J.; Morino, K.; Moore, I.K.; Hannah, J.Y.; Liu, Z.-X.; Dong, J.; Mustard, K.J.; Hawley, S.A. Aging-associated reductions in AMP-activated protein kinase activity and mitochondrial biogenesis. *Cell Metab.* **2007**, *5*, 151–156. [CrossRef]

24. Martín, D.; Salinas, M.; Fujita, N.; Tsuruo, T.; Cuadrado, A. Ceramide and reactive oxygen species generated by H_2O_2 induce caspase-3-independent degradation of Akt/protein kinase B. *J. Biol. Chem.* **2002**, *277*, 42943–42952. [CrossRef]

25. Kahn, B.B.; Alquier, T.; Carling, D.; Hardie, D.G. AMP-activated protein kinase: Ancient energy gauge provides clues to modern understanding of metabolism. *Cell Metab.* **2005**, *1*, 15–25. [CrossRef] [PubMed]

26. Miller, R.A.; Birnbaum, M.J. An energetic tale of AMPK-independent effects of metformin. *J. Clin. Investig.* **2010**, *120*, 2267–2270. [CrossRef] [PubMed]

27. Raffetin, A.; Bruneel, F.; Roussel, C.; Thellier, M.; Buffet, P.; Caumes, E.; Jauréguiberry, S. Use of artesunate in non-malarial indications. *Med. Mal. Infect.* **2018**, *48*, 238–249. [CrossRef]

28. Wang, Z.; Ye, Z.; Huang, G.; Wang, N.; Wang, E.; Guo, Q. Sevoflurane post-conditioning enhanced hippocampal neuron resistance to global cerebral ischemia induced by cardiac arrest in rats through PI3K/Akt survival pathway. *Front. Cell. Neurosci.* **2016**, *10*, 271. [CrossRef]

29. Chong, C.-M.; Zheng, W. Artemisinin protects human retinal pigment epithelial cells from hydrogen peroxide-induced oxidative damage through activation of ERK/CREB signaling. *Redox Biol.* **2016**, *9*, 50–56. [CrossRef]

30. Cavallucci, V.; D'Amelio, M.; Cecconi, F. Aβ toxicity in Alzheimer's disease. *Mol. Neurobiol.* **2012**, *45*, 366–378. [CrossRef]

31. Golpich, M.; Amini, E.; Mohamed, Z.; Azman Ali, R.; Mohamed Ibrahim, N.; Ahmadiani, A. Mitochondrial dysfunction and biogenesis in neurodegenerative diseases: Pathogenesis and treatment. *CNS Neurosci. Ther.* **2017**, *23*, 5–22. [CrossRef] [PubMed]

32. Bhatti, J.S.; Kumar, S.; Vijayan, M.; Bhatti, G.K.; Reddy, P.H. Therapeutic strategies for mitochondrial dysfunction and oxidative stress in age-related metabolic disorders. *Prog. Mol. Biol. Transl. Sci.* **2017**, *146*, 13–46. [PubMed]

33. Han, X.; Tai, H.; Wang, X.; Wang, Z.; Zhou, J.; Wei, X.; Ding, Y.; Gong, H.; Mo, C.; Zhang, J. AMPK activation protects cells from oxidative stress-induced senescence via autophagic flux restoration and intracellular NAD+ elevation. *Aging Cell* **2016**, *15*, 416–427. [CrossRef] [PubMed]

34. Adegoke, O.O.; Qiao, F.; Liu, Y.; Longley, K.; Feng, S.; Wang, H. Overexpression of Ubiquilin-1 Alleviates Alzheimer's Disease-Caused Cognitive and Motor Deficits and Reduces Amyloid-beta Accumulation in Mice. *J. Alzheimer's Dis. JAD* **2017**, *59*, 575–590. [CrossRef] [PubMed]

35. Lin, S.-P.; Li, W.; Liu, R.; Yang, S. Artemisinin Prevents Glutamate-Induced Neuronal Cell Death Via Akt Pathway Activation. *Front. Cell. Neurosci.* **2018**, *12*, 108. [CrossRef]

36. Karbwang, J.; Sukontason, K.; Rimchala, W.; Namsiripongpun, W.; Tin, T.; Auprayoon, P.; Tumsupapong, S.; Bunnag, D.; Harinasuta, T. Preliminary report: A comparative clinical trial of artemether and quinine in severe falciparum malaria. *Southeast Asian J. Trop. Med. Public Health* **1992**, *23*, 768.

International Journal of
Molecular Sciences

Review

Causative Links between Protein Aggregation and Oxidative Stress: A Review

Elise Lévy [1,2], **Nadine El Banna** [2], **Dorothée Baïlle** [2], **Amélie Heneman-Masurel** [2], **Sandrine Truchet** [1], **Human Rezaei** [1], **Meng-Er Huang** [2], **Vincent Béringue** [1], **Davy Martin** [1,*] and **Laurence Vernis** [2,*]

[1] Molecular Virology and Immunology Unit (VIM-UR892), INRA, Université Paris-Saclay, 78352 Jouy-en-Josas, France
[2] Institut Curie, PSL Research University, CNRS UMR3348, Université Paris-Sud, Université Paris-Saclay, 91400 Orsay, France
* Correspondence: davy.martin@inra.fr (D.M.); laurence.vernis@curie.fr (L.V.)

Received: 17 July 2019; Accepted: 1 August 2019; Published: 9 August 2019

Abstract: Compelling evidence supports a tight link between oxidative stress and protein aggregation processes, which are noticeably involved in the development of proteinopathies, such as Alzheimer's disease, Parkinson's disease, and prion disease. The literature is tremendously rich in studies that establish a functional link between both processes, revealing that oxidative stress can be either causative, or consecutive, to protein aggregation. Because oxidative stress monitoring is highly challenging and may often lead to artefactual results, cutting-edge technical tools have been developed recently in the redox field, improving the ability to measure oxidative perturbations in biological systems. This review aims at providing an update of the previously known functional links between oxidative stress and protein aggregation, thereby revisiting the long-established relationship between both processes.

Keywords: protein aggregation; redox; oxidative stress; proteinopathy

---

1. Introduction

Protein aggregation consists of any association of proteins into larger structures with non-native conformation [1]. Aggregates can have either an amorphous or a highly ordered structure (amyloid). The presence of aggregates is generally indicative of proteostasis imbalance, either due to insufficient proteostasis or to disrupted chaperone capacity. Aggregates can also be a response to cellular stress related to environmental changes [2] (Figure 1).

The intrinsic parameters of protein aggregation are not fully understood, but the ability of certain proteins to aggregate more readily compared to others has been known for a while. Intrinsic aggregation capacity can be obvious, as in the case of peptide poly(Q) tracts that form high molecular weight aggregates in some neurodegenerative diseases (including Huntington's disease). Such aggregates exhibit a fibrillar or ribbon-like morphology, reminiscent of prion rods and amyloid-β (Aβ) fibrils in Alzheimer's disease (AD) [3]. Aggregation is controlled in vitro, not only by poly(Q) tracts length (51–122 glutamines causing huntingtin's aggregation in vitro), but also by protein concentration and reaction time [4]. In addition, most proteins contain one or more aggregation prone-regions (APR), which are protected from aggregation by protein interactions or by specific structural features (e.g., burying into a hydrophobic core) in physiological conditions [5].

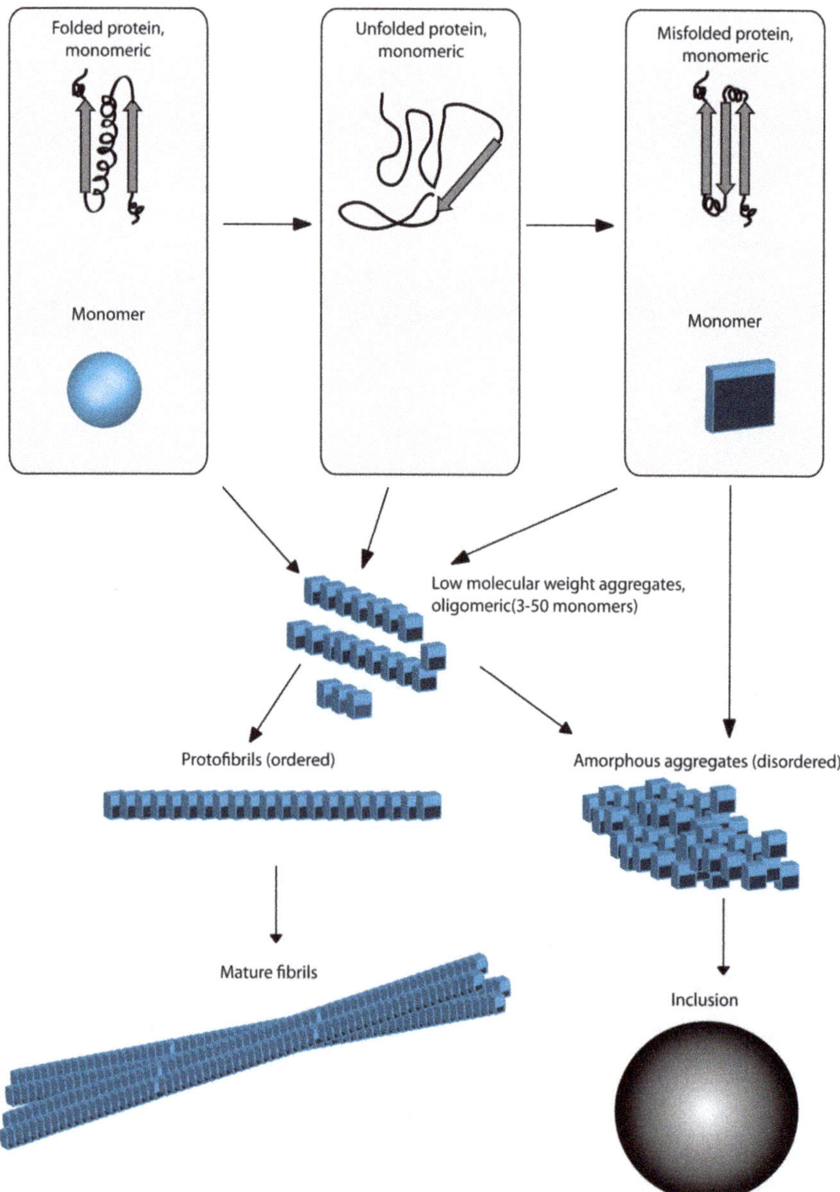

Figure 1. General picture of the protein aggregation process. The unfolded and misfolded monomer structures are aggregation prone. Folded monomers can also aggregate from native-like conformations without going through the unfolded step. The association of several monomers gives rise to oligomeric aggregates with low molecular weight. The addition of oligomers in an ordered manner permits the growth of oligomers to protofibrils and mature fibrils. Amorphous aggregates can arise from the precipitation of monomers or oligomers, possibly leading to protein inclusions.

However, the aggregation of proteins without any obvious features is intriguing. An interesting study used three different stress agents: Arsenite, a toxic metalloid, hydrogen peroxide (H_2O_2), a ubiquitous oxidative stress agent, and azetidine-2-carboxylic acid (AZC), a proline analog whose incorporation into proteins provokes conformation alterations, misfolding, and aggregation [6,7]. The authors identified that despite distinct stress conditions with distinct mechanisms of action, some of the proteins that aggregate are of similar types, suggesting that proteins within aggregates are intrinsically aggregation-prone [7]. Indeed, it appears that aggregation prone proteins like PrP^C (the prion protein) or Shadoo (also a member of the prion protein family) exhibit large intrinsically disordered domains (IDDs). The resulting conformational plasticity is likely to offer a wide range of interacting possibilities for these proteins, thereby regulating their localization and aggregation propensity [8–10]. Several computational approaches have tried to question the intrinsic determinisms of aggregation. Noticeably, the Waltz algorithm was developed to try to predict amylogenic regions in protein sequences, based on a scoring matrix deduced from the biophysical and structural analysis of previously characterized hexapeptides with amyloid properties [11]. Based on the accuracy of their pWALTZ prion prediction method derived from Waltz, Sabate and co-workers recently suggested a model for prion formation that depends on the presence of specific short sequence elements, embedded in intrinsically Q/N-rich regions, with high amyloid propensity [12].

Interestingly, the aggregation of globular proteins into amyloids from a native or native-like state has also been described. In such cases, aggregation-prone states can be reached from native-like conformations after small temperature or pH changes, or stress modulations in general, without the need to cross the energy barrier to unfold [13]. Several examples in the literature illustrate this fact, such as the aggregation of one of the acylphosphatases from the *Drosophila melanogaster* (AcPDro2) [13], the human lysozyme aggregating through a nucleation-dependent growth process [14], the globular acylphosphatase from *Sulfolobus solfataricus* (forming aggregates in which the monomers maintain their native-like topology [15]), and the transthyretin-like domain of human carboxypeptidase D (h-TTL), a monomeric protein with homology to human transthyretin that aggregates under close to physiological conditions [16]. Another example is the src tyrosine kinase SH3 domain, whose aggregation-prone state favors a domain swap that allows amyloid formation [17]. In such cases, the free energy gap between the native and the aggregation-prone state is an essential determinant of the aggregation propensity of the proteins [14,17]. Moreover, even though unfolded proteins have a high capacity to aggregate, the residual aggregation potential of proteins within a folded state may be physiologically relevant and play a role in several clinical situations [13,14,16].

Protein aggregation is obviously related to several human pathological situations. Amyloidoses, for example, are a group of diseases originating from the aggregation of oligomers into amyloid fibrils that deposit in tissues and are, in turn, toxic to the cells. Which of the oligomers or the fibrils are the most toxic to cells is still a matter of debate. Further, well-known neurodegenerative diseases, including Alzheimer's (AD), Parkinson's (PD), and motor neuron diseases, appear to be a consequence of protein aggregation in neurons, leading to toxicity and neuronal cell death. Although pathological protein aggregation may have different causes, the chronic disturbance of cellular homeostasis (due to exogenous pollutants, or aging for example) is likely to play a major role by modifying the physico-chemical equilibrium and influencing protein folding and aggregation. Redox perturbations, in particular, have long been linked to protein aggregation diseases.

Reactive oxygen species (ROS) are normal by-products arising from various cellular reactions, mostly during electron transport in mitochondria or chloroplasts. Intracellular ROS levels are maintained low within cells, ensuring redox homeostasis for proper cellular chemical reactions. Oxidative stress occurs when the ROS concentration is excessive regarding the antioxidant capacities of the cell, leading to the oxidation of cellular molecules and their alteration [18]. Proteins appear to be a major target for oxidation due to their elevated quantities compared to other cell components and also due to their high reactivity with ROS [19]. Proteins are susceptible to ROS modifications of amino acid side chains that alter their structure.

We provide here a review of recent data on the functional links between oxidative stress and protein aggregation.

2. Oxidative Stress Can Produce Aggregation: A Mechanistic View

2.1. Oxidation of Critical Amino Acids Induces Structural Changes within Proteins, Leading to Aggregation

2.1.1. Cysteine Oxidation

Among the amino acids, cysteine (Cys) possesses a thiol group, which is highly nucleophilic. This structural feature makes cysteine particularly prone to oxidation by ROS. Amorphous aggregation of γ-cristallins is the cause of cataracts, a widespread disease among seniors. Internal disulfide bond formation between Cys32 and Cys41 due to oxidation was found necessary and sufficient to provoke aggregation under physiological conditions, likely by stabilizing an unfolded intermediate prone to protein–protein interaction between an extruded hairpin and a distal β-sheet in the γD-crystallin [20].

Being essential for serotonin synthesis, Tryptophan hydroxylase 2 (TPH2) is considered a phenotypic marker for serotonin neurons. Known to be extremely labile to oxidation, TPH2 aggregates through both intra- and inter-molecular disulfide cross-linking upon oxidation of cysteines [21]. Accordingly, in a systematic cysteine-mutagenesis approach, a single cysteine out of 13 was found sufficient for aggregation, whereas only cysteine-less mutants were found resistant to aggregation upon oxidation, thereby indicating that cysteines are necessary for the responses of TPH2 to oxidation. These results led the authors to hypothesize that redox homeostasis changes occurring during Parkinson's disease might be involved in disulfide links in TPH2, causing TPH2 to shift from a soluble compartment to large inclusion bodies, consequently losing its catalytic function [21].

Alternatively, disulfide bonds can occur in proteins' native structures in physiological conditions. Disulfide-rich domains (DRDs) are peptide domains whose native structures are stabilized by covalent disulfide bonds through an oxidative folding reaction. The authors questioned whether this specific folding might be associated with an increased aggregation propensity of DRDs. Among the 97 DRDs analyzed in silico, a majority were intrinsically disordered, but remained more soluble and had fewer aggregating regions than those of other globular domains [22]. This work suggests that DRDs might have evolved to avoid aggregation before proteins acquire their covalently linked native structures or after oxidative stress.

2.1.2. Other Aminoacids Involved

Caseins are very abundant in milk as they represent more than 80% of total milk protein content [23]. Inter- and intra- covalent di-tyrosine (di-Tyr) and di-tryptophan (di-Trp) cross-links have been shown to provoke α- and β-casein aggregation, due to the Trp-or Tyr- derived radicals produced by photo-oxidation mediated by riboflavin, a photosensitizing vitamin [24]. Oxygen was found to strongly modulate this phenomenon and to increase protein aggregation by decreasing the overall cross-link formation, but allowing the formation of oxidized Trp, Tyr, Methionine (Met), and Histidine (His) residues [24].

Glyceraldehyde-3-phosphate dehydrogenase (GAPDH) is involved in energy production and has been shown to convert from its native soluble state into a non-native high molecular weight, which is insoluble, and an aggregated state, during the course of several diseases, including Alzheimer's disease [25–27]. Interestingly, methionine, rather than cysteine oxidation, was shown to be a primary cause of GAPDH aggregation, as mutating methionine 46 to leucine rendered GAPDH highly resistant to aggregation after exposure to (3E)-4-ethyl-2-hydroxyimino-5-nitro-3-hexenamide (NOR3), a potent oxidative agent driving GAPDH aggregation [25,28,29]. In this case, the authors propose that methionine oxidation represents a "linchpin", a permissive event for subsequent misfolding and aggregation [28].

It appears that protein residue oxidation, including the resulting disulfide bonds between cysteines, is not directly responsible for aggregation, per se, but in most cases induces limited or extensive unfolding of the surrounding environment of the protein. This further favors protein–protein interactions and aggregation, as summarized in Table 1.

Table 1. Residue oxidation causing protein aggregation.

Protein(s) Involved	Mechanism	Residue Involved	Related Disease/Pathway	Reference
γD-crystallin	Cys32–Cys41 internal disulfide bond formation leads to the stabilization of a partially unfolded domain, which is prone to further intermolecular interactions. Internal disulfide bond formation provokes aggregation under physiological conditions.	Cysteine	Cataracts in older people	[20]
Tryptophan hydroxylase 2 (TPH2)	Intra- and inter-molecular disulfide bonds responsible for high molecular weight aggregates.	Cysteine	Parkinson's disease	[21]
Milk caseins	Oxidation of Tryptophan, Tyrosine, Methionine, and Histidine residues decreases di-Tyr and di-Trp formation but allows increased protein aggregation.	Tryptophan, Tyrosine, Methionine, Histidine	None	[24]
Glyceraldehyde-3-phosphate dehydrogenase (GAPDH)	Oxidation of Methionine allows subsequent misfolding and further aggregation of GAPDH.	Methionine	None	[28]

2.2. Role of Metals

Metals' contributions to Fenton's or Haber-Weiss' reactions have been described for a number of years, as they lead to highly reactive ROS that can be deleterious to cell components. In particular, they are responsible for amino acid oxidation. The involvement of metals in various pathological situations has also been recognized [30].

Parkinson's disease pathology involves misfolding and aggregation of the presynaptic protein α-synuclein, together with an alteration of the homeostasis of brain metals, including iron. A recent analysis of iron's role in the aggregation and secondary structure of N-terminally acetylated α-synuclein (^{NAc}αS), the pathologically relevant form in PD, revealed the importance of oxidation in the phenomenon [31]. The sole addition of iron(II) in presence of oxygen was shown to induce an antiparallel right-twisted conformation of ^{NAc}αS, and generate the oligomer-locked ^{NAc}αS–Fe^{II}/O$_2$ conformation, thus initiating oligomer formation but preventing further processing into fibrils. This was not observed in the absence of oxygen. In contrast, the addition of iron(III) led to the formation of fibrils in the presence of oxygen. Thus, the iron oxidation status in the presence of oxygen differentially controls aggregation. This may have physiological or pathological implications [31].

Particular mutations in the Apolipoprotein A-I (APOA1) gene provoke hereditary amyloidosis. Interestingly, Fe(II) can reduce the formation of fibrillar APOA1 species, as opposed to Fe(III), which enhances their formation in vitro [32]. The increased levels in Fe(III) compared to Fe(II) under particular oxidative physiological conditions, e.g., in older patients or in presence of air pollutants, might thus contribute to the development of amyloidosis and possibly other diseases involving protein aggregation [32].

Several studies have demonstrated amyloid-β-mediated ROS production in the presence of Cu ions, and mechanisms that can generate a superoxide anion (O$_2$• −), hydrogen peroxide (H$_2$O$_2$), and a hydroxyl radical (HO•) have been proposed [33,34]. The Amyloid-β peptide (Aβ) is produced after cleavage of the transmembrane Amyloid Precursor Protein (APP) by β- and γ-secretases. Aβ is thus released in the extracellular space. Its accumulation and aggregation are a hallmark of Alzheimer's disease (See Figure 2). A recent work identified that Cu(I) or Cu(II) cations predominantly bind the M1 site located in domain E2 of APP, with a comparably high affinity (picomolar) being liganded by four histidine residues. Cu(II) binding to M1 was found to stabilize E2 but also alters its structural conformation into a more open state [35]. The authors demonstrated the aerobic catalytic oxidation of ascorbate, an abundant antioxidant molecule in the central nervous system, by the Cu-E2 complex, with an experimental set up close to physiological conditions. APPβ might thus play the role of redox catalyst in vivo and cause protein damage, as observed in brain tissues from Alzheimer's patients [35].

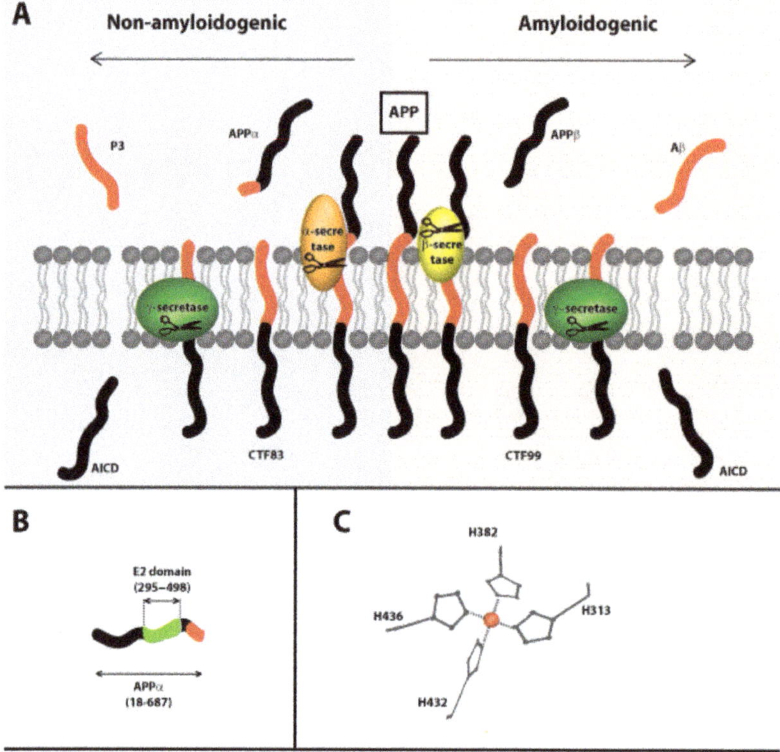

Figure 2. Amyloid Precursor Protein (APP) processing. (**A**) Cleavage of APP occurs through two pathways. The non-amyloidogenic pathway (grey, left) involves two cleavages by α- and γ-secretases and produces a long APPα fragment, which is secreted. C-Terminal Fragment (CTF)83, 3-kd peptide, (P3), and APP intracellular domain (AICD) fragments are also released. In parallel, the amyloidogenic pathway (pink, right) involves two cleavages by β- and γ-secretases, producing a long APPβ, which is secreted. CTF99, AICD, and Aβ fragments are also produced. In pathological conditions, Aβ peptides accumulate and can ultimately aggregate and form oligomers and fibrils that are toxic for the cells (adapted from [36]). (**B**) Schematic representation of the APPα fragment, produced after cleavage of the APP by α-secretase. The E2 domain is highlighted (green). (**C**) Zooming in on the four cysteine residues involved in the tetra-His M1 site, included in the E2 domain. The Cu cation is pictured as a red ball (adapted from [37] with permission).

2.3. Carbonylation Leads to Aggregation

Carbonylation is a particular type of oxidation involving the irreversible addition of a carbonyl group (CO) into proteins. The ROS-mediated carbonylation of proteins mainly affects lysine, arginine, threonine, and proline residues [38] and was frequently reported in chronic inflammatory diseases [39,40]. Protein carbonylation is commonly considered a standard marker of oxidative stress. While carbonylation is reported as having a role in protein quality control, in tagging damaged proteins for degradation with the 20S proteasome (via an unresolved mechanism), excessive carbonylation has been shown to be responsible for protein aggregation, especially when proteasome activity is impaired [41]. However, few publications provide mechanistic insight into how protein carbonylation causes aggregation. A recent report indicated that these specific modifications are also associated with physiological aging and favor the formation of denser and more compact aggregates, including several proteins previously reported for their aggregation propensity [42]. Inducing carbonyl stress in young mice increased protein aggregation, similar to the aggregation naturally observed in old mice, indicating that post-translational oxidative alterations are responsible for increased protein aggregation [42]. Despite evidence that carbonylation leads to aggregation, to our knowledge, no direct mechanism has been explicitly reported.

Recent attention has been paid to reactive dicarbonyl species, such as methylglyoxal (MGO) and glyoxal (GO), considered to be side-products of several metabolic pathways, as they produce specific oxidative modifications of proteins, mainly reacting with lysine and arginine [43]. These oxidative modifications, including carbonylation, are subsequently responsible for secondary and tertiary structure alterations and the formation of high molecular mass protein aggregates and may have an underestimated role in several proteinopathies, including Alzheimer's and Parkinson's diseases.

2.4. Protein Oxidation Influences Aggregation by Modulating Chaperone Protein Activity

Oxidation-induced structural changes in proteins may also prevent specific interactions with partners. A very interesting case has been described recently, with the nucleotide exchanger and co-chaperone Mge1 within the mitochondria. Mge1 is actually modified by persistent oxidative stress, and methionine 155, in particular, can be reversibly oxidized into methionine sulfoxide. As a result, due to local structural changes in Mge1 binding to its partner, the heat shock protein 70 (Hsp70) is defective. Hsp70 is a central heat-shock protein involved in controlling protein folding and plays an important role in ensuring proteostasis. If not reduced by the endogenous methionine sulfoxide reductase, the oxidized Mge1 aggregates into amyloid-type particles. Interestingly, the authors noticed that highly oxidized Mge1 actually increased the binding capacity of Hsp70 to a denatured protein, suggesting that the oxidation-induced defective binding of Meg1 to Hsp70 and MgeI aggregation may protect the cells from the aggregation of other proteins in an oxidative stress context [44]. It is interesting to note that some isoforms of Hsp70 are stress-modulated. For instance, oxidants like methylene blue and hydrogen peroxide inactivate Hsp72, possibly due to the oxidation of two specific cysteine residues resulting in structural changes within the nucleotide-binding domain. Noticeably, this oxidation-induced inactivation of Hsp72 is associated with decreased levels of tau in several Alzheimer's disease models. The authors suggest that Hsp72 inactivation could clear the cytosol from misfolded Hsp72 substrates like tau, even though the mechanism remains unclear [45].

Tsa1 is a surprising protein that exhibits a double role, acting as a peroxidase but also as a chaperone for aggregated proteins—specifically, by chaperoning misassembled ribosomal proteins, thereby preventing toxicity to arise from aggregation [46]. It is noticeable that the absence of Tsa1 both provokes an increase in [H_2O_2] and the accumulation of aggregated proteins in the meantime [47]. This dual role might argue for a strong functional connection between oxidative stress and protein aggregation.

2.5. Protein Oxidation Influences Aggregation by Perturbing the Translational Process

Yeast Sup35 is a translation termination factor well known for its capacity to form a prion (i.e., a self-perpetuating amyloid aggregate), in response to environmental or cellular factors. This prion is called [*PSI*+] [48]. When aggregated into [*PSI*+], Sup35 loses its function in translation termination, and an elevated read-through of the stop codons occurs, thereby generating C-terminally extended polypeptides. Following exposure to H_2O_2, [*PSI*$^+$] formation was shown to occur with increased frequency, possibly linked to the oxidation of methionine residues [49]. As a consequence, exposure to oxidative stress perturbs the translational process through titration of Sup35.

Defective mRNAs are normally handled by mRNA quality control systems within cells, among which the non-stop decay (NSD) pathway and the no-go decay (NGD) pathways prevent the production of abnormal, potentially aggregation-prone proteins. Interestingly, NSD components (e.g., Ski7) and NGD components (e.g., Dom34/Hbs1) were recently shown to be required during oxidative stress [50]. Moreover, overexpression of Sup35 actually decreases stop codon read-through and improves the tolerance of yeast cells to oxidative stress, thus providing an unanticipated link between oxidative stress tolerance and NSD, as NSD substrates noticeably accumulate as a consequence of [*PSI*$^+$] formation after oxidative stress [50].

2.6. Oxidative Stress Contributes to Aggregation by Modulating the Proteasome and Autophagy Capacity

Proteasomes are central players of the Ubiquitin–Proteasome System (UPS), which ensures the quality control of proteins. The 20S proteasome, constituting 28 subunits, is considered the "core" proteasome and is involved in unfolded, misfolded, or intrinsically disordered or oxidized protein removal by proteolytic degradation. Ubiquitination is not needed for degradation by the 20S proteasome, unlike the 26S proteasome. Several other proteasome components regarded as regulators have been characterized, including the 19S proteasome. Obviously, increasing 20S proteasome activity in pathogenic conditions due to protein aggregation might be of interest. The small, imidazoline-derivative molecule TCH-165 was shown to increase 20S proteasome levels to the detriment of the 26S proteasome, resulting in the enhanced proteolysis of intrinsically disordered proteins. These include aggregation-prone proteins such as α-synuclein and tau, whose aggregation are hallmarks of Parkinson's and Alzheimer's diseases, respectively, and proto-oncogenes, such as ornithine decarboxylase and c-Fos. Noticeably, high concentrations of this molecule resulted in the accumulation of ubiquitinated substrates [51].

Similarly, proteasome impairment has been shown to favor the accumulation and aggregation of aggregation-prone proteins [52–55]. UPS disturbances have been correlated with the spreading of diseases involving protein aggregation, including Alzheimer's, Parkinson's, and Huntington's diseases, as well as amyotrophic lateral sclerosis (ALS). ALS, in particular, is characterized by ubiquitinated proteic inclusions.

For our current focus, it is worth noticing that the UPS is actually redox regulated, and S-glutathionylation, specifically, has been involved in the post-translational control of proteasome activity. For example, 20S glutathionylation was shown to be responsible for increased proteolytic activity [56], favoring the removal of oxidized or unstructured proteins in stressful situations. After an acute oxidative event, a coordinated response involving poly[ADP-ribose] polymerase 1 (PARP-1) activation and a ubiquitination block, followed by the inactivation of the proteolytic capacity, occurs, before massive de novo synthesis of proteasome players and consequently increased proteolytic activity [57]. In contrast to 20S, 26S proteolytic activity was reduced in oxidative conditions. Indeed, the S-gluthathionylation of a regulatory subunit of the 19S proteasome, resulting in diminished 26S proteasome activity, was observed in the presence of chronically increased hydrogen peroxide concentrations both in vitro and in cellulo [58]. These data might reveal differential redox regulations for the proteasome's components. Under normal physiological conditions, the oxidized proteasomes (ubiquitin-dependent proteolytic 20S core and regulatory 19S) are S-glutathionylated, as evidenced in several cellular contexts [56,58–60].

It is also interesting to report here that the anti-malaria drug dihydroartemisinin, an artemisinin derivative used in clinics, acts by inhibiting the parasite's proteasome, possibly through oxidative damage to the proteasome [61].

Thus, chronic oxidative stress favors protein aggregation through by impairing proteasome capacity (Figure 3). In addition, the slow and gradual accumulation of aggregates during aging has been observed. These aggregates are composed of oxidized proteins (including carbonylated aggregates) due to proteins escaping the UPS and are suggested to bind the proteasome and inhibit its function [62,63], thereby fostering this aggregation loop.

Autophagy is a cellular process that allows the recycling of cellular components, such as proteins or organelles, through a lysosome-mediated catabolic pathway. Particular types of autophagy are primarily distinguished (macroautophagy, chaperone-mediated autophagy, microautophagy) by how target substrates are recognized and delivered to lysosomes. More specifically, autophagy is central in eliminating the substrates that will ultimately give rise to amyloids and fibrils in several neurodegenerative diseases (Aβ peptide and tau protein in AD [64,65], alpha-synuclein in PD [66], mutated huntingtin in Huntington disease [67], and superoxide dismutase 1 (SOD1) in ALS [68]). Accordingly, mice deficient in autophagy show neurodegenerative disorders, and defects in autophagy have been associated with various human neurodegenerative diseases [69–73].

Redox regulation of autophagy has been suggested, as antioxidant treatments actually prevent autophagy [74]. This type of regulation was proposed by several authors [75–79]. Interestingly, recent work identified that the absence of the glutathione reductase *gsr-1* gene leads to major redox homeostasis unbalance in the model organism, *Caenorhabditis elegans*, and is also responsible for autophagy impairment by preventing the nuclear translocation of a key transcription factor, HLH-30/TFEB. In addition, the aggregation of both homologous and heterologous proteins (Aβ peptide (AD), α-synuclein:: YFP (PD), and Q40::YFP (Huntington disease) expressed in *C. elegans* was increased. This study reveals a glutathione-dependent regulation of autophagy, allowing the control of protein aggregation, a process also conserved from lower eukaryotes to mammals [80]. Recently, several drugs were developed to modulate autophagy in search of therapeutic improvement in neurodegenerative disorders. Noticeably, Rapamycin targets mTor, a master regulator of cell growth and metabolism. Rapamycin activates autophagy and lysosomal biogenesis [81] and, furthermore, was shown capable of reducing Aβ accumulation and improving cognitive impairments in a transgenic mouse model by increasing autophagy [82,83]. Autophagy activation by rapamycin in neurons was also shown to favor the clearing of intracellular aggregates of misfolded prion proteins and to reduce neurotoxicity [84]. The rapamycin derivative, Rilmenidine, which is protective against oxidative cytotoxicity [85], was also shown to increase autophagy, despite failing to decrease the accumulation and aggregation of SOD1 in a mouse ALS model [86]. Another rapamycin derivative, biolimus, was also shown to be capable of activating autophagy efficiently in smooth muscle cells [87]. Most of these drugs have been shown to play on the oxidative balance within cells. Nevertheless, whether the effect of these drugs on protein aggregation is related to redox modulation has not been consistently characterized.

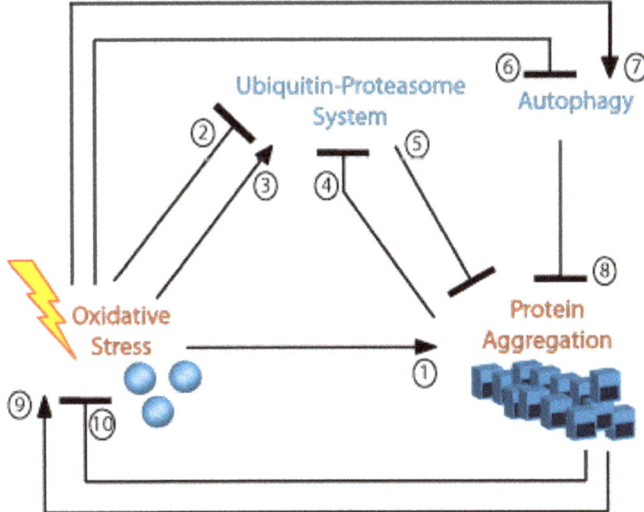

Figure 3. General diagram showing the regulation of protein aggregation. Arrows indicate the activation relationship, and blunt-ended arrows indicate the repression relationship. References supporting this diagram within this review are as follows: 1: [20,21,24,28,31,32,35,41–47,49,50]; 2: [58,61]; 3: [57]; 4: [62,63]; 5: [51–55]; 6: [80]; 7: [74–80]; 8: [64–73,82–84]; 9: [47,88–99]; 10: [100–107].

3. Aggregation Can, in Turn, Produce Oxidative Stress, or Protect Against Oxidative Insults

3.1. Pro-Oxidative Effects

As noticed previously, the yeast peroxiredoxin Tsa1 acts both as a chaperone and an ROS scavenger. The proline analogue azetidine-2-carboxylic acid (AZC) induces aggregation in yeast cells, and yeast mutants lacking *TSA1* are highly sensitive to AZC-induced misfolding. The toxicity of AZC is actually related to ROS accumulation, as decreasing ROS levels prevents sensitivity to AZC. Interestingly, inhibiting nascent protein synthesis with cycloheximide rescues the *tsa1* mutant sensitivity to AZC, confirming that aggregation in this case causes ROS production [47].

ROS production due to mitochondrial dysfunction was often found to be associated with protein aggregation, as in the case of Huntington's [88,89] or Alzheimer's disease [90]. Alpha–synuclein aggregation is a hallmark of Parkinson's disease, as previously explained, and is known to influence mitochondrial morphology, interrupting ER–mitochondria communication [108–110] and modulating mitochondrial fragmentation [111], even though the molecular determinants of these changes are not very clear. The preferential binding of pathological alpha–synuclein aggregates to mitochondria was recently identified in neurons [91], and this process was accompanied by cellular respiration defects, suggesting mitochondrial dysfunction and leading to the hypothesis of a direct mitochondrial impairment by alpha-synuclein aggregates, inducing ROS production.

RNA aptamers are synthetic nontoxic and non-immunogenic RNA oligonucleotides able to bind a specific target and can thus be used as therapeutic weapons. For instance, specific RNA aptamers can efficiently inhibit aggregation of mutant huntingtin, with a pathogenic polyglutamine stretch, by stabilizing the monomer both in vitro and in cellulo. Oxidative stress is a hallmark of Huntington's disease and has been shown to occur in cells exposed to mutant huntingtin aggregates. Whether it is a cause or a consequence of the disease is not yet clear. However, it is interesting to note that RNA aptamers inhibiting mutant huntingtin's aggregation lead to reduced oxidative stress levels in cellular models [92]. The authors smartly engineered cells expressing RNA aptamers under the control of an oxidative stress-inducible promoter. A nine-fold increase in aptamer expression occurred in mutant huntingtin-expressing cells, which consequently reduced mutant huntingtin aggregation in these cells. As a consequence, oxidative stress levels were also reduced, so RNA aptamer expression was also, in turn, reduced [93]. This smart design unambiguously demonstrates that protein aggregation is responsible for intracellular oxidative stress production in this context. Accordingly, mitochondrial dysfunction was also reduced using this particular experimental design, indicating that ROS production is likely related to mitochondrial impairment [93].

Human Amylin (hA) is a 4 kDa pancreatic hormone, synthesized and secreted along with insulin by islet beta cells. Like other amyloid proteins, it is prone to aggregation and remains a hallmark of type-2 diabetes mellitus [96]. Previous studies indicated that either exposure of beta cells to hA or hA-overexpression in cells results in intracellular ROS accumulation, supporting the hypothesis that hA aggregation might be a cause of oxidative stress [94], possibly through mitochondrial dysfunction [95]. Recent work identified that a misfolded amylin actually activates an upstream apoptosis signal regulating kinase-1 (ASK1), with a concomitant decrease in the intracellular levels of reduced glutathione [96]. Moreover, the pro-oxidative activity and expression of a plasma membrane bound NADPH oxidase (NOX) and its regulatory subunits were stimulated, suggesting that NOX1 and ASK1 mediate the cytotoxic effect of aggregated hA in pancreatic beta-cells [96]. Interestingly, NOX had also been previously identified as a main ROS producer in cultured neurons exposed to amyloid-β peptides [97,98].

Stabilizing α-synuclein monomers also proved to decrease the growth of misfolded cytotoxic aggregates and consequently reduced oxidative damage to the cells by limiting the binding of aggregates to the cell membrane [99]. Thus, a membrane's structure seems to play an important role in oxidative stress production in response to protein aggregation, possibly involving membrane bound oxidases for ROS production, as a consequence of aggregates binding to the membrane.

3.2. Anti-Oxidative Effects

Even though the role of aggregates in ROS production was unambiguously demonstrated through obvious examples, other findings describe how aggregates can also "buffer" oxidative effects. Surprisingly, 40-aminoacids, as well as 28-aminoacids amyloid-β peptides (Aβ$_{1-40}$ or Aβ$_{1-28}$), either soluble or aggregated, revealed potent anti-oxidant activity in cell-free systems, despite their accumulation and aggregation being a hallmark of Alzheimer's disease, which is also characterized by chronic oxidative stress [100,101]. The authors suggest that peptides might first chelate metal ions, including Zn(II), Cu(II), and Fe(II), as previously suggested by other authors [102–104], which would inhibit Fenton's reaction, and then scavenge radicals by oxidation of His and Tyr residues [100,101]. Whether a peptide with similar His and Tyr residue content are also capable of "buffering" free radicals would be worth testing. A more recent study reported similar anti-oxidant activity of amyloid-β aggregates. Overexpression of 21 variants of the amyloid-β 42 peptide fused with GFP in yeast (covering a broad range of intrinsic aggregation propensities) led to various levels of intrinsic oxidative stress production [105]. A striking correlation was observed—the more aggregation-prone the mutated GFP-Abeta is, the less oxidative stress is produced, suggesting that large insoluble aggregates might act to limit cellular oxidative stress. An aggregation propensity threshold could be defined, above which proteins with a high aggregation propensity accumulate into foci. The authors then suggested that protein foci formation is an active ATP-dependent process, which might serve as a protective mechanism against oxidative stress damage, despite its high energetic price [106].

Similarly, it is noteable that [*PSI+*] formation has been reported as protective against oxidative stress, as antioxidant enzyme-lacking yeasts become more sensitive to H$_2$O$_2$ when [*PSI+*] is eliminated [107].

Finally, Carija et al. propose that different kinds of aggregates have different effects on ROS levels, as small particles called "diffuse" aggregates are associated with increased intracellular oxidative stress, whereas larger protein inclusions are not [105]. Similar conclusions were previously reached, that hydrogen peroxide is generated during the early stages of protein aggregation in the course of Alzheimer's disease pathogenesis, possibly by an early form of protein aggregation, in the absence of a mature amyloid fibril [112].

4. Aggregation and Oxidation can be Parts of a Vicious Circle

In about 12% of familiar and 1.5% of sporadic cases of ALS, mutations in the superoxide dismutase 1 (SOD1), a major player in ROS detoxification, are found. The misfolding and aggregating of SOD1 are considered a hallmark of the disease [113]. A mutated version of SOD1 named SOD1^{A4V} is particularly aggregation-prone. Its expression is associated with a profound disturbance of free ubiquitin distribution within the cells, and with UPS dysfunction. These data suggest that protein aggregation, per se, might cause UPS disturbance and participate in a vicious circle that eventually prevents the elimination of aggregates [114]. Because SOD1 is a major player in ROS detoxification, and its absence is responsible for increased oxidative stress, the latter being suggested to play a key role in ALS progression [115,116], it is a possibility that oxidative stress plays a role in UPS disturbance in this particular case. Both would participate a vicious circle involving SOD1 aggregation and oxidative stress.

In light of the mechanistic insights described previously, multifunctional modulators that are able to chelate metals and prevent ROS generation and protein aggregation have been designed for several years, in an attempt to develop new therapeutic tools [117,118]. TGR86 is composed of both the metal chelating clioquinol and the antioxidant, epigallocatechin gallate [119]. TGR86 was shown to be capable of both interacting with amyloid-β and efficiently modulating its aggregation and complexing with Cu(II), as well as preventing ROS generation. Another bifunctional molecule, composed of a nitroxide spin label linked to an amyloidophilic fluorine (spin-labeled fluorine, SLF), was recently developed [120]. Using super-resolution confocal microscopy imaging, the authors identified that assembling those two modules within SLF creates a synergistic effect in cultured neurons, preventing the intracellular accumulation of amyloid-β on the one hand, together with the reduction and the scavenging of amyloid-β-induced ROS on the other. Further studies are now needed to evaluate the therapeutic potential of these molecules in Alzheimer's disease.

Author Contributions: All co-authors conceived, designed, and drafted the review. All authors approved the submitted version.

Funding: This resarch was funded by regular funding from the Centre National de la Recherche Scientifique (CNRS) and Institut Curie. E.L. is supported by a doctoral research fellowship from the Doctoral School ABIES, AgroParisTech. N.E.B. is supported by a doctoral research fellowship from the Doctoral School SDSV, Université Paris-Saclay, and from "La Fondation pour la Recherche Médicale". S.T., H.R., V.B., and D.M. are supported by Institut National de la Recherche Agronomique (INRA). L.V. is supported by Inserm.

Conflicts of Interest: The authors declare no conflict of interest. The funding sponsors had no role in the writing of the manuscript.

References

1. Hipp, M.S.; Park, S.H.; Hartl, F.U. Proteostasis impairment in protein-misfolding and -aggregation diseases. *Trends Cell Biol.* **2014**, *24*, 506–514. [CrossRef] [PubMed]

2. Li, J.; Labbadia, J.; Morimoto, R.I. Rethinking HSF1 in Stress, Development, and Organismal Health. *Trends Cell Biol.* **2017**, *27*, 895–905. [CrossRef] [PubMed]

3. Scherzinger, E.; Lurz, R.; Turmaine, M.; Mangiarini, L.; Hollenbach, B.; Hasenbank, R.; Bates, G.P.; Davies, S.W.; Lehrach, H.; Wanker, E.E. Huntingtin-encoded polyglutamine expansions form amyloid-like protein aggregates *in vitro* and *in vivo*. *Cell* **1997**, *90*, 549–558. [CrossRef]

4. Scherzinger, E.; Sittler, A.; Schweiger, K.; Heiser, V.; Lurz, R.; Hasenbank, R.; Bates, G.P.; Lehrach, H.; Wanker, E.E. Self-assembly of polyglutamine-containing huntingtin fragments into amyloid-like fibrils: Implications for Huntington's disease pathology. *Proc. Natl. Acad. Sci. USA* **1999**, *96*, 4604–4609. [CrossRef] [PubMed]

5. Beerten, J.; Schymkowitz, J.; Rousseau, F. Aggregation prone regions and gatekeeping residues in protein sequences. *Curr. Top. Med. Chem.* **2012**, *12*, 2470–2478. [CrossRef] [PubMed]

6. Trotter, E.W.; Berenfeld, L.; Krause, S.A.; Petsko, G.A.; Gray, J.V. Protein misfolding and temperature up-shift cause G1 arrest via a common mechanism dependent on heat shock factor in Saccharomycescerevisiae. *Proc. Natl. Acad. Sci. USA* **2001**, *98*, 7313–7318. [CrossRef] [PubMed]

7. Weids, A.J.; Ibstedt, S.; Tamas, M.J.; Grant, C.M. Distinct stress conditions result in aggregation of proteins with similar properties. *Sci. Rep.* **2016**, *6*, 24554. [CrossRef]

8. Ciric, D.; Richard, C.A.; Moudjou, M.; Chapuis, J.; Sibille, P.; Daude, N.; Westaway, D.; Adrover, M.; Beringue, V.; Martin, D.; et al. Interaction between Shadoo and PrP Affects the PrP-Folding Pathway. *J. Virol.* **2015**, *89*, 6287–6293. [CrossRef]

9. Pepe, A.; Avolio, R.; Matassa, D.S.; Esposito, F.; Nitsch, L.; Zurzolo, C.; Paladino, S.; Sarnataro, D. Regulation of sub-compartmental targeting and folding properties of the Prion-like protein Shadoo. *Sci. Rep.* **2017**, *7*, 3731. [CrossRef]

10. Sarnataro, D. Attempt to Untangle the Prion-Like Misfolding Mechanism for Neurodegenerative Diseases. *Int. J. Mol. Sci.* **2018**, *19*, 3081. [CrossRef]

11. Maurer-Stroh, S.; Debulpaep, M.; Kuemmerer, N.; Lopez de la Paz, M.; Martins, I.C.; Reumers, J.; Morris, K.L.; Copland, A.; Serpell, L.; Serrano, L.; et al. Exploring the sequence determinants of amyloid structure using position-specific scoring matrices. *Nat. Methods* **2010**, *7*, 237–242. [CrossRef] [PubMed]

12. Sabate, R.; Rousseau, F.; Schymkowitz, J.; Ventura, S. What makes a protein sequence a prion? *PLoS Comput. Biol.* **2015**, *11*, e1004013. [CrossRef] [PubMed]

13. Soldi, G.; Bemporad, F.; Torrassa, S.; Relini, A.; Ramazzotti, M.; Taddei, N.; Chiti, F. Amyloid formation of a protein in the absence of initial unfolding and destabilization of the native state. *Biophys. J.* **2005**, *89*, 4234–4244. [CrossRef] [PubMed]

14. Dumoulin, M.; Kumita, J.R.; Dobson, C.M. Normal and aberrant biological self-assembly: Insights from studies of human lysozyme and its amyloidogenic variants. *Acc. Chem. Res.* **2006**, *39*, 603–610. [CrossRef] [PubMed]

15. Bemporad, F.; Vannocci, T.; Varela, L.; Azuaga, A.I.; Chiti, F. A model for the aggregation of the acylphosphatase from Sulfolobus solfataricus in its native-like state. *Biochim. Biophys. Acta* **2008**, *1784*, 1986–1996. [CrossRef] [PubMed]

16. Garcia-Pardo, J.; Grana-Montes, R.; Fernandez-Mendez, M.; Ruyra, A.; Roher, N.; Aviles, F.X.; Lorenzo, J.; Ventura, S. Amyloid formation by human carboxypeptidase D transthyretin-like domain under physiological conditions. *J. Biol. Chem.* **2014**, *289*, 33783–33796. [CrossRef] [PubMed]

17. Zhuravlev, P.I.; Reddy, G.; Straub, J.E.; Thirumalai, D. Propensity to form amyloid fibrils is encoded as excitations in the free energy landscape of monomeric proteins. *J. Mol. Biol.* **2014**, *426*, 2653–2666. [CrossRef] [PubMed]

18. Kim, G.H.; Kim, J.E.; Rhie, S.J.; Yoon, S. The Role of Oxidative Stress in Neurodegenerative Diseases. *Exp. Neurobiol.* **2015**, *24*, 325–340. [CrossRef]

19. Ahmad, S.; Khan, H.; Shahab, U.; Rehman, S.; Rafi, Z.; Khan, M.Y.; Ansari, A.; Siddiqui, Z.; Ashraf, J.M.; Abdullah, S.M.; et al. Protein oxidation: An overview of metabolism of sulphur containing amino acid, cysteine. *Front. Biosci.* **2017**, *9*, 71–87. [CrossRef]

20. Serebryany, E.; Woodard, J.C.; Adkar, B.V.; Shabab, M.; King, J.A.; Shakhnovich, E.I. An Internal Disulfide Locks a Misfolded Aggregation-prone Intermediate in Cataract-linked Mutants of Human gammaD-Crystallin. *J. Biol. Chem.* **2016**, *291*, 19172–19183. [CrossRef]

21. Kuhn, D.M.; Sykes, C.E.; Geddes, T.J.; Jaunarajs, K.L.; Bishop, C. Tryptophan hydroxylase 2 aggregates through disulfide cross-linking upon oxidation: Possible link to serotonin deficits and non-motor symptoms in Parkinson's disease. *J. Neurochem.* **2011**, *116*, 426–437. [CrossRef] [PubMed]

22. Fraga, H.; Grana-Montes, R.; Illa, R.; Covaleda, G.; Ventura, S. Association between foldability and aggregation propensity in small disulfide-rich proteins. *Antioxid. Redox Signal.* **2014**, *21*, 368–383. [CrossRef] [PubMed]

23. Jenness, R. Comparative aspects of milk proteins. *J. Dairy Res.* **1979**, *46*, 197–210. [CrossRef] [PubMed]

24. Fuentes-Lemus, E.; Silva, E.; Leinisch, F.; Dorta, E.; Lorentzen, L.G.; Davies, M.J.; Lopez-Alarcon, C. Alpha- and beta-casein aggregation induced by riboflavin-sensitized photo-oxidation occurs via di-tyrosine cross-links and is oxygen concentration dependent. *Food Chem.* **2018**, *256*, 119–128. [CrossRef] [PubMed]

25. Nakajima, H.; Amano, W.; Kubo, T.; Fukuhara, A.; Ihara, H.; Azuma, Y.T.; Tajima, H.; Inui, T.; Sawa, A.; Takeuchi, T. Glyceraldehyde-3-phosphate dehydrogenase aggregate formation participates in oxidative stress-induced cell death. *J. Biol. Chem.* **2009**, *284*, 34331–34341. [CrossRef] [PubMed]

26. Cumming, R.C.; Schubert, D. Amyloid-beta induces disulfide bonding and aggregation of GAPDH in Alzheimer's disease. *FASEB J.* **2005**, *19*, 2060–2062. [CrossRef] [PubMed]

27. Tsuchiya, K.; Tajima, H.; Kuwae, T.; Takeshima, T.; Nakano, T.; Tanaka, M.; Sunaga, K.; Fukuhara, Y.; Nakashima, K.; Ohama, E.; et al. Pro-apoptotic protein glyceraldehyde-3-phosphate dehydrogenase promotes the formation of Lewy body-like inclusions. *Eur. J. Neurosci.* **2005**, *21*, 317–326. [CrossRef] [PubMed]

28. Samson, A.L.; Knaupp, A.S.; Kass, I.; Kleifeld, O.; Marijanovic, E.M.; Hughes, V.A.; Lupton, C.J.; Buckle, A.M.; Bottomley, S.P.; Medcalf, R.L. Oxidation of an exposed methionine instigates the aggregation of glyceraldehyde-3-phosphate dehydrogenase. *J. Biol. Chem.* **2014**, *289*, 26922–26936. [CrossRef] [PubMed]

29. Nakajima, H.; Amano, W.; Fujita, A.; Fukuhara, A.; Azuma, Y.T.; Hata, F.; Inui, T.; Takeuchi, T. The active site cysteine of the proapoptotic protein glyceraldehyde-3-phosphate dehydrogenase is essential in oxidative stress-induced aggregation and cell death. *J. Biol. Chem.* **2007**, *282*, 26562–26574. [CrossRef]

30. Sayre, L.M.; Perry, G.; Atwood, C.S.; Smith, M.A. The role of metals in neurodegenerative diseases. *Cell. Mol. Biol.* **2000**, *46*, 731–741.

31. Abeyawardhane, D.L.; Fernandez, R.D.; Murgas, C.J.; Heitger, D.R.; Forney, A.K.; Crozier, M.K.; Lucas, H.R. Iron Redox Chemistry Promotes Antiparallel Oligomerization of alpha-Synuclein. *J. Am. Chem. Soc.* **2018**, *140*, 5028–5032. [CrossRef] [PubMed]

32. Del Giudice, R.; Pesce, A.; Cozzolino, F.; Monti, M.; Relini, A.; Piccoli, R.; Arciello, A.; Monti, D.M. Effects of iron on the aggregation propensity of the N-terminal fibrillogenic polypeptide of human apolipoprotein AI. *BioMetals* **2018**, *31*, 551–559. [CrossRef] [PubMed]

33. Cheignon, C.; Jones, M.; Atrian-Blasco, E.; Kieffer, I.; Faller, P.; Collin, F.; Hureau, C. Identification of key structural features of the elusive Cu-Abeta complex that generates ROS in Alzheimer's disease. *Chem. Sci.* **2017**, *8*, 5107–5118. [CrossRef] [PubMed]

34. Cheignon, C.; Tomas, M.; Bonnefont-Rousselot, D.; Faller, P.; Hureau, C.; Collin, F. Oxidative stress and the amyloid beta peptide in Alzheimer's disease. *Redox Biol.* **2018**, *14*, 450–464. [CrossRef] [PubMed]

35. Young, T.R.; Pukala, T.L.; Cappai, R.; Wedd, A.G.; Xiao, Z. The Human Amyloid Precursor Protein Binds Copper Ions Dominated by a Picomolar-Affinity Site in the Helix-Rich E2 Domain. *Biochemistry* **2018**, *57*, 4165–4176. [CrossRef]

36. Thinakaran, G.; Koo, E.H. Amyloid precursor protein trafficking, processing, and function. *J. Biol. Chem.* **2008**, *283*, 29615–29619. [CrossRef] [PubMed]

37. Dienemann, C.; Coburger, I.; Mehmedbasic, A.; Andersen, O.M.; Than, M.E. Mutants of metal binding site M1 in APP E2 show metal specific differences in binding of heparin but not of sorLA. *Biochemistry* **2015**, *54*, 2490–2499. [CrossRef]

38. Weng, S.L.; Huang, K.Y.; Kaunang, F.J.; Huang, C.H.; Kao, H.J.; Chang, T.H.; Wang, H.Y.; Lu, J.J.; Lee, T.Y. Investigation and identification of protein carbonylation sites based on position-specific amino acid composition and physicochemical features. *MC Bioinform.* **2017**, *18*, 1105. [CrossRef]

39. Arena, S.; Salzano, A.M.; Renzone, G.; D'Ambrosio, C.; Scaloni, A. Non-enzymatic glycation and glycoxidation protein products in foods and diseases: An interconnected, complex scenario fully open to innovative proteomic studies. *Mass Spectrom. Rev.* **2014**, *33*, 49–77. [CrossRef]

40. Haigis, M.C.; Yankner, B.A. The aging stress response. *Mol. Cell* **2010**, *40*, 333–344. [CrossRef]

41. Castro, J.P.; Ott, C.; Jung, T.; Grune, T.; Almeida, H. Carbonylation of the cytoskeletal protein actin leads to aggregate formation. *Free Radic. Biol. Med.* **2012**, *53*, 916–925. [CrossRef] [PubMed]

42. Tanase, M.; Urbanska, A.M.; Zolla, V.; Clement, C.C.; Huang, L.; Morozova, K.; Follo, C.; Goldberg, M.; Roda, B.; Reschiglian, P.; et al. Role of Carbonyl Modifications on Aging-Associated Protein Aggregation. *Sci. Rep.* **2016**, *6*, 19311. [CrossRef] [PubMed]

43. Bhat, S.A.; Bhat, W.F.; Afsar, M.; Khan, M.S.; Al-Bagmi, M.S.; Bano, B. Modification of chickpea cystatin by reactive dicarbonyl species: Glycation, oxidation and aggregation. *Arch. Biochem. Biophys.* **2018**, *650*, 103–115. [CrossRef] [PubMed]

44. Karri, S.; Singh, S.; Paripati, A.K.; Marada, A.; Krishnamoorthy, T.; Guruprasad, L.; Balasubramanian, D.; Sepuri, N.B.V. Adaptation of Mge1 to oxidative stress by local unfolding and altered Interaction with mitochondrial Hsp70 and Mxr2. *Mitochondrion* **2018**, *46*, 140–148. [CrossRef] [PubMed]

45. Miyata, Y.; Rauch, J.N.; Jinwal, U.K.; Thompson, A.D.; Srinivasan, S.; Dickey, C.A.; Gestwicki, J.E. Cysteine reactivity distinguishes redox sensing by the heat-inducible and constitutive forms of heat shock protein 70. *Chem. Biol.* **2012**, *19*, 1391–1399. [CrossRef] [PubMed]

46. Rand, J.D.; Grant, C.M. The thioredoxin system protects ribosomes against stress-induced aggregation. *Mol. Biol. Cell* **2006**, *17*, 387–401. [CrossRef] [PubMed]

47. Weids, A.J.; Grant, C.M. The yeast peroxiredoxin Tsa1 protects against protein-aggregate-induced oxidative stress. *J. Cell Sci.* **2014**, *127*, 1327–1335. [CrossRef]

48. Wickner, R.B. [URE3] as an altered URE2 protein: Evidence for a prion analog in Saccharomyces cerevisiae. *Science* **1994**, *264*, 566–569. [CrossRef]

49. Doronina, V.A.; Staniforth, G.L.; Speldewinde, S.H.; Tuite, M.F.; Grant, C.M. Oxidative stress conditions increase the frequency of de novo formation of the yeast [PSI+] prion. *Mol. Microbiol.* **2015**, *96*, 163–174. [CrossRef]

50. Jamar, N.H.; Kritsiligkou, P.; Grant, C.M. The non-stop decay mRNA surveillance pathway is required for oxidative stress tolerance. *Nucleic Acids Res.* **2017**, *45*, 6881–6893. [CrossRef]

51. Njomen, E.; Osmulski, P.A.; Jones, C.L.; Gaczynska, M.; Tepe, J.J. Small Molecule Modulation of Proteasome Assembly. *Biochemistry* **2018**, *57*, 4214–4224. [CrossRef] [PubMed]

52. Matsui, H.; Ito, H.; Taniguchi, Y.; Inoue, H.; Takeda, S.; Takahashi, R. Proteasome inhibition in medaka brain induces the features of Parkinson's disease. *J. Neurochem.* **2010**, *115*, 178–187. [CrossRef] [PubMed]

53. Kitajima, Y.; Tashiro, Y.; Suzuki, N.; Warita, H.; Kato, M.; Tateyama, M.; Ando, R.; Izumi, R.; Yamazaki, M.; Abe, M.; et al. Proteasome dysfunction induces muscle growth defects and protein aggregation. *J. Cell Sci.* **2014**, *127*, 5204–5217. [CrossRef] [PubMed]

54. Wang, R.; Zhao, J.; Zhang, J.; Liu, W.; Zhao, M.; Li, J.; Lv, J.; Li, Y. Effect of lysosomal and ubiquitin-proteasome system dysfunction on the abnormal aggregation of alpha-synuclein in PC12 cells. *Exp. Ther. Med.* **2015**, *9*, 2088–2094. [CrossRef] [PubMed]

55. Kumar, V.; Singh, D.; Singh, B.K.; Singh, S.; Mittra, N.; Jha, R.R.; Patel, D.K.; Singh, C. Alpha-synuclein aggregation, Ubiquitin proteasome system impairment, and L-Dopa response in zinc-induced Parkinsonism: Resemblance to sporadic Parkinson's disease. *Mol. Cell. Biochem.* **2018**, *444*, 149–160. [CrossRef] [PubMed]

56. Silva, G.M.; Netto, L.E.S.; Simões, V.; Santos, L.F.A.; Gozzo, F.C.; Demasi, M.A.A.; Oliveira, C.L.P.; Bicev, R.N.; Klitzke, C.F.; Sogayar, M.C.; et al. Redox Control of 20S Proteasome Gating. *Antioxid. Redox Signal.* **2012**, *16*, 1183–1194. [CrossRef] [PubMed]

57. Korovila, I.; Hugo, M.; Castro, J.P.; Weber, D.; Hohn, A.; Grune, T.; Jung, T. Proteostasis, oxidative stress and aging. *Redox Biol.* **2017**, *13*, 550–567. [CrossRef]

58. Zmijewski, J.W.; Banerjee, S.; Abraham, E. S-glutathionylation of the Rpn2 regulatory subunit inhibits 26 S proteasomal function. *J. Biol. Chem.* **2009**, *284*, 22213–22221. [CrossRef] [PubMed]

59. Demasi, M.; Shringarpure, R.; Davies, K.J. Glutathiolation of the proteasome is enhanced by proteolytic inhibitors. *Arch. Biochem. Biophys.* **2001**, *389*, 254–263. [CrossRef]

60. Silva, G.M.; Netto, L.E.; Discola, K.F.; Piassa-Filho, G.M.; Pimenta, D.C.; Barcena, J.A.; Demasi, M. Role of glutaredoxin 2 and cytosolic thioredoxins in cysteinyl-based redox modification of the 20S proteasome. *FEBS J.* **2008**, *275*, 2942–2955. [CrossRef]

61. Bridgford, J.L.; Xie, S.C.; Cobbold, S.A.; Pasaje, C.F.A.; Herrmann, S.; Yang, T.; Gillett, D.L.; Dick, L.R.; Ralph, S.A.; Dogovski, C.; et al. Artemisinin kills malaria parasites by damaging proteins and inhibiting the proteasome. *Nat. Commun.* **2018**, *9*, 3801. [CrossRef] [PubMed]

62. Nystrom, T. Role of oxidative carbonylation in protein quality control and senescence. *EMBO J.* **2005**, *24*, 1311–1317. [CrossRef] [PubMed]

63. Grune, T.; Jung, T.; Merker, K.; Davies, K.J. Decreased proteolysis caused by protein aggregates, inclusion bodies, plaques, lipofuscin, ceroid, and 'aggresomes' during oxidative stress, aging, and disease. *Int. J. Biochem. Cell Biol.* **2004**, *36*, 2519–2530. [CrossRef] [PubMed]

64. Cho, M.-H.; Cho, K.; Kang, H.-J.; Jeon, E.-Y.; Kim, H.-S.; Kwon, H.-J.; Kim, H.-M.; Kim, D.-H.; Yoon, S.-Y. Autophagy in microglia degrades extracellular β-amyloid fibrils and regulates the NLRP3 inflammasome. *Autophagy* **2014**, *10*, 1761–1775. [CrossRef]

65. Caccamo, A.; Ferreira, E.; Branca, C.; Oddo, S. p62 improves AD-like pathology by increasing autophagy. *Mol. Psychiatry* **2016**, *22*, 865. [CrossRef]

66. Cuervo, A.M.; Stefanis, L.; Fredenburg, R.; Lansbury, P.T.; Sulzer, D. Impaired Degradation of Mutant α-Synuclein by Chaperone-Mediated Autophagy. *Science* **2004**, *305*, 1292. [CrossRef]

67. Ravikumar, B.; Duden, R.; Rubinsztein, D.C. Aggregate-prone proteins with polyglutamine and polyalanine expansions are degraded by autophagy. *Hum. Mol. Genet.* **2002**, *11*, 1107–1117. [CrossRef]

68. Yung, C.; Sha, D.; Li, L.; Chin, L.-S. Parkin Protects Against Misfolded SOD1 Toxicity by Promoting Its Aggresome Formation and Autophagic Clearance. *Mol. Neurobiol.* **2016**, *53*, 6270–6287. [CrossRef]

69. Bordi, M.; Berg, M.J.; Mohan, P.S.; Peterhoff, C.M.; Alldred, M.J.; Che, S.; Ginsberg, S.D.; Nixon, R.A. Autophagy flux in CA1 neurons of Alzheimer hippocampus: Increased induction overburdens failing lysosomes to propel neuritic dystrophy. *Autophagy* **2016**, *12*, 2467–2483. [CrossRef]

70. Nixon, R.A.; Yang, D.-S. Autophagy failure in Alzheimer's disease—Locating the primary defect. Neurobiol. Dis. **2011**, *43*, 38–45. [CrossRef]

71. Ramirez, A.; Heimbach, A.; Grundemann, J.; Stiller, B.; Hampshire, D.; Cid, L.P.; Goebel, I.; Mubaidin, A.F.; Wriekat, A.L.; Roeper, J.; et al. Hereditary parkinsonism with dementia is caused by mutations in ATP13A2, encoding a lysosomal type 5 P-type ATPase. *Nat. Genet.* **2006**, *38*, 1184–1191. [CrossRef] [PubMed]

72. Zavodszky, E.; Seaman, M.N.; Moreau, K.; Jimenez-Sanchez, M.; Breusegem, S.Y.; Harbour, M.E.; Rubinsztein, D.C. Mutation in VPS35 associated with Parkinson's disease impairs WASH complex association and inhibits autophagy. *Nat. Commun.* **2014**, *5*, 3828. [CrossRef] [PubMed]

73. Chen, Y.; Liu, H.; Guan, Y.; Wang, Q.; Zhou, F.; Jie, L.; Ju, J.; Pu, L.; Du, H.; Wang, X. The altered autophagy mediated by TFEB in animal and cell models of amyotrophic lateral sclerosis. *Am. J. Transl. Res.* **2015**, *7*, 1574–1587. [PubMed]

74. Levonen, A.L.; Hill, B.G.; Kansanen, E.; Zhang, J.; Darley-Usmar, V.M. Redox regulation of antioxidants, autophagy, and the response to stress: Implications for electrophile therapeutics. *Free Radic. Biol. Med.* **2014**, *71*, 196–207. [CrossRef] [PubMed]

75. Scherz-Shouval, R.; Shvets, E.; Fass, E.; Shorer, H.; Gil, L.; Elazar, Z. Reactive oxygen species are essential for autophagy and specifically regulate the activity of Atg4. *EMBO J.* **2007**, *26*, 1749–1760. [CrossRef] [PubMed]

76. Bensaad, K.; Cheung, E.C.; Vousden, K.H. Modulation of intracellular ROS levels by TIGAR controls autophagy. *EMBO J.* **2009**, *28*, 3015–3026. [CrossRef]

77. Chen, Y.; Azad, M.B.; Gibson, S.B. Superoxide is the major reactive oxygen species regulating autophagy. *Cell Death Differ.* **2009**, *16*, 1040. [CrossRef]

78. Dodson, M.; Darley-Usmar, V.; Zhang, J. Cellular metabolic and autophagic pathways: Traffic control by redox signaling. *Free Radic. Biol. Med.* **2013**, *63*, 207–221. [CrossRef]

79. Lee, J.; Giordano, S.; Zhang, J. Autophagy, mitochondria and oxidative stress: Cross-talk and redox signalling. *Biochem. J.* **2012**, *441*, 523–540. [CrossRef]

80. Guerrero-Gomez, D.; Mora-Lorca, J.A.; Saenz-Narciso, B.; Naranjo-Galindo, F.J.; Munoz-Lobato, F.; Parrado-Fernandez, C.; Goikolea, J.; Cedazo-Minguez, Á.; Link, C.D.; Neri, C.; et al. Loss of glutathione redox homeostasis impairs proteostasis by inhibiting autophagy-dependent protein degradation. *Cell Death Differ.* **2019**. Available online: https://www.nature.com/articles/s41418-018-0270-9 (accessed on 15 February 2019). [CrossRef]

81. Civiletto, G.; Dogan, S.A.; Cerutti, R.; Fagiolari, G.; Moggio, M.; Lamperti, C.; Beninca, C.; Viscomi, C.; Zeviani, M. Rapamycin rescues mitochondrial myopathy via coordinated activation of autophagy and lysosomal biogenesis. *EMBO Mol. Med.* **2018**, *10*, e8799. [CrossRef] [PubMed]

82. Richardson, A.; Galvan, V.; Lin, A.-L.; Oddo, S. How longevity research can lead to therapies for Alzheimer's disease: The rapamycin story. *Exp. Gerontol.* **2015**, *68*, 51–58. [CrossRef] [PubMed]

83. Zhang, L.; Wang, L.; Wang, R.; Gao, Y.; Che, H.; Pan, Y.; Fu, P. Evaluating the Effectiveness of GTM-1, Rapamycin, and Carbamazepine on Autophagy and Alzheimer Disease. *Med. Sci. Monit.* **2017**, *23*, 801–808. [CrossRef] [PubMed]

84. Thellung, S.; Scoti, B.; Corsaro, A.; Villa, V.; Nizzari, M.; Gagliani, M.C.; Porcile, C.; Russo, C.; Pagano, A.; Tacchetti, C.; et al. Pharmacological activation of autophagy favors the clearing of intracellular aggregates of misfolded prion protein peptide to prevent neuronal death. *Cell Death Dis.* **2018**, *9*, 166. [CrossRef] [PubMed]

85. Choi, D.H.; Kim, D.H.; Park, Y.G.; Chun, B.G.; Choi, S.H. Protective effects of rilmenidine and AGN 192403 on oxidative cytotoxicity and mitochondrial inhibitor-induced cytotoxicity in astrocytes. *Free Radic. Biol. Med.* **2002**, *33*, 1321–1333. [CrossRef]

86. Perera, N.D.; Sheean, R.K.; Lau, C.L.; Shin, Y.S.; Beart, P.M.; Horne, M.K.; Turner, B.J. Rilmenidine promotes MTOR-independent autophagy in the mutant SOD1 mouse model of amyotrophic lateral sclerosis without slowing disease progression. *Autophagy* **2017**, *14*, 534–551. [CrossRef]

87. Kim, Y.; Park, J.K.; Seo, J.H.; Ryu, H.S.; Lim, K.S.; Jeong, M.H.; Kang, D.H.; Kang, S.W. A rapamycin derivative, biolimus, preferentially activates autophagy in vascular smooth muscle cells. *Sci. Rep.* **2018**, *8*, 16551. [CrossRef]

88. Bossy-Wetzel, E.; Petrilli, A.; Knott, A.B. Mutant huntingtin and mitochondrial dysfunction. *Trends Neurosci.* **2008**, *31*, 609–616. [CrossRef]

89. Gruber, A.; Hornburg, D.; Antonin, M.; Krahmer, N.; Collado, J.; Schaffer, M.; Zubaite, G.; Luchtenborg, C.; Sachsenheimer, T.; Brugger, B.; et al. Molecular and structural architecture of polyQ aggregates in yeast. *Proc. Natl. Acad. Sci. USA* **2018**, *115*, E3446–E3453. [CrossRef]

90. Cenini, G.; Rub, C.; Bruderek, M.; Voos, W. Amyloid beta-peptides interfere with mitochondrial preprotein import competence by a coaggregation process. *Mol. Biol. Cell* **2016**, *27*, 3257–3272. [CrossRef]

91. Wang, X.; Becker, K.; Levine, N.; Zhang, M.; Lieberman, A.P.; Moore, D.J.; Ma, J. Pathogenic alpha-synuclein aggregates preferentially bind to mitochondria and affect cellular respiration. *Acta Neuropathol. Commun.* **2019**, *7*, 41. [CrossRef]

92. Chaudhary, R.K.; Patel, K.A.; Patel, M.K.; Joshi, R.H.; Roy, I. Inhibition of Aggregation of Mutant Huntingtin by Nucleic Acid Aptamers *In vitro* and in a Yeast Model of Huntington's Disease. *Mol. Ther.* **2015**, *23*, 1912–1926. [CrossRef] [PubMed]

93. Patel, K.A.; Kolluri, T.; Jain, S.; Roy, I. Designing Aptamers which Respond to Intracellular Oxidative Stress and Inhibit Aggregation of Mutant Huntingtin. *Free Radic. Biol. Med.* **2018**, *120*, 311–316. [CrossRef]

94. Mattson, M.P.; Goodman, Y. Different amyloidogenic peptides share a similar mechanism of neurotoxicity involving reactive oxygen species and calcium. *Brain Res.* **1995**, *676*, 219–224. [CrossRef]

95. Lim, Y.A.; Rhein, V.; Baysang, G.; Meier, F.; Poljak, A.; Raftery, M.J.; Guilhaus, M.; Ittner, L.M.; Eckert, A.; Gotz, J. Abeta and human amylin share a common toxicity pathway via mitochondrial dysfunction. *Proteomics* **2010**, *10*, 1621–1633. [CrossRef] [PubMed]

96. Singh, S.; Bhowmick, D.C.; Pany, S.; Joe, M.; Zaghlula, N.; Jeremic, A.M. Apoptosis signal regulating kinase-1 and NADPH oxidase mediate human amylin evoked redox stress and apoptosis in pancreatic beta-cells. *Biochim. Biophys. Acta* **2018**, *1860*, 1721–1733. [CrossRef]

97. Abramov, A.Y.; Duchen, M.R. The role of an astrocytic NADPH oxidase in the neurotoxicity of amyloid beta peptides. *Philos. Trans. R. Soc. B: Biol. Sci.* **2005**, *360*, 2309–2314. [CrossRef]

98. Abeti, R.; Abramov, A.Y.; Duchen, M.R. Beta-amyloid activates PARP causing astrocytic metabolic failure and neuronal death. *Brain* **2011**, *134*, 1658–1672. [CrossRef]

99. Palazzi, L.; Bruzzone, E.; Bisello, G.; Leri, M.; Stefani, M.; Bucciantini, M.; Polverino de Laureto, P. Oleuropein aglycone stabilizes the monomeric alpha-synuclein and favours the growth of non-toxic aggregates. *Sci. Rep.* **2018**, *8*, 8337. [CrossRef]

100. Baruch-Suchodolsky, R.; Fischer, B. Soluble amyloid beta1-28-copper(I)/copper(II)/Iron(II) complexes are potent antioxidants in cell-free systems. *Biochemistry* **2008**, *47*, 7796–7806. [CrossRef]

101. Baruch-Suchodolsky, R.; Fischer, B. Abeta40, either soluble or aggregated, is a remarkably potent antioxidant in cell-free oxidative systems. *Biochemistry* **2009**, *48*, 4354–4370. [CrossRef] [PubMed]

102. Garzon-Rodriguez, W.; Yatsimirsky, A.K.; Glabe, C.G. Binding of Zn(II), Cu(II), and Fe(II) ions to Alzheimer's A beta peptide studied by fluorescence. *Bioorganic Med. Chem. Lett.* **1999**, *9*, 2243–2248. [CrossRef]

103. Khan, A.; Dobson, J.P.; Exley, C. Redox cycling of iron by Abeta42. *Free Radic. Biol. Med.* **2006**, *40*, 557–569. [CrossRef] [PubMed]

104. Yang, E.Y.; Guo-Ross, S.X.; Bondy, S.C. The stabilization of ferrous iron by a toxic beta-amyloid fragment and by an aluminum salt. *Brain Res.* **1999**, *839*, 221–226. [CrossRef]

105. Carija, A.; Navarro, S.; de Groot, N.S.; Ventura, S. Protein aggregation into insoluble deposits protects from oxidative stress. *Redox Biol.* **2017**, *12*, 699–711. [CrossRef] [PubMed]

106. Sanchez de Groot, N.; Gomes, R.A.; Villar-Pique, A.; Babu, M.M.; Coelho, A.V.; Ventura, S. Proteome response at the edge of protein aggregation. *Open Biol.* **2015**, *5*, 140221. [CrossRef] [PubMed]

107. Sideri, T.C.; Stojanovski, K.; Tuite, M.F.; Grant, C.M. Ribosome-associated peroxiredoxins suppress oxidative stress-induced de novo formation of the [PSI+] prion in yeast. *Proc. Natl. Acad. Sci. USA* **2010**, *107*, 6394–6399. [CrossRef] [PubMed]

108. Cali, T.; Ottolini, D.; Negro, A.; Brini, M. alpha-Synuclein controls mitochondrial calcium homeostasis by enhancing endoplasmic reticulum-mitochondria interactions. *J. Biol. Chem.* **2012**, *287*, 17914–17929. [CrossRef] [PubMed]

109. Guardia-Laguarta, C.; Area-Gomez, E.; Rub, C.; Liu, Y.; Magrane, J.; Becker, D.; Voos, W.; Schon, E.A.; Przedborski, S. alpha-Synuclein is localized to mitochondria-associated ER membranes. *J. Neurosci.* **2014**, *34*, 249–259. [CrossRef]

110. Paillusson, S.; Gomez-Suaga, P.; Stoica, R.; Little, D.; Gissen, P.; Devine, M.J.; Noble, W.; Hanger, D.P.; Miller, C.C.J. Alpha-Synuclein binds to the ER-mitochondria tethering protein VAPB to disrupt Ca(2+) homeostasis and mitochondrial ATP production. *Acta Neuropathol.* **2017**, *134*, 129–149. [CrossRef]

111. Pozo Devoto, V.M.; Dimopoulos, N.; Alloatti, M.; Pardi, M.B.; Saez, T.M.; Otero, M.G.; Cromberg, L.E.; Marín-Burgin, A.; Scassa, M.E.; Stokin, G.B.; et al. αSynuclein control of mitochondrial homeostasis in human-derived neurons is disrupted by mutations associated with Parkinson's disease. *Sci. Rep.* **2017**, *7*, 5042. [CrossRef] [PubMed]

112. Tabner, B.J.; El-Agnaf, O.M.; German, M.J.; Fullwood, N.J.; Allsop, D. Protein aggregation, metals and oxidative stress in neurodegenerative diseases. *Biochem. Soc. Trans.* **2005**, *33*, 1082–1086. [CrossRef] [PubMed]

113. Zhu, C.; Beck, M.V.; Griffith, J.D.; Deshmukh, M.; Dokholyan, N.V. Large SOD1 aggregates, unlike trimeric SOD1, do not impact cell viability in a model of amyotrophic lateral sclerosis. *Proc. Natl. Acad. Sci.* **2018**, *115*, 4661–4665. [CrossRef] [PubMed]

114. Farrawell, N.E.; Lambert-Smith, I.; Mitchell, K.; McKenna, J.; McAlary, L.; Ciryam, P.; Vine, K.L.; Saunders, D.N.; Yerbury, J.J. SOD1(A4V) aggregation alters ubiquitin homeostasis in a cell model of ALS. *J. Cell Sci.* **2018**, *131*, jcs209122. [CrossRef] [PubMed]

115. D'Amico, E.; Factor-Litvak, P.; Santella, R.M.; Mitsumoto, H. Clinical perspective on oxidative stress in sporadic amyotrophic lateral sclerosis. *Free Radic. Biol. Med.* **2013**, *65*, 509–527. [CrossRef]

116. Petrov, D.; Daura, X.; Zagrovic, B. Effect of Oxidative Damage on the Stability and Dimerization of Superoxide Dismutase 1. *Biophys. J.* **2016**, *110*, 1499–1509. [CrossRef] [PubMed]

117. Dedeoglu, A.; Cormier, K.; Payton, S.; Tseitlin, K.A.; Kremsky, J.N.; Lai, L.; Li, X.; Moir, R.D.; Tanzi, R.E.; Bush, A.I.; et al. Preliminary studies of a novel bifunctional metal chelator targeting Alzheimer's amyloidogenesis. *Exp. Gerontol.* **2004**, *39*, 1641–1649. [CrossRef]

118. Sharma, A.K.; Pavlova, S.T.; Kim, J.; Finkelstein, D.; Hawco, N.J.; Rath, N.P.; Kim, J.; Mirica, L.M. Bifunctional Compounds for Controlling Metal-Mediated Aggregation of the Aβ42 Peptide. *J. Am. Chem. Soc.* **2012**, *134*, 6625–6636. [CrossRef]

119. Rajasekhar, K.; Mehta, K.; Govindaraju, T. Hybrid Multifunctional Modulators Inhibit Multifaceted Abeta Toxicity and Prevent Mitochondrial Damage. *ACS Chem. Neurosci.* **2018**, *9*, 1432–1440. [CrossRef]
120. Hilt, S.; Altman, R.; Kalai, T.; Maezawa, I.; Gong, Q.; Wachsmann-Hogiu, S.; Jin, L.W.; Voss, J.C. A Bifunctional Anti-Amyloid Blocks Oxidative Stress and the Accumulation of Intraneuronal Amyloid-Beta. *Molecules* **2018**, *23*, 2010. [CrossRef]

International Journal of
Molecular Sciences

Review

Dual Role of Reactive Oxygen Species in Muscle Function: Can Antioxidant Dietary Supplements Counteract Age-Related Sarcopenia?

Simona Damiano [†], Espedita Muscariello [†], Giuliana La Rosa, Martina Di Maro, Paolo Mondola and Mariarosaria Santillo *

Dipartimento di Medicina Clinica e Chirurgia, Università di Napoli "Federico II", Via S. Pansini, 5, 80131 Naples, Italy
* Correspondence: marsanti@unina.it; Tel.: +39-0817463233
† These authors contributed equally.

Received: 3 July 2019; Accepted: 1 August 2019; Published: 5 August 2019

Abstract: Sarcopenia is characterized by the progressive loss of skeletal muscle mass and strength. In older people, malnutrition and physical inactivity are often associated with sarcopenia, and, therefore, dietary interventions and exercise must be considered to prevent, delay, or treat it. Among the pathophysiological mechanisms leading to sarcopenia, a key role is played by an increase in reactive oxygen and nitrogen species (ROS/RNS) levels and a decrease in enzymatic antioxidant protection leading to oxidative stress. Many studies have evaluated, in addition to the effects of exercise, the effects of antioxidant dietary supplements in limiting age-related muscle mass and performance, but the data which have been reported are conflicting. In skeletal muscle, ROS/RNS have a dual function: at low levels they increase muscle force and adaptation to exercise, while at high levels they lead to a decline of muscle performance. Controversial results obtained with antioxidant supplementation in older persons could in part reflect the lack of univocal effects of ROS on muscle mass and function. The purpose of this review is to examine the molecular mechanisms underlying the dual effects of ROS in skeletal muscle function and the analysis of literature data on dietary antioxidant supplementation associated with exercise in normal and sarcopenic subjects.

Keywords: sarcopenia; reactive oxygen species; redox signaling; antioxidant supplementation; exercise

---

1. Age-Related Sarcopenia

Aging is characterized by a progressive decline in muscle mass and strength. In 1988 Irwin Rosenberg proposed the term sarcopenia (from the Greek "sarx" = flesh and "penia" = loss) to describe muscle size decrease that occurs in the elderly [1].

Roubenoff tried to differentiate sarcopenia from other processes leading to muscle mass loss, such as wasting and cachexia [2]. Wasting was considered an unintentional weight loss, including loss of both fat and lean body mass, due to inadequate caloric intake. Cachexia, on the other hand, was defined as "the loss of fat free mass with no significant weight loss" as a result of hypermetabolism and hypercatabolism mediated by cytokines. Conversely, sarcopenia was regarded as a process that could take place even without malnutrition or disease and in this sense could be considered a natural process in aging. The currently most used definition in clinical practice and diagnostic criteria consensus for age-related sarcopenia has been developed by the European Working Group on Sarcopenia in Older People (EWGSOP) [3], which claims that a subject with low muscle strength and low muscle mass or quality must be diagnosed with sarcopenia. Hence, sarcopenia can occur either acutely or chronically. Diagnosis of sarcopenia relies on combined measurement of (1) muscle mass, which is assessed for example by dual energy X-ray absorptiometry (DXA) or bioelectrical impedance analysis

(BIA); (2) muscle strength, assessed by functional tests like grip strength; and (3) physical performance, including assessment of mobility, strength, and balance [4].

Sarcopenia can be considered a multifactorial event which is characterized by inflammation, oxidative stress, motor neuron loss, and changing of endocrine function [5]. Loss of strength and mobility, together with balance disorders induced by sarcopenia, increase the rate of falls and fractures in old age, leading to immobilization, which in turn contributes to worsening of sarcopenia. In addition, old age is often associated with appetite loss, protein-energy malnutrition, and weight loss, which concur with sarcopenia. Therefore, an adequate nutrient intake is recommended in sarcopenic patients to preserve muscle mass. Often physical inactivity and malnutrition reinforce each other; indeed, reduced mobility leads to a decline in nutrition capability, worsening malnutrition and sarcopenia. However, all these factors contribute to skeletal muscle atrophy and weakness, accompanied by disability increase, frailty, and life quality impairment [5].

Several studies have focused on the pathophysiology of age-related sarcopenia; nonetheless, the biological mechanisms underlying decline in muscle strength and mass with age are not completely understood. Loss in muscle mass and impairment of muscle force has been associated with the disruption of excitation–contraction coupling. In single human skeletal myocytes obtained by needle biopsy of the vastus lateralis, a significant reduction of dihydropyridine (DHP)-sensitive Ca^{2+} currents have been recorded in fibers from old people compared to those in young people. Moreover, a reduced peak of Ca^{2+} transient and a decrease in the voltage-/Ca^{2+}-dependent Ca^{2+} release ratio has been registered in old fibers compared to young ones, suggesting that the decrease of Ca^{2+} available for mechanical responses in aged skeletal muscle is due to DHP receptor (DHPR)-ryanodine receptor (RyR) uncoupling [6]. Similar results have been obtained in experiments conducted in mice [7]. Age-associated muscle mass and strength decline may also be explained by motor unit remodeling, which has been observed in old mice, in which denervated muscle fibers are reinnervated by axonal sprouting of adjacent motor units [8]. Moreover, fibers from aged rats contract more slowly than those from young rats; in an in vitro motility assay with isolated soleus muscle fibers an age-related alteration in myosin with a decrease of the maximum shortening velocity has been evidenced [9]. A replacement of muscle mass by fat and connective tissue with aging results in a gradual decrease of muscle size/volume. Moreover, muscle biopsies from old and young subjects have shown that number and size of muscle fibers, mainly of type II (fast-twitch) are reduced in the elderly [10]. Indeed, a loss of muscle mass explains only in part muscle decline, and other possible mechanisms are related to alterations in muscle fiber quality and changes in fiber type [11]. It has been also hypothesized that a reduction and/or dysregulation of satellite cells involved in skeletal muscle regeneration may contribute to the loss of skeletal muscle mass observed in aging [12]. Among the mechanisms underlying sarcopenia, an emerging role is being played by endoplasmic reticulum (ER) stress due to accumulation of unfolded or misfolded proteins within the ER and its adaptive responses involving reactive oxygen species (ROS) signaling [13]. Finally, recent studies on the role of gut microbiota in inflammatory diseases indicate a relevant contribution of gut microbial changes and activity that occur with aging to the types of inflammatory molecules present in the environment surrounding muscles [14].

The molecular mechanisms leading to sarcopenia are complex and have not been completely clarified. Redox signaling and oxidative damage are among the most accepted mechanisms underlying the decline of muscle mass and strength with aging and they will therefore be discussed in more detail in separate sections of this review.

2. Endogenous Sources of Reactive Oxygen Species and Antioxidant Systems

ROS are generated through various pathways; one of the main sites of ROS production is the mitochondrial electron transport chain, where the transfer of a single electron to molecular oxygen gives rise to a monovalent reduction of oxygen, which leads to the formation of superoxide ions. The superoxide production can also occur enzymatically through NADPH oxidase enzymes or the xanthine/xanthine oxidase system. NADPH oxidase was discovered first in phagocytes, where it generates high levels of superoxide as a microbicidal mechanism in host defense [15]. NADPH oxidase is an enzymatic complex consisting of various cytosolic units (p40-, p47-, and p67phox), and the membrane-anchored components p22phox and cytochrome b558 (gp91phox); the latter is the site of the catalytic activity of the complex. The activation of the complex also requires the involvement of small GTP binding proteins Rap1A and Rac. In mammalians there are seven genes encoding distinct catalytic subunits, namely, NOXs 1-5 and DUOX1-2 [16]. Xanthine oxidase enzyme catalyzes the hydroxylation of hypoxanthine to xanthine and xanthine to uric acid. In both steps, molecular oxygen is reduced, forming the superoxide anion [17].

Among the antioxidant enzymes, superoxide dismutases are responsible for the dismutation of O_2^- in molecular oxygen and hydrogen peroxide [18,19], which is converted into water by catalase or glutathione-peroxidase. There are three SOD isoenzymes, namely, the cytosolic dimeric Cu,Zn SOD (SOD1) [20], the mitochondrial manganese-containing SOD (MnSOD or SOD2) [21], and the tetrameric extracellular CuZn SOD (EcSOD or SOD3) [22].

In the reaction catalyzed by glutathione-peroxidase, glutathione is oxidized to glutathione disulfate, which can be converted to glutathione by glutathione reductase in a "NADPH-consuming" process (NADPH → NADP$^+$) [23]. Catalase is a high molecular weight tetrameric enzyme containing porphyrin in the active site [24]. In the presence of transition metal ions (e.g., $Fe^{2+\backslash 3+}$, $Cu^{+\backslash 2+}$), hydrogen peroxide produces the highly reactive oxygen species hydroxyl radical (OH·) and hydroxyl ion (OH$^-$) according to the Fenton reaction [25].

Endoplasmic reticulum (ER) is involved in the control of the redox state of the cells as well [26]. The accumulation of unfolded or misfolded proteins within the ER leads to ER stress and to the activation of the signaling pathways of the unfolded protein response (UPR) as adaptive response. Aiming to re-establish ER proteostasis, the UPR pathway reduces ER protein load and enhances ER quality control and autophagy. The adaptive response of the UPR pathway induces antioxidant gene transcription through the activation of nuclear factor E2-related factor 2 (Nrf 2). However, the UPR pathway can even enhance ROS production; indeed, the increased protein folding during ER stress causes the peroxide levels to rise through the pancreatic ER kinase (PERK)/C/EBP homologous protein (CHOP) pathway, thus enhancing the expression of Ero1, encoding for an ER peroxidase. This pathway leads to oxidative stress, and, eventually, apoptosis [27].

The levels of ROS/reactive nitrogen species (RNS) inside the cells are strictly controlled by the balance between the rate of synthesis by ROS/RNS generating systems and the rate of removal through the non-enzymatic and enzymatic antioxidant systems. An accurate control of ROS levels guarantees the maintenance of physiological levels of ROS necessary for cell functions. Excessive ROS/RNS production or the impairment of antioxidant status may disturb the cellular redox balance, inducing oxidative stress in cells or tissues.

3. The Role of Reactive and Nitrogen Species in Muscle Functions

Many studies suggest that high levels of free radicals can damage biological molecules while low levels play physiological roles such as regulation of cell signaling [28–36]. ROS, including mainly superoxide anions and hydrogen peroxide and hydroxyl radicals, are continuously produced by muscle cells in resting condition, and their levels increase during contraction modulating force production [37]. However, while at low levels a ROS increase produces an enhancement in the development of muscle force up to a maximum peak, a further ROS increase induces a dramatic decline in the force [38]. Moreover, intense and prolonged exercise has often been associated with increase of ROS production and oxidative damage to cellular constituents.

ROS are produced in muscle cells by various sources in different cell compartments. Mitochondria are an important site of ROS production, and during exercise, ROS increases due to high oxygen consumption by increased mitochondrial activity [39,40]. Another source of ROS in skeletal muscle is NOX enzymes [41]. NOX1, NOX2, and NOX4 are three NOX isoforms expressed in skeletal muscle [42]. NOXs are located in the sarcoplasmic reticulum (SR), transverse tubule, and plasma membrane; in particular, the superoxide anion generated by NOX2 in the SR can stimulate the correct release of calcium from intracellular stores through the oxidation of RyR1 [43]. Thus, this enzyme plays an important functional role in excitation contraction coupling. Although mitochondria have long been considered the principal site of muscle ROS production [35], much evidence suggests that NOXs are the main source of ROS induced by skeletal muscle contractions. Indeed, during contractions, the increase in cytosolic ROS comes first and is greater than the rise in mitochondrial ROS [44].

ROS-mediated ROS release due to crosstalk between NOXs and mitochondria has been described for different cells. The uncoupling of electron transfer can result from oxidation of mitochondrial electron transfer chain complexes [45]. Moreover, NOX-derived ROS can induce mitochondrial ROS production through the opening of mitochondrial ATP-sensitive K+ channels (mito-KATP) [46] and consequent potassium influx into the matrix that reduces the mitochondrial membrane potential, opening permeability transition pores. Another mechanism that can explain the crosstalk between NOXs and mitochondrial ROS production is the rise of intracellular Ca^{2+} levels by NOX-derived ROS, which, increasing mitochondrial Ca^{2+} load, induce ROS production by these organelles [47]. The existence of NOX-mitochondria redox crosstalk has also been demonstrated insmooth and cardiac muscle cells. Although there is no direct evidence of the existence of ROS-mediated ROS release in skeletal muscle cells, it is plausible that these mechanisms are also active in these cell types.

Another striking example of ROS-mediated ROS release is the crosstalk between ER and mitochondria. ER oxidative stress can be transmitted to mitochondria through Ca^{++} influx. Across mitochondrial-associated ER membranes (MAMs), Ca^{++} ions and other metabolites are transferred to mitochondria, leading to the opening of permeability transition pores and increased mitochondrial ROS levels [26].

RNS as well as ROS may be messenger molecules that activate muscle adaptive responses through the induction of redox-sensitive signaling to maintain cellular oxidant-antioxidant homeostasis during exercise. RNS arise from several sources and the levels increase with contractile activity. Nitric oxide (NO) is formed from L-arginine in a reaction catalyzed by the nitric oxide synthase (NOS) enzyme. In skeletal muscle there are all the different isoforms of this enzyme: nNOS, eNOS, and iNOS. A calmodulin (CaM)-binding domain is present in nNOS and eNOS, and, therefore, these Ca^{2+}sensitive isoforms are responsive to contractile activity [48]. Moreover, NO generated by NOSs readily reacts with superoxide to form peroxynitrite ($NO^+ O_2^-/ONOO^-$), which decreases the bioavailability of NO and superoxide, modifying the redox balance in the myocyte [49].

Regular physical exercise modulates ROS in a bell-shaped hormesis curve due to the activation by ROS of adaptive response resulting in increasing activity of repair enzymes and low degree of oxidative stress [50]. Overall, it seems that with the exception of high intensity and long duration exercise, physical activity cannot result in harmful oxidative damage.

4. ROS-Mediated Mechanisms in the Development of Age-Related Sarcopenia

Cumulative damage to skeletal muscle and nerve cells in sarcopenia may result from oxidative stress. Oxidative stress causes damage in many tissues, including the loss of muscle mass and strength, which is associated with impairment of neurotransmitter release and neuronal degeneration.

Sarcopenia could be caused by an increase of endogenous ROS formation in skeletal muscle but the source of ROS in sarcopenic muscle is still relatively unknown; however, an age-associated increase of ROS levels in muscle mass, as a consequence of an upregulation of NOX2 enzyme, has been reported [51]. Moreover, a study by Sullivan-Gunn and Lewandowski [52] has highlighted the role of NOX2 enzyme in a healthy mouse model of aging, suggesting that elevated levels of H_2O_2 from NOX2, as well as the lack of antioxidant protection from catalase and glutathione peroxidase (GPx), carry out a key role in the onset of sarcopenia. The lack of SOD1 also causes a reduction of skeletal muscle mass, impairment of neurotransmitter release, and neuronal degeneration in mice [53].

O_2^- radicals induce neuromuscular degeneration and mitochondrial dysfunction in SOD1-deficient mice [54]; moreover, the reduction of cellular antioxidant capacity by disruption of the SOD1 gene in mice (*Sod1*$^{-/-}$) causes severe oxidative stress and oxidative damage associated with an acceleration of age-related loss of skeletal muscle mass accompanied by neuromuscular junction (NMJ) morphologic changes, increased denervation, and an elevated production of superoxide and hydrogen peroxide by muscle mitochondria [54]. Elevated ROS have been shown to interfere with synaptic vesicle axonal transport and formation of new vesicles in the trans-Golgi network [55]. This, in turn, may lead to the accumulation of fewer synaptic vesicles at NMJs, resulting in reduced neurotransmitter release [56]. In addition, *Sod1* gene ablation in adult mice causes physiological changes at the NMJ, similar to that occurring in old wild types [57]. Moreover, the increase of cytosolic oxidative stress caused either by the deletion of SOD1 (*Sod1*$^{-/-}$ mice) or by introduction of mutations of SOD1, (e.g., the SOD1^{G93A} amyotrophic lateral sclerosis mutant mouse model) increases ROS levels, causing muscle atrophy and weakness in mice which are phenotypically similar to muscle changes observed in older animals [54,58].

The use of SOD1^{G93A} and other SOD1 mutant models to decipher the mechanistic aspects of oxidative stress in muscle atrophy could be confounded by the toxic gain of function that results in the formation of SOD1 protein aggregates [59]. Indeed, recently, it has been shown that a gain of function of the mutated SOD1^{G93A} could be associated with an increase of oxidative stress, intracellular calcium concentration, and proapoptotic effect in both human neuroblastoma SK-N-BE and mouse motor-neuron-like NSC-34 cells. These effects are carried out through an activation of extracellular signal regulated kinases (ERK) 1–2, serine threonine kinase (Akt), and intracellular calcium levels mediated by the activation of muscarinic M1 receptor [28]. Systemic administration of endogenous nitric oxide donor *S*-nitrosoglutathione enhances extracellular SOD3 expression and the antioxidant activity protecting structural and functional integrity of skeletal muscle [60]. However, it is noteworthy to underline that SOD1 is secreted as well [61], and, therefore, the elevated extracellular oxygen radical, associated with sarcopenia, could also be scavenged by cytosolic SOD1.

ER stress and UPR response also play an important role in age-related sarcopenia [13]. ER stress can directly impact muscle mass since sustained ER stress leads to cell death of muscle cells [62], which is mediated by increased ER ROS. ER stress can also inhibit rapamycin complex 1 (mTORC 1) that mediates the response to anabolic stimulus of nutrients and contractile activity, thus inducing anabolic resistance and reduced regenerative potential of skeletal muscle observed during aging. Moreover, it has been shown that in the chronic kidney disease uremic toxin-accumulated sarcopenia model, ER stress and UPR pathways account for the inhibition of myoblast differentiation and myotubular atrophy induction through the activation of a ROS-eIF2α axis [63].

Aging is also characterized by mitochondrial dysfunction in skeletal muscle with accumulation of mitochondrial damage and oxidative stress [64]. Mitochondrial dynamics are controlled by fusion and fission proteins. Mitofusin 1 and 2 (Mnf 1 and 2) are involved in the outer mitochondrial membrane fusion while optic atrophy 1 (OPA1) mediates fusion of the inner mitochondrial membrane. Aging is associated with a progressive reduction in Mnf2, and Mfn2 deficiency in mouse skeletal muscle reduces mitophagy, leading to the accumulation of damaged mitochondria [65]. On the other hand, OPA1, which is also a sensor of physical activity, is downregulated during aging-related sarcopenia. Interestingly, in adult mice, acute, muscle-specific deletion of OPA1 leads to ER stress, which through UPR pathways, ROS, and FoxOs induces a catabolic program, muscle loss, and systemic inflammation [66].

In conclusion, age-related ROS overproduction generates oxidative damage of muscle but it also plays a role in regulating intracellular signal transduction pathways that are directly or indirectly involved in skeletal muscle atrophy, motoneuronal degeneration, and impairment of muscle contractility.

5. Redox Signaling in Exercise Adaptation in Age-Related Sarcopenia

Muscle has meaningful plastic properties. Indeed, regular physical exercise ameliorates skeletal muscle performance along with other multiple body functions. During exercise the great oxygen flux required for ATP production leads to ROS generation at different rates and from different sources depending on type, intensity, and duration of exercise. ROS generated during exercise regulate signal transduction pathways responsible for muscle remodeling and for the adaptive response necessary to limit oxidative stress.

Nuclear factor (NF) κB, mitogen-activated protein kinases (MAPKs), and peroxisome proliferator-activated receptor γ co-activator 1 α (PGC1α) are among the main redox-sensitive pathways activated during muscle activity involved in the adaptive response to oxidative stress [67].

NFκB, induced by hydrogen peroxide and pro-inflammatory cytokines, increases the expression of proteins and enzymes that require consensus binding of κB, including SOD2, glutamyl-cysteine synthetase (GCS), iNOS, and cyclooxygenase 2 (COX2), among others [68].

MAPK pathways, comprising c-Jun N-terminal kinases (JNK), ERK 1–2, and p38 MAPK, are activated by a variety of physiological events associated with exercise like ROS, hormones, calcium influx, and neural or mechanical stimuli [69]. Through p38 MAPK activation/phosphorylation, ROS increases glucose uptake by muscle cells during exercise [70].

PGC1α plays a pivotal role in the regulation of mitochondria biogenesis, antioxidant enzyme expression, and regulation of anti-inflammatory cytokine expression [67]. Moreover, PGC1α promotes mitochondrial oxidative metabolism [71] and plays a role in fiber-type specificity inducing slow phenotype specification [72]. PGC1α is regulated at transcriptional and post-transcriptional levels by pathways activated during muscle contraction like AMP-activated protein kinase (AMPK), sirtuin 1 (SIRT1), protein kinase C, changes in intracellular Ca^{2+} concentration, p38 MAPK, NO, ROS, and hypoxia-inducible factor-1 (HIF-1) [73]. An important anabolic pathway inducing protein synthesis involves activation of the phosphatidylinositol 3-kinase (PI3K)/Akt, which stimulates mammalian target of rapamycin (mTOR), which is essential for muscle growth during development and regeneration; this pathway has a role also in the regulation of muscle mass which strictly depends on protein synthesis [74]. In most cases, the effects of ROS on these signaling molecules differ in dependence of the levels. For example, JNK phosphorylation levels depend on the relative activity of specific kinases and phosphatases. Low levels of ROS induce JNK phosphorylation without affecting phosphatase, leading to a transient activation of JNK, while higher ROS levels may activate the JNK pathway and inactivate phosphatases, resulting in a prolonged activation of JNK [75]. In addition, ROS activate the PI3K/Akt pathway, either by directly activating PI3K or inactivating phosphatase and tensin homolog (PTEN) which inhibits the activation of Akt through cysteine residues oxidation [76]. At lower levels, ROS also oxidize the disulfide bridges in Akt leading to a short-term activation of Akt signaling [77].

Muscle adaptation to different training conditions is associated with the activation of different pathways. The main muscle adaptive response to non-exhaustive endurance training is mitochondria biogenesis, which increases muscle oxidative capability [78]. The underlined mechanism relies on the production of reactive oxygen species like ubisemiquinone, NO, superoxide, and H_2O_2 [79,80].

A key modulator of mitochondria biogenesis is PGC1α, which guarantees the balance between the production and scavenging of reactive oxygen species by regulating both mitochondrial biogenesis and the expression of antioxidant enzymes like SOD1, GPx, and catalase [81]. Unlike endurance training, resistance training induces skeletal muscle hypertrophy. Indeed, resistance training is associated with an increase in protein biosynthesis with respect to protein breakdown. ROS produced during resistance exercise activate multiple pathways, including the insulin/IGF-1-IP3K, MAPKs, and Ca-calmodulin pathways [82] involved in protein biosynthesis.

Finally, short-term anaerobic exercise, such as sprinting, is associated with high levels of ROS and oxidative stress. During sprinting exercise, the main ROS sources are NOXs [83] and the xanthine/xanthine oxidase system [84], while mitochondria play only a minor role. Additionally, exhaustive endurance and resistance exercise are related to increased levels of skeletal muscle ROS, oxidative stress, and cortisol, leading to transitory immunosuppression [85]. In the elderly, protein anabolic pathways are reduced [86] while protein catabolic pathways are activated [87], contributing to muscle atrophy. In particular, with aging a reduction in the mitochondrial protein synthesis rate in muscle is correlated with a decrease in mitochondrial enzyme activity and oxidative capability [88]. In addition, a significant decline in mitochondria biogenesis, lower levels of testosterone, and a PGC1α, Akt, and mTOR expression decrease contribute to the loss of muscle mass and strength, which are hallmarks of sarcopenia [89,90]. Much evidence highlights the role of reactive oxygen species in age-related neuromuscular deficit [91]. Impaired mitochondrial electron transport chain has been involved in the increase of ROS levels in aged skeletal muscle [92]; moreover, in aging, skeletal muscles produce higher levels of ROS during an acute bout of exercise, while chronic exercise has a protective effect against oxidative damage [93]. In addition, physical inactivity increases the levels of oxidative stress contributing to the onset of sarcopenia [94].

Much evidence highlights the positive role of exercise in elderly skeletal muscle. Elderly subjects who regularly exercise show oxidative stress levels comparable to that of younger individuals who do not perform physical activity [95].

Both redox-sensitive mitochondria biogenesis and PGC1α levels are increased by exercise training and increased expression of PGC1α in old mice is associated with high mitochondria biogenesis and lower oxidative stress, inflammation, and apoptosis [89].

Overall, ROS/RNS signaling plays a pivotal role in skeletal muscle contractile function, hypertrophy, mitochondrial biogenesis, and glucose uptake as an adaptive response to exercise. Increasing knowledge of redox signaling involved in responses to exercise will lead to the development of new approaches to regulate muscle metabolism and function to prevent loss of muscle mass and performance in age-related diseases.

6. Dietary Intervention in Age-Related Sarcopenia

It is well known that old age is often associated with appetite loss, which contributes to decreased food intake, protein-energy malnutrition, and weight loss. Therefore, it is important to evaluate different dietary interventions to assure an adequate nutrient intake in sarcopenic patients to preserve muscle mass.

In the elderly, muscle protein homeostasis is impaired because of a reduced synthesis and increased rate of degradation. In addition, reduction in muscle mass is facilitated by physical inactivity and decreased dietary protein intake [96]. Several authors suggest the consumption of good sources of proteins low in fat, including lean meat, poultry, and fish, and high protein intake (1.2–1.4 g/Kg/die) [97] for the treatment of sarcopenia. In vitro experiments, performed in cardiomyoblast cell lines, have shown that serum from vegan subjects induces oxidative stress and cell death compared to vegetarian and omnivorous sera [98], suggesting a mechanistic link between deficient protein intake, oxidative stress, and loss of muscle mass.

The list of natural compounds with antioxidant activity is very long. However, we have focused our attention on the most known nutritional antioxidants, including L-ascorbic acid (vitamin C), tocopherols (vitamin E), carotenoids, flavonoids, and polyphenols [99–105], whose function on muscle have been more extensively studied.

Vitamin C is the primary water-soluble and non-enzymatic antioxidant in plasma and tissues. Humans, unlike most mammals and other animals, do not have the ability to synthesize vitamin C due to lack of the last enzyme in the biosynthetic process, and it must therefore be obtained by dietary intake. The principal sources of vitamin C are kiwifruit, strawberries, broccoli, kale, tomatoes, and sweet red pepper [106]. Important enzymatic reactions requiring vitamin C as an essential cofactor are the biosynthesis of collagen, carnitine, and neuropeptides, and the regulation of gene expression [107]. Cohort studies also indicate that higher vitamin C status, assessed by measuring circulating vitamin C levels, is associated with lower risks of hypertension, coronary heart disease, and stroke [108]. This vitamin is also involved in the regeneration of fat-soluble vitamin E [109], which is particularly able to inhibit lipid peroxidation.

Naturally occurring vitamin E includes eight fat-soluble isoforms, but in the human body α-tocopherol is the most common isoform. Plant seeds, especially sunflower seeds, almonds, and hazelnuts are rich sources of α-tocopherol; moreover, many vegetable oils (e.g., olive oil and canola oil) and tomato, avocado, spinach, asparagus, Swiss chard, and broccoli also contain this vitamin. α-tocopherol is uniquely suited to intercept peroxyl radicals and thus prevent lipid peroxidation and the detrimental effects of free radicals in membranes and plasma lipoproteins [110]. α-tocopherol is also likely to be involved in cell-mediated immunity [111,112]. In addition to its direct antioxidant properties, vitamin E modulates signal transduction and gene expression in a redox-dependent and redox-independent manner, regulating cellular functions relevant for its action and for preventing a number of diseases, including cancer, atherosclerosis, inflammation, and neurodegenerative diseases [113]. Carotenoids are another important type of dietary antioxidant which play a protective role in many diseases. They are organic pigments present in fruits and vegetables like pumpkins, carrots, corn, and tomatoes [114]. They have important antioxidant functions such as singlet oxygen quenching and radical scavenging [115]. Recently, the benefits of carotenoids against oxidative stress in human have been reviewed [116]. Some carotenoids like beta-carotene are dietary precursors of the fat-soluble vitamin A, or retinol, with important antioxidant activity and liver protection functions [117]. In addition, vitamin A is converted in retinoic acid and functions as a ligand, regulating the expression of genes involved in cell metabolism [118]. Some evidence suggests that carotenoids beside their antioxidant activity also exert signaling functions. Indeed, carotenoids or their metabolites may up-regulate the expression of antioxidant or detoxifying enzymes via the activation of the Nrf2-dependent pathway [119].

Polyphenols are phytochemicals with antioxidant properties which occur in vegetal food [105]. Quercetin, a natural flavonoid found mainly in nuts, grapes, onions, broccoli, apples, and black tea, has shown important antioxidant activity and protective effects on the intestinal mucosal barrier [120], as well as the ability to reduce inflammation by suppressing the expression of pro-inflammatory mediators [121,122].

Resveratrol is a natural polyphenolic compound occurring in several plants and in food, including in red wine, peanuts, blueberries, raspberries, and mulberries [123]. In preclinical studies, it has been observed that resveratrol is important in the prevention and/or treatment of cancer, cardiovascular disease, and neurodegenerative diseases [124]. Another biologically active polyphenolic compound is curcumin, which is found in turmeric, a spice derived from the rhizomes of the plant *Curcuma longa* Linn. Many preclinical studies show that curcumin modulates numerous molecular targets and exerts antioxidant, anti-inflammatory, anticancer, and neuroprotective activities [125]. Furthermore, curcumin is important to prevent and treat Type 2 diabetes mellitus disease [126]. The effects of polyphenols cannot be explained solely on the basis of their antioxidant action; indeed, the health benefits of these substances may rely on their effects on enzyme, membrane, or nuclear receptors and intracellular transduction mechanisms [127]. Functions and sources of the main dietary antioxidants are summarized in Table 1.

In the elderly there is an impairment of the endogenous antioxidant defense system [141] and a decline of mitochondrial function associated with inadequate antioxidant dietary intake. Great interest has been devoted to antioxidant supplementation as a potential intervention in sarcopenia [142] since oxidative damage is considered to be one of the mechanisms leading to the loss of muscle mass and function.

Studies conducted in animal models in which supplementation of diet with antioxidants and physical activity have been combined in many cases seem to support the use of antioxidants to ameliorate exercise performance. Resveratrol, which is able to promote mitochondrial adaptive response and strength of upper limbs in mice after 12 weeks of treadmill exercise training suggests that this natural antioxidant could be used as a performance enhancer [138]. Resveratrol also exerts beneficial effects on muscle performance in animal models of aging. The dietary administration of resveratrol, in combination with habitual exercise in mice, has been shown to improve mitochondrial function and aging-related decline in physical performance [139]. In addition, curcumin supplementation ameliorates exercise performance in rats increasing time of run to exhaustion compared to control animals [137]. On the contrary, vitamin A administration in exercise training rats has been shown to induce lipid peroxidation and protein damage, decrease SOD1 levels, and attenuated exercise-dependent increase of SOD2 in the skeletal muscle [135].

Table 1. Functions and sources of principal nutritional antioxidants.

Nutritional Antioxidants	Functions	Sources	References Related to Effects on Skeletal Muscle
Vitamin E (α–tocopherol)	• Boosts antioxidant defense • Protects cell membranes • Enhances immune functions • Lipid peroxidation inhibitor	Vegetable oil, nuts, avocados	[128–130]
Vitamin C (L-ascorbic acid)	• Principal hydrophilic antioxidant • Scavenger of free radicals • Enhances immune function • Needed to make collagen, carnitine and neurotransmitters, e.g., serotonin and norepinephrine	Kiwifruit, strawberries, broccoli, kale, tomatoes, sweet red pepper	[128,129,131]
Quercetin	• Antioxidant activity and protective effects on intestinal mucosal barrier • Reduced exercise-induced lipid peroxidation • Reduced oxidative stress and inflammatory biomarkers	Nuts, grapes, onions, broccoli, apples, black tea	[132–134]
Carotenoids	• Scavenger of free radicals and singlet oxygen quenching • Cancer prevention • Prevention of liver diseases	Pumpkins, carrots, corn, tomatoes	[135,136]
Curcumin	• Scavenger of reactive oxygen species (ROS) and reactive nitrogen species • Neuroprotective activity • Disease prevention (Type 2 diabetes mellitus, cancer)	*Curcuma longa*	[137]
Resveratrol	• Prevention of cancer and cardiovascular disease • Inhibition of neuroinflammation • Increment of aerobic capacity	Skin of grapes, blueberries, raspberries, mulberries	[138–140]

However, the positive findings obtained in animals have not always been confirmed in human trials. In young football athletes, antioxidant supplementation has been shown to not counteract muscle damage or soreness induced by acute exercise and also to not ameliorate physical performance, although it has been seen to reduce oxidative stress [128]. Moreover, even where a decrease of oxidative stress or inflammatory markers and reduced onset of age-related pathophysiological disorders has been shown in human subjects assuming quercetin supplementation accompanied by constant and moderate physical exercise, an amelioration of muscle performance has not been clearly evidenced [132–134]. A recent review by Beaudart et al. [143] summarizing the results of 37 randomized clinical trials in which exercise and nutritional interventions, including the administration of natural antioxidants, were combined, evidenced a beneficial effect of exercise on muscle mass and strength in subjects over 65 years, while the effects of dietary antioxidant supplementation were more controversial. Moreover, importantly, some studies have highlighted that prolonged antioxidant supplementation can lead to undesirable effects like disruption of endogenous antioxidant levels, thus failing to counteract exercise-induced oxidative stress, and interfering with muscle adaptation to exercise [144–146]. Moreover, long-term administration of vitamin C has been observed to prevent mitochondrial biogenesis, decreasing the expression of endogenous antioxidant enzymes [131].

However, other literature data have suggested beneficial effects of antioxidant supplementation in muscle performance in normal young adult and sarcopenic subjects. A study conducted in subjects aged 65–80 years assuming resveratrol combined with 12 weeks of exercise has indicated a novel anabolic role of resveratrol in exercise-induced adaptations of older persons, and this suggests that this compound combined with exercise is likely to better counteract sarcopenia than exercise alone [140]. Moreover, the strongest evidence supporting the beneficial effects of antioxidant supplementation in physical performance has been obtained with vitamin E alone and in combination with vitamin C. Recently, He et al. [129] showed a decrease of markers of muscle damage, an amelioration of antioxidant status, and a delayed onset of muscle soreness after repeated downhill runs in moderately-trained males assuming vitamin C and E.

Several studies have reported the association between vitamin E and sarcopenia. In particular, Cesari et al. [130], in their study "Invecchiare in Chianti", showed that vitamin E daily intake level positively correlated with knee extension strength and total physical performance. Moreover, it seems that vitamin C, by regenerating vitamin E, is responsible for muscle protection [147]. Flavonoids have many health benefits and improve exercise performance in athletes and in subjects not necessarily in constant training, such as aged people [148]. Finally, sarcopenia has also been associated with low levels of carotenoids even if further studies are necessary to better ascertain the existence of a direct link between low carotenoids status and muscle decline [136].

Several reasons could underlie these contradictory results. The lack of beneficial effects of antioxidants in sarcopenia could be due to the fact that supplemented antioxidants, which are non-enzymatic antioxidants, may not be able to make up for the enzymatic antioxidant deficiency that characterizes sarcopenia. It must also be considered that some antioxidant enzymes, such as SOD1 [149] and SOD2 [150], are up-regulated by ROS and that the scavenging effects of antioxidants could further down-regulate them, worsening muscle oxidative damage and performance.

Moreover, the dual role of ROS in muscle performance could explain the conflicting data obtained with different nutritional protocols and exercise settings. Antioxidant treatment associated with exercise could eliminate the adaptive response, and, therefore, it has been suggested that if antioxidants are administered before exercise-induced ROS levels reach their peak they can prevent the physiological function of ROS, while they can exert beneficial effects if administered after the bell-shaped curve of ROS has reached its summit [50].

The search for the right association of antioxidant supplementation in sarcopenic subjects who perform physical activity still has a long way to go. In experimental protocols considering the contemporary administration of antioxidant supplementation and physical training in elderly subjects, the type, strength, and duration of exercise, as well as the research design and the timing and extent of favorable effects of ROS in muscle adaptation to exercise [151] must be carefully taken into account in order to better clarify the issue.

7. Conclusions

Western societies are characterized by a progressive aging population which results in increasing prevalence of sarcopenia with subsequent increasing healthcare costs. Aging is associated with skeletal muscle oxidative stress due to increased ROS generation and impairment of antioxidant enzymes systems. On the other hand, muscle disuse and malnutrition are often associated with aging and also contribute to increased oxidative damage of skeletal muscle, leading to a decline in muscle mass and strength (Figure 1). From this complex scenario, it is clear that both physical activity and nutritional interventions must be contemplated for sarcopenic individuals.

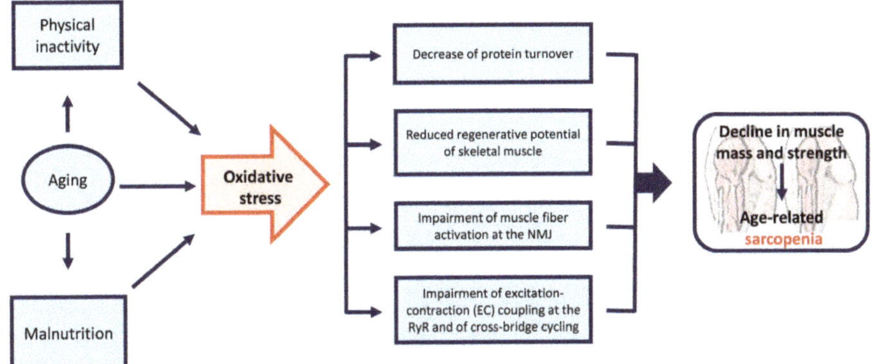

Figure 1. Aging, physical inactivity, and malnutrition lead to oxidative stress, contributing to sarcopenia.

Due to the central role played by oxidative stress in the onset of sarcopenia, a great effort has been focused on the understanding of the underlined cellular and molecular mechanisms involving ROS signaling in the attempt to identify the right strategies based on antioxidant supplementation for prevention and treatment of this condition.

Unfortunately, despite the fact that redox mechanisms leading to muscle mass and strength loss in sarcopenia have in part been unraveled, clinical data with antioxidant supplementation are far from clear and as a result of these conflicting reports, antioxidant supplementation cannot yet be considered as a nutritional intervention to prevent and treat sarcopenia. A more intense research effort must be made to clarify the right association between diet and exercise effective to counteract the onset and extent of age-related sarcopenia.

Author Contributions: Conceptualization, supervision, and writing, M.S.; writing—review and editing, S.D. and E.M.; writing, G.L.R., M.D.M. and P.M.

Funding: This research received no external funding

Acknowledgments: We thank Ilaria Di Gregorio for helping with the editing of the manuscript.

Conflicts of Interest: The authors declare no conflict of interest.

References

1. Rosenberg, I.H. Sarcopenia: Origins and clinical relevance. *J. Nutr.* **1997**, *127*, 990–991. [CrossRef] [PubMed]
2. Roubenoff, R.; Heymsfield, S.B.; Kehayias, J.J.; Cannon, J.G.; Rosenberg, I.H. Standardization of nomenclature of body composition in weight loss. *Am. J. Clin. Nutr.* **1997**, *66*, 192–196. [CrossRef] [PubMed]
3. Cruz-Jentoft, A.J.; Baeyens, J.P.; Bauer, J.M.; Boirie, Y.; Cederholm, T.; Landi, F.; Martin, F.C.; Michel, J.P.; Rolland, Y.; Schneider, S.M.; et al. Sarcopenia: European consensus on definition and diagnosis: Report of the European Working Group on Sarcopenia in Older People. *Age Ageing* **2010**, *39*, 412–423. [CrossRef] [PubMed]
4. Cruz-Jentoft, A.J.; Sayer, A.A. Sarcopenia. *Lancet* **2019**, *393*, 2636–2646. [CrossRef]
5. Marques, A.; Queirós, C. Frailty, Sarcopenia and Falls. In *Fragility Fracture Nursing: Holistic Care and Management of the Orthogeriatric Patient*; Hertz, K., Santy-Tomlinson, J., Eds.; Springer: Cham, Switzerland, 2018; Chapter 2.
6. Delbono, O.; O'Rourke, K.S.; Ettinger, W.H. Excitation-calcium release uncoupling in aged single human skeletal muscle fibers. *J. Membr. Biol.* **1995**, *148*, 211–222. [CrossRef] [PubMed]
7. Wang, Z.-M.; Messi, M.L.; Delbono, O. Sustained overexpression of IGF-1 prevents age-dependent decrease in charge movement and intracellular calcium in mouse skeletal muscle. *Biophys. J.* **2002**, *82*, 1338–1344. [CrossRef]

8. Brooks, S.V.; Faulkner, J.A. Skeletal muscle weakness in old age: Underlying mechanisms. *Med. Sci. Sports Exerc.* **1994**, *26*, 432–439. [CrossRef]

9. Hook, P.; Li, X.; Sleep, J.; Hughes, S.; Larsson, L. In vitro motility speed of slow myosin extracted from single soleus fibres from young and old rats. *J. Physiol.* **1999**, *520*, 463–471. [CrossRef]

10. Lexell, J. Human aging, muscle mass, and fiber type composition. *J. Gerontol. A Biol. Sci. Med. Sci.* **1995**, *50*, 11–16. [CrossRef]

11. Frontera, W.R.; Suh, D.; Krivickas, L.S.; Hughes, V.A.; Goldstein, R.; Roubenoff, R. Skeletal muscle fiber quality in older men and women. *Am. J. Physiol. Cell Physiol.* **2000**, *279*, C611–C618. [CrossRef]

12. Joanisse, S.; Nederveen, J.P.; Snijders, T.; McKay, B.R.; Parise, G. Skeletal Muscle Regeneration, Repair and Remodelling in Aging: The Importance of Muscle Stem Cells and Vascularization. *Gerontology* **2017**, *63*, 91–100. [CrossRef]

13. Bohnert, K.R.; McMillan, J.D.; Kumar, A. Emerging roles of ER stress and unfolded protein response pathways in skeletal muscle health and disease. *Cell Physiol.* **2018**, *233*, 67–78. [CrossRef]

14. Picca, A.; Fanelli, F.; Calvani, R.; Mulè, G.; Pesce, V.; Sisto, A.; Pantanelli, C.; Bernabei, R.; Landi, F.; Marzetti, E. Gut Dysbiosis and muscle Aging: Searching for Novel Targets against Sarcopenia. *Mediat. Inflamm.* **2018**, *2018*, 7026198. [CrossRef]

15. Dahlgren, C.; Karlsson, A.; Bylund, J. Intracellular Neutrophil Oxidants: From Laboratory Curiosity to Clinical Reality. *J. Immunol.* **2019**, *202*, 3127–3134. [CrossRef]

16. Santillo, M.; Colantuoni, A.; Mondola, P.; Guida, B.; Damiano, S. NOX signaling in molecular cardiovascular mechanisms involved in the blood pressure homeostasis. *Front. Physiol.* **2015**, *6*, 194. [CrossRef]

17. Xu, H.; Li, C.; Mozziconacci, O.; Zhu, R.; Xu, Y.; Tang, Y.; Chen, R.; Huang, Y.; Holzbeierlein, J.M.; Schöneich, C.; et al. Xanthine oxidase-mediated oxidative stress promotes cancer cell-specific apoptosis. *Free Radic. Biol. Med.* **2019**, *139*, 70–79. [CrossRef]

18. Deby, C.; Goutier, R. New perspectives on the biochemistry of superoxide anion and the efficiency of superoxide dismutase. *Biochem. Pharmacol.* **1990**, *39*, 399–405. [CrossRef]

19. Fridovich, I. The biology of oxygen radicals. *Science* **1978**, *201*, 875–880. [CrossRef]

20. McCord, J.M.; Fridovich, I. Superoxide dismutase. An enzymic function for erythrocuprein (hemocuprein). *J. Biol. Chem.* **1996**, *244*, 6049–6055.

21. Weisiger, R.A.; Fridovich, I. Superoxide dismutase. Organelle specificity. *J. Biol. Chem.* **1973**, *248*, 3582–3592.

22. Marklund, S.L. Human copper-containing superoxide dismutase of high molecular weight. *Proc. Natl. Acad. Sci. USA* **1982**, *79*, 7634–76388. [CrossRef]

23. Oldford, C.; Kuksal, N.; Gill, R.; Young, A.; Mailloux, R.J. Estimation of the hydrogen peroxide producing capacities of liver and cardiac mitochondria isolated from C57BL/6N and C57BL/6J mice. *Free Radic. Biol. Med.* **2019**, *135*, 15–27. [CrossRef]

24. Del Río, L.A.; López-Huertas, E. ROS Generation in Peroxisomes and its Role in Cell Signaling. *Plant Cell Physiol.* **2016**, *57*, 1364–1376. [CrossRef]

25. Formanowicz, D.; Radom, M.; Rybarczyk, A.; Formanowicz, P. The role of Fenton reaction in ROS-induced toxicity underlying atherosclerosis—modeled and analyzed using a Petri net-based approach. *Biosystems* **2018**, *165*, 71–87. [CrossRef]

26. Amodio, G.; Moltedo, O.; Faraonio, R.; Remondelli, P. Targeting the Endoplasmic Reticulum Unfolded Protein Response to Counteract the Oxidative Stress-Induced Endothelial Dysfunction. *Oxid. Med. Cell. Longev.* **2018**, *2018*, 4946289. [CrossRef]

27. Marciniak, S.J.; Yun, C.Y.; Oyadomari, S.; Novoa, I.; Zhang, Y.; Jungreis, R.; Nagata, K.; Harding, H.P.; Ron, D. CHOP induces death by promoting protein synthesis and oxidation in the stressed endoplasmic reticulum. *Genes Dev.* **2004**, *18*, 3066–3077. [CrossRef]

28. Damiano, S.; Sasso, A.; Accetta, R.; Monda, M.; De Luca, B.; Pavone, L.M.; Belfiore, A.; Santillo, M.; Mondola, P. Effect of Mutated Cu, Zn Superoxide Dismutase (SOD1G93A) on Modulation of Transductional Pathway Mediated by M1 Muscarinic Receptor in SK-N-BE and NSC-34 Cells. *Front. Physiol.* **2018**, *9*, 611. [CrossRef]

29. Potenza, N.; Mosca, N.; Mondola, P.; Damiano, S.; Russo, A.; De Felice, B. Human miR-26a-5p regulates the glutamate transporter SLC1A1 (EAAT3) expression. Relevance in multiple sclerosis. *Biochim. Biophys. Acta Mol. Basis Dis.* **2018**, *1864*, 317–323. [CrossRef]

30. Mondola, P.; Damiano, S.; Sasso, A.; Santillo, M. The Cu, Zn Superoxide Dismutase: Not Only a Dismutase Enzyme. *Front. Physiol.* **2016**, *7*, 594. [CrossRef]

31. Accetta, R.; Damiano, S.; Morano, A.; Mondola, P.; Paternò, R.; Avvedimento, E.V.; Santillo, M. Reactive Oxygen Species Derived from NOX3 and NOX5 Drive Differentiation of Human Oligodendrocytes. *Front. Cell. Neurosci.* **2016**, *10*, 146. [CrossRef]

32. Damiano, S.; Sasso, A.; De Felice, B.; Terrazzano, G.; Bresciamorra, V.; Carotenuto, A.; Orefice, N.S.; Orefice, G.; Vacca, G.; Belfiore, A.; et al. The IFN-β 1b effect on Cu Zn superoxide dismutase (SOD1) in peripheral mononuclear blood cells of relapsing-remitting multiple sclerosis patients and in neuroblastoma SK-N-BE cells. *Brain Res. Bull.* **2015**, *118*, 1–6. [CrossRef]

33. Damiano, S.; Morano, A.; Ucci, V.; Accetta, R.; Mondola, P.; Paternò, R.; Avvedimento, V.E.; Santillo, M. Dual oxidase 2 generated reactive oxygen species selectively mediate the induction of mucins by epidermal growth factor in enterocytes. *Int. J. Biochem. Cell Biol.* **2015**, *60*, 8–18. [CrossRef]

34. Damiano, S.; Petrozziello, T.; Ucci, V.; Amente, S.; Santillo, M.; Mondola, P. Cu-Zn superoxide dismutase activates muscarinic acetylcholine M1 receptor pathway in neuroblastoma cells. *Mol. Cell. Neurosci.* **2013**, *52*, 31–37. [CrossRef]

35. Damiano, S.; Fusco, R.; Morano, A.; De Mizio, M.; Paternò, R.; De Rosa, A.; Spinelli, R.; Amente, S.; Frunzio, R.; Mondola, P.; et al. Reactive oxygen species regulate the levels of dual oxidase (Duox1-2) in human neuroblastoma cells. *PLoS ONE* **2012**, *7*, e34405. [CrossRef]

36. Terrazzano, G.; Rubino, V.; Damiano, S.; Sasso, A.; Petrozziello, T.; Ucci, V.; Palatucci, A.T.; Giovazzino, A.; Santillo, M.; De Felice, B.; et al. T cellactivation induces CuZnsuperoxide dismutase (SOD)-1 intracellular re-localization, production and secretion. *Biochim. Biophys. Acta* **2014**, *1843*, 265–274. [CrossRef]

37. Smith, M.A.; Reid, M.B. Redox modulation of contractile function in respiratory and limb skeletal muscle. *Respir. Physiol. Neurobiol.* **2006**, *151*, 229–241. [CrossRef]

38. Debold, E.P. Potential molecular mechanisms underlying muscle fatigue mediated by reactive oxygen and nitrogen species. *Front. Physiol.* **2015**, *6*, 239. [CrossRef]

39. Davies, K.J.; Quintanilha, A.T.; Brooks, G.A.; Packer, L. Free radicals and tissue damage produced by exercise. *Biochem. Biophys. Res. Commun.* **1982**, *107*, 1198–1205. [CrossRef]

40. Urso, M.L.; Clarkson, P.M. Oxidative stress, exercise, and antioxidant supplementation. *Toxicology* **2003**, *189*, 41–54. [CrossRef]

41. Powers, S.K.; Jackson, M.J. Exercise-induced oxidative stress: Cellular mechanisms and impact on muscle force production. *Physiol. Rev.* **2008**, *88*, 1243–1276. [CrossRef]

42. Ferreira, L.F.; Laitano, O. Regulation of NADPH oxidases in skeletal muscle. *Free Radic. Biol. Med.* **2016**, *98*, 18–28. [CrossRef] [PubMed]

43. Cherednichenko, G.; Zima, A.V.; Feng, W.; Schaefer, S.; Blatter, L.A.; Pessah, I.N. NADH oxidase activity of rat cardiac sarcoplasmic reticulum regulates calcium-induced calcium release. *Circ. Res.* **2004**, *94*, 478–486. [CrossRef] [PubMed]

44. Pearson, T.; Kabayo, T.; Ng, R.; Chamberlain, J.; McArdle, A.; Jackson, M.J. Skeletal muscle contractions induce acute changes in cytosolic superoxide, but slower responses in mitochondrial superoxide and cellular hydrogen peroxide. *PLoS ONE* **2014**, *9*, e96378. [CrossRef] [PubMed]

45. Brandes, R.P. Triggering mitochondrial radical release: A new function for NADPH oxidases. *Hypertension* **2005**, *45*, 847–848. [CrossRef] [PubMed]

46. Zhang, D.X.; Chen, Y.F.; Campbell, W.B.; Zou, A.P.; Gross, G.J.; Li, P.L. Characteristics and superoxide-induced activation of reconstituted myocardial mitochondrial ATP-sensitive potassium channels. *Circ. Res.* **2001**, *89*, 1177–1183. [CrossRef] [PubMed]

47. Gorlach, A.; Bertram, K.; Hudecova, S.; Krizanova, O. Calcium and ROS: A mutual interplay. *Redox Biol.* **2015**, *6*, 260–271. [CrossRef]

48. Stamler, J.S.; Meissner, G. Physiology of nitric oxide in skeletal muscle. *Physiol. Rev.* **2001**, *81*, 209–237. [CrossRef] [PubMed]

49. Ward, C.W.; Prosser, B.L.; Lederer, W.J. Mechanical stretch-induced activation of ROS/RNS signaling in striated muscle. *Antioxid. Redox Signal.* **2014**, *20*, 929–936. [CrossRef]

50. Radak, Z.; Ishiharaa, K.; Tekusb, E.; Vargac, C.; Posac, A.; Baloghd, L.; Boldoghe, I.; Koltaia, E. Exercise, oxidants, and antioxidants change the shape of the bell-shaped hormesis curve. *Redox Biol.* **2017**, *12*, 285–290. [CrossRef]

51. Oudot, A.; Martin, C.; Busseuilla, D.; Vergelya, C.; Demaisonc, L.; Rochette, L. NADPH oxidases are in part responsible for increased cardiovascular superoxide production during aging. *Free Radic. Biol. Med.* **2006**, *40*, 2214–2222. [CrossRef]

52. Sullivan-Gunn, M.J.; Lewandowski, P.A. Elevated hydrogen peroxide and decreased catalase and glutathione peroxidase protection are associated with aging sarcopenia. *BMC Geriatr.* **2013**, *13*, 104. [CrossRef]

53. Shi, Y.; Ivanikkov, M.V.; Walsh, M.E.; Liu, Y.; Zhang, Y.; Jaramillo, C.A.; Macleod, G.T.; Van Remmen, H. The lack of CuZnSOD leads to impaired neurotransmitter release, neuromuscular junction destabilization and reduced muscle strength in mice. *PLoS ONE* **2014**, *9*, e100834. [CrossRef]

54. Muller, F.L.; Song, W.; Liu, Y.; Chaudhuri, A.; Pieke-Dahl, S.; Strong, R.; Huang, T.T.; Epstein, C.J.; Roberts, L.J.; Csete, M.; et al. Absence of CuZn superoxide dismutase leads to elevated oxidative stress and acceleration of age-dependent skeletal muscle atrophy. *Free Radic. Biol. Med.* **2006**, *40*, 1993–2004. [CrossRef]

55. Fang, C.; Bourdette, D.; Banker, G. Oxidative stress inhibits axonal transport: Implications for neurodegenerative diseases. *Mol. Neurodegener.* **2012**, *7*, 29. [CrossRef]

56. Liu, Y.; Chang, A. Heat shock response relieves ER stress. *EMBO J.* **2008**, *27*, 1049–1059. [CrossRef]

57. Ivannikov, M.V.; Van Remmen, H. Sod1 gene ablation in adult mice leads to physiological changes at the neuromuscular junction similar to changes that occur in old wild type mice. *Free Radic. Biol. Med.* **2015**, *84*, 254–262. [CrossRef]

58. Muller, F.L.; Song, W.; Jang, Y.C.; Liu, Y.; Sabia, M.; Richardson, A.; Van Remmen, H. Denervation-induced skeletal muscle atrophy is associated with increased mitochondrial ROS production. *Am. J. Physiol. Regul. Integr. Comp. Physiol.* **2007**, *293*, R1159–R1168. [CrossRef]

59. Watanabe, M.; Dykes–Hoberg, M.; Culotta, V.; Price, D.L.; Wong, P.C.; Rothstein, J.D. Histological evidence of protein aggregation in mutant SOD1 transgenic mice and in amyotrophic lateral sclerosis neural tissues. *Neurobiol. Dis.* **2001**, *8*, 933–941. [CrossRef]

60. Okutsu, M.; Call, J.A.; Lira, V.A.; Zhang, M.; Donet, J.A.; French, B.A.; Martin, K.S.; Peirce-Cottler, S.M.; Rembold, C.M.; Annex, B.H.; et al. Extracellular Superoxide Dismutase Ameliorates Skeletal Muscle Abnormalities, Cachexia and Exercise Intolerance in Mice with Congestive Heart Failure. *Circ. Heart Fail.* **2014**, *7*, 519–530. [CrossRef]

61. Mondola, P.; Ruggiero, G.; Serù, R.; Damiano, S.; Grimaldi, S.; Garbi, C.; Monda, M.; Greco, D.; Santillo, M. The Cu, Zn superoxide dismutase in neuroblastoma SK-N-BE cells is exported by a microvesicles dependent pathway. *Brain Res. Mol. Brain Res.* **2003**, *110*, 45–51. [CrossRef]

62. Deldicque, L. Endoplasmic reticulum stress in human skeletal muscle: Any contribution to sarcopenia? *Front. Physiol.* **2013**, *4*, 236. [CrossRef] [PubMed]

63. Jheng, J.R.; Chen, Y.S.; Ao, U.I.; Chan, D.C.; Huang, J.W.; Hung, K.Y.; Tarng, D.C.; Chiang, C.K. The double-edged sword of endoplasmic reticulum stress in uremic sarcopenia through myogenesis perturbation. *J. Cachexia Sarcopenia Muscle* **2018**, *9*, 570–584. [CrossRef] [PubMed]

64. Bratic, A.; Larsson, N.G. The role of mitochondria in aging. *J. Clin. Investig.* **2013**, *123*, 951–957. [CrossRef] [PubMed]

65. Sebastián, D.; Sorianello, E.; Segalés, J.; Irazoki, A.; Ruiz-Bonilla, V.; Sala, D.; Planet, E.; Berenguer-Llergo, A.; Muñoz, J.P.; Sánchez-Feutrie, M.; et al. Mfn2 deficiency links age-related sarcopenia and impaired autophagy to activation of an adaptive mitophagy pathway. *EMBO J.* **2016**, *35*, 1677–1693. [CrossRef]

66. Tezze, C.; Romanello, V.; Desbats, M.A.; Fadini, G.P.; Albiero, M.; Favaro, G.; Ciciliot, S.; Soriano, M.E.; Morbidoni, V.; Cerqua, C.; et al. Age-Associated Loss of OPA1 in Muscle Impacts Muscle Mass, Metabolic Homeostasis, Systemic Inflammation, and Epithelial Senescence. *Cell Metab.* **2017**, *25*, 1374–1389. [CrossRef] [PubMed]

67. Ji, L.L.; Zhang, Y. Antioxidant and anti-inflammatory effects of exercise: Role of redox signaling. *Free Radic. Res.* **2014**, *48*, 3–11. [CrossRef] [PubMed]

68. Ghosh, S.; Karin, M. Missing pieces in the NF-kappaB puzzle. *Cell* **2002**, *109*, S81–S96. [CrossRef]

69. Nader, G.; Esser, K. Intracellular signaling specificity in skeletal muscle in response to different modes of exercise. *J. Appl. Physiol.* **2001**, *90*, 1936–1942. [CrossRef]

70. Chambers, M.A.; Moylan, J.S.; Smith, J.D.; Goodyear, L.J.; Reid, M.B. Stretch-stimulated glucose uptake in skeletal muscle is mediated by reactive oxygen species and p38 MAP-kinase. *J. Physiol.* **2009**, *587*, 3363–3373. [CrossRef]

71. Thirupathi, A.; de Souza, C.T. Multi-regulatory network of ROS: The interconnection of ROS, PGC-1 alpha, and AMPK-SIRT1 during exercise. *J. Physiol. Biochem.* **2017**, *73*, 487–494. [CrossRef]

72. Lin, J.; Wu, H.; Tarr, P.T.; Zhang, C.Y.; Wu, Z.; Boss, O.; Michael, L.F.; Puigserver, P.; Isotani, E.; Olson, E.N.; et al. Transcriptional coactivator PGC-1 alpha drives the formation of slow-twitch muscle fibres. *Nature* **2002**, *418*, 797–801. [CrossRef]

73. Gundersen, K. Excitation-transcription coupling in skeletal muscle: The molecular pathways of exercise. *Biol. Rev. Camb. Philos. Soc.* **2011**, *86*, 564–600. [CrossRef]

74. Schiaffino, S.; Mammucari, C. Regulation of skeletal muscle growth by the IGF1-Akt/PKB pathway: Insights from genetic models. *Skelet. Muscle* **2011**, *1*, 4. [CrossRef]

75. Zhang, J.; Wang, X.; Vikash, V.; Ye, Q.; Wu, D.; Liu, Y.; Dong, W. ROS and ROS-Mediated Cellular Signaling. *Oxid. Med. Cell. Longev.* **2016**, *2016*, 4350965. [CrossRef]

76. Leslie, N.R.; Downes, C.P. PTEN: The down side of PI 3-kinase signaling. *Cell. Signal.* **2002**, *14*, 285–295. [CrossRef]

77. Murata, H.; Ihara, Y.; Nakamura, H.; Yodoi, J.; Sumikawa, K.; Kondo, T. Glutaredoxin exerts an antiapoptotic effect by regulating the redox state of Akt. *J. Biol. Chem.* **2003**, *278*, 50226–50233. [CrossRef]

78. Bassel-Duby, R.; Olson, E.N. Signaling pathways in skeletal muscle remodeling. *Annu. Rev. Biochem.* **2006**, *75*, 19–37. [CrossRef]

79. Davies, K.J.A. Cardiovascular Adaptive Homeostasis in Exercise. *Front. Physiol.* **2018**, *9*, 369. [CrossRef]

80. Santillo, M.; Pagliaro, P. Editorial: Redox and Nitrosative Signaling in Cardiovascular System: From Physiological Response to Disease. *Front. Physiol.* **2018**, *9*, 1538. [CrossRef]

81. St-Pierre, J.; Drori, S.; Uldry, M.; Silvaggi, J.M.; Rhee, J.; Jäger, S.; Handschin, C.; Zheng, K.; Lin, J.; Yang, W.; et al. Suppression of reactive oxygen species and neurodegeneration by the PGC-1 transcriptional coactivators. *Cell* **2006**, *127*, 397–408. [CrossRef]

82. Mason, S.A.; Morrison, D.; Mc Conell, G.K.; Wadley, G.D. Muscle redox signaling pathways in exercise. Role of antioxidants. *Free Radic. Biol. Med.* **2016**, *98*, 29–45. [CrossRef]

83. Sakellariou, G.K.; Vasilaki, A.; Palomero, J.; Kayani, A.; Zibrik, L.; McArdle, A.; Jackson, M. Studies of mitochondrial and nonmitochondrial sources implicate nicotinamide adenine dinucleotide phosphate oxidase(s) in the increased skeletal muscle superoxide generation that occurs during contractile activity. *Antioxid. Redox Signal.* **2013**, *18*, 603–621. [CrossRef]

84. Kang, C.; O'Moore, K.M.; Dickman, J.R.; Ji, L.L. Exercise activation of muscle peroxisome proliferator-activated receptor-gamma coactivator-1alpha signaling is redox sensitive. *Free Radic. Biol. Med.* **2009**, *47*, 1394–1400. [CrossRef]

85. Ji, L.L. Redox signaling in skeletal muscle: Role of aging and exercise. *Adv. Physiol. Educ.* **2015**, *39*, 352–359. [CrossRef]

86. Ebner, N.; Sliziuk, V.; Scherbakov, N.; Sandek, A. Muscle wasting in ageing and chronic illness. *ESC Heart Fail.* **2015**, *2*, 58–68. [CrossRef]

87. Mosoni, L.; Malmezat, T.; Valluy, M.C.; Houlier, M.L.; Attaix, D.; Mirand, P.P. Lower recovery of muscle protein lost during starvation in old rats despite a stimulation of protein synthesis. *Am. J. Physiol.* **1999**, *277*, E608–E616. [CrossRef]

88. Rooyackers, O.E.; Adey, D.B.; Ades, P.A.; Nair, K.S. Effect of age on in vivo rates of mitochondrial protein synthesis in human skeletal muscle. *Proc. Natl. Acad. Sci. USA* **1996**, *93*, 15364–15369. [CrossRef]

89. Wenz, T.; Rossi, S.G.; Rotundo, R.L.; Spiegelman, B.M.; Moraes, C.T. Increased muscle PGC-1alpha expression protects from sarcopenia and metabolic disease during aging. *Proc. Natl. Acad. Sci. USA* **2009**, *93*, 15364–15369. [CrossRef]

90. Calvani, R.; Joseph, A.M.; Adhihetty, P.J.; Miccheli, A.; Bossola, M.; Leeuwenburgh, C.; Bernabei, R.; Marzetti, E. Mitochondrial pathways in sarcopenia of aging and disuse muscle atrophy. *Biol. Chem.* **2013**, *394*, 393–414. [CrossRef]

91. Jackson, M.J.; McArdle, A. Role of reactive oxygen species in age-related neuromuscular deficits. *J. Physiol.* **2016**, *594*, 1979–1988. [CrossRef]

92. Ji, L.L. Exercise at old age: Does it increase or alleviate oxidative stress? *Ann. N. Y. Acad. Sci.* **2001**, *928*, 236–247. [CrossRef]

93. Bejma, J.; Ji, L.L. Aging and acute exercise enhances free radical generation and oxidative damage in skeletal muscle. *J. Appl. Physiol.* **1999**, *87*, 465–470. [CrossRef]

94. Derbré, F.; Gratas-Delamarche, A.; Gómez-Cabrera, M.C.; Viña, J. Inactivity-induced oxidative stress: A central role in age-related sarcopenia? *Eur. J. Sport Sci.* **2014**, *14*, S98–S108. [CrossRef]

95. Bouzid, M.A.; Filaire, E.; Matran, R.; Robin, S.; Fabre, C. Lifelong Voluntary Exercise Modulates Age-Related Changes in Oxidative Stress. *Int. J. Sports Med.* **2018**, *39*, 21–28. [CrossRef]

96. Tipton, K.D. Muscle protein metabolism in the elderly: Influence of exercise and nutrition. *Can. J. Appl. Physiol.* **2001**, *26*, 588–606. [CrossRef]

97. Muscariello, E.; Nasti, G.; Siervo, M.; Di Maro, M.; Lapi, D.; D'Addio, G.; Colantuoni, A. Dietary protein intake in sarcopenic obese older women. *Clin. Interv. Aging* **2016**, *11*, 133–140. [CrossRef]

98. Vanacore, D.; Messina, G.; Lama, S.; Bitti, G.; Ambrosio, P.; Tenore, G.; Messina, A.; Monda, V.; Zappavigna, S.; Boccellino, M.; et al. Effect of restriction vegan diet's on muscle mass, oxidative status, and myocytes differentiation: A pilot study. *J. Cell. Physiol.* **2018**, *233*, 9345–9353. [CrossRef]

99. Monroy, A.; Lithgow, G.J.; Alavez, S. Curcumin and neurodegenerative diseases. *Biofactors* **2013**, *39*, 122–132. [CrossRef]

100. Pellavio, G.; Rui, M.; Caliogna, L.; Martino, E.; Gastaldi, G.; Collina, S.; Laforenza, U. Regulation of Aquaporin Functional Properties Mediated by the Antioxidant Effects of Natural Compounds. *Int. J. Mol. Sci.* **2017**, *18*, 2665. [CrossRef]

101. Boots, A.W.; Haenen, G.R.; Bast, A. Health effects of quercetin: From antioxidant to nutraceutical. *Eur. J. Pharmacol.* **2008**, *585*, 325–337. [CrossRef]

102. Wu, Q.; Wang, X.; Nepovimova, E.; Wang, Y.; Yang, H.; Li, L.; Zhang, X.; Kuca, K. Antioxidant agents against trichothecenes: New hints for oxidative stress treatment. *Oncotarget* **2017**, *8*, 110708–110726. [CrossRef]

103. Damiano, S.; Sasso, A.; De Felice, B.; Di Gregorio, I.; La Rosa, G.; Lupoli, G.A.; Belfiore, A.; Mondola, P.; Santillo, M. Quercetin Increases MUC2 and MUC5AC Gene Expression and Secretion in Intestinal Goblet Cell-Like LS174T via PLC/PKCα/ERK1-2 Pathway. *Front. Physiol.* **2018**, *9*, 357. [CrossRef]

104. Nieman, D.C.; Laupheimer, M.W.; Ranchordas, M.K.; Burke, L.M.; Stear, S.J.; Castell, L.M. A-Z of nutritional supplements: Dietary supplements, sports nutrition foods and ergogenic aids for health and performance. Part 33. *Br. J. Sport Med.* **2012**, *46*, 618–620. [CrossRef]

105. Tresserra-Rimbau, A.; Arranz, S.; Vallverdu-Queralt, A. New Insights into the Benefits of Polyphenols in Chronic Diseases. *Oxid. Med. Cell. Longev.* **2017**, *2017*, 1432071. [CrossRef]

106. Fenech, M.; Amaya, I.; Valpuesta, V.; Botella, M.A. Vitamin C Content in Fruits: Biosynthesis and Regulation. *Front. Plant Sci.* **2018**, *9*, 2006. [CrossRef]

107. Levine, M.; Padayatty, S.J. *Modern Nutrition in Health and Disease*; Ross, A.C., Caballero, B., Cousins, R.J., Tucker, K.L., Ziegler, T.R., Eds.; Wolters Kluwer Health, Lippincott Williams & Wilkins: Baltimore, MD, USA, 2014; pp. 416–426.

108. Hill, A.; Wendt, S.; Benstoem, C.; Neubauer, C.; Meybohm, P.; Langlois, P.; Adhikari, N.K.J.; Heyland, D.K.; Stoppe, C. Vitamin C to Improve Organ Dysfunction in Cardiac Surgery Patients—Review and Pragmatic Approach. *Nutrients* **2018**, *10*, 974. [CrossRef]

109. Cerullo, F.; Gambassi, G.; Cesari, M. Rationale for Antioxidant Supplementation in Sarcopenia. *J. Aging Res.* **2012**, *2012*, 316943. [CrossRef]

110. Szymańska, R.; Nowicka, B.; Kruk, J. Vitamin E—Occurrence, Biosynthesis by Plants and Functions in Human Nutrition. *Mini Rev. Med. Chem.* **2017**, *17*, 1039–1052. [CrossRef]

111. Marko, M.G.; Ahmed, T.; Bunnell, S.C.; Wu, D.; Chung, H.; Huber, B.T.; Meydani, S.N. Age-associated decline in effective immune synapse formation of CD4(+) T cells is reversed by vitamin E supplementation. *J. Immunol.* **2007**, *178*, 1443–1449. [CrossRef]

112. Molano, A.; Meydani, S.N. Vitamin E, signalosomes and gene expression in T cells. *Mol. Asp. Med.* **2012**, *33*, 55–62. [CrossRef]

113. Zingg, J.M. Vitamin E: Regulatory Role on Signal Transduction. *IUBMB Life* **2019**, *71*, 456–478. [CrossRef]

114. Elvira-Torales, L.I.; García-Alonso, J.; Periago-Castón, M.J. Nutritional Importance of Carotenoids and Their Effect on Liver Health: A Review. *Antioxidants* **2019**, *8*, 229. [CrossRef]

115. Sandmann, G. Antioxidant Protection from UV- and Light-Stress Related to Carotenoid Structures. *Antioxidants* **2019**, *8*, 219. [CrossRef]

116. Bohn, T. Carotenoids and markers of oxidative stress in human observational studies and intervention trials: Implications for chronic diseases. *Antioxidants* **2019**, *8*, 179. [CrossRef]

117. Yilmaz, B.; Sahin, K.; Bilen, H.; Bahcecioglu, I.H.; Bilir, B.; Ashraf, S.; Kucuk, O. Carotenoids and non-alcoholic fatty liver disease. *Hepatobiliary Surg. Nutr.* **2015**, *4*, 161–171. [CrossRef]

118. Seif El-Din, S.H.; El-Lakkany, N.M.; El-Naggar, A.A.; Hammam, O.A.; Abd El-Latif, H.A.; Ain-Shoka, A.A.; Ebeid, F.A. Effects of rosuvastatin and/or β-carotene on non-alcoholic fatty liver in rats. *Res. Pharm. Sci.* **2015**, *10*, 275–287.

119. Kaulmann, A.; Bohn, T. Carotenoids, inflammation, and oxidative stress—Implications of cellular signaling pathways and relation to chronic disease prevention. *Nutr. Res.* **2014**, *34*, 907–929. [CrossRef]

120. Kumar, S.; Pandey, A.K. Chemistry and biological activities of flavonoids: An overview. *Sci. World J.* **2013**, *2013*, 162750. [CrossRef]

121. Espley, R.V.; Butts, C.A.; Laing, W.A.; Martell, S.; Smith, H.; McGhie, T.K.; Zhang, J.; Paturi, G.; Hedderley, D.; Bovy, A.; et al. Dietary flavonoids from modified apple reduce inflammation markers and modulate gut microbiota in mice. *J. Nutr.* **2014**, *144*, 146–154. [CrossRef]

122. Lee, S.G.; Kim, B.; Yang, Y.; Pham, T.X.; Park, Y.K.; Manatou, J.; Koo, S.I.; Chun, O.K.; Lee, J.Y. Berry anthocyanins suppress the expression and secretion of proinflammatory mediators in macrophages by inhibiting nuclear translocation of NF-kappaB independent of NRF2-mediated mechanism. *J. Nutr. Biochem.* **2014**, *25*, 404–411. [CrossRef]

123. Romero-Perez, A.I.; Ibern-Gomez, M.; Lamuela-Raventos, R.M.; de La Torre-Boronat, M.C. Piceid, the major resveratrol derivative in grape juices. *J. Agric. Food Chem.* **1999**, *47*, 1533–1536. [CrossRef]

124. Aggarwal, B.B.; Bhardwaj, A.; Aggarwal, R.S.; Seeram, N.P.; Shishodia, S.; Takada, Y. Role of resveratrol in prevention and therapy of cancer: Preclinical and clinical studies. *Anticancer Res.* **2004**, *24*, 2783–2840.

125. Kuttan, G.; Kumar, K.B.; Guruvayoorappan, C.; Kuttan, R. Antitumor, anti-invasion, and antimetastatic effects of curcumin. *Adv. Exp. Med. Biol.* **2007**, *595*, 173–184. [CrossRef]

126. Chuengsamarn, S.; Rattanamongkolgul, S.; Luechapudiporn, R.; Phisalaphong, C.; Jirawatnotai, S. Curcumin extract for prevention of type 2 diabetes. *Diabetes Care* **2012**, *35*, 2121–2127. [CrossRef]

127. Cipolletti, M.; Solar Fernandez, V.; Montalesi, E.; Marino, M.; Fiocchetti, M. Beyond the Antioxidant Activity of Dietary Polyphenols in Cancer: The Modulation of Estrogen Receptors (ERs) Signaling. *Int. J. Mol. Sci.* **2018**, *19*, 2624. [CrossRef]

128. De Oliveira, D.C.X.; Rosa, F.T.; Simões-Ambrósio, L.; Jordao, A.A.; Deminice, R. Antioxidant vitamin supplementation prevents oxidative stress but does not enhance performance in young football athletes. *Nutrition* **2019**, *63–64*, 29–35. [CrossRef]

129. He, F.; Hockemeyer, J.A.; Sedlock, D. Does combined antioxidant vitamin supplementation blunt repeated bout effect? *Int. J. Sports Med.* **2015**, *36*, 407–413. [CrossRef]

130. Cesari, M.; Pahor, M.; Bartali, B.; Cherubini, A.; Penninx, B.W.; Williams, G.R.; Atkinson, H.; Martin, A.; Guralnik, J.M.; Ferrucci, L. Antioxidants and physical performance in elderly persons: The Invecchiare in Chianti (InCHIANTI) study. *Am. J. Clin. Nutr.* **2004**, *79*, 289–294. [CrossRef]

131. Gomez-Cabrera, M.C.; Domenech, E.; Romagnoli, M.; Arduini, A.; Borras, C.; Pallardo, F.V.; Sastre, J.; Vina, J. Oral administration of vitamin C decreases muscle mitochondrial biogenesis and hampers training-induced adaptations in endurance performance. *Am. J. Clin. Nutr.* **2008**, *87*, 142–149. [CrossRef]

132. Simioni, C.; Zauli, G.; Martelli, A.M.; Vitale, M.; Sacchetti, G.; Gonelli, A.; Neri, L.M. Oxidative stress: Role of physical exercise and antioxidant nutraceuticals in adulthood and aging. *Oncotarget* **2018**, *9*, 17181–17198. [CrossRef]

133. Scholten, S.D.; Sergeev, I.N. Long-term quercetin supplementation reduces lipid peroxidation but does not improve performance in endurance runners. *Open Access J. Sports Med.* **2013**, *4*, 53–61. [CrossRef]

134. McAnulty, L.S.; Miller, L.E.; Hosick, P.A.; Utter, A.C.; Quindry, J.C.; McAnulty, S.R. Effect of resveratrol and quercetin supplementation on redox status and inflammation after exercise. *Appl. Physiol. Nutr. Metab.* **2013**, *38*, 760–765. [CrossRef]

135. Petiz, L.L.; Girardi, C.S.; Bortolin, R.C.; Kunzler, A.; Gasparotto, J.; Rabelo, T.K.; Matté, C.; Moreira, J.C.; Gelain, D.P. Vitamin A Oral Supplementation Induces Oxidative Stress and Suppresses IL-10 and HSP70 in Skeletal Muscle of Trained Rats. *Arch. Biochem. Biophys.* **2007**, *458*, 141–145. [CrossRef]

136. Semba, R.D.; Lauretani, F.; Ferrucci, L. Carotenoids as protection against sarcopenia in older adults. *Arch. Biochem. Biophys.* **2007**, *458*, 141–145. [CrossRef]

137. Sahin, K.; Pala, R.; Tuzcu, M.; Ozdemir, O.; Orhan, C.; Sahin, N.; Juturu, V. Curcumin prevents muscle damage by regulating NF-kappaB and Nrf2 pathways and improves performance: An in vivo model. *J. Inflamm. Res.* **2016**, *9*, 147–154. [CrossRef]

138. Menzies, K.J.; Singh, K.; Saleem, A.; Hood, D.A. Sirtuin 1-mediated effects of exercise and resveratrol on mitochondrial biogenesis. *J. Biol. Chem.* **2013**, *288*, 6968–6979. [CrossRef]

139. Murase, T.; Haramizu, S.; Ota, N.; Hase, T. Suppression of the aging associated decline in physical performance by a combination of resveratrol intake and habitual exercise in senescence accelerated mice. *Biogerontology* **2009**, *10*, 423–434. [CrossRef]

140. Always, S.E.; McCrory, J.L.; Kearcher, K.; Vickers, A.; Frear, B.; Gilleland, D.L.; Bonner, D.E.; Thomas, J.M.; Donley, D.A.; Lively, M.W.; et al. Resveratrol Enhances Exercise-Induced Cellular and Functional Adaptations of Skeletal Muscle in Older Men and Women. *J. Gerontol. A Biol. Sci. Med. Sci.* **2017**, *72*, 1595–1606. [CrossRef]

141. Kregel, K.C.; Zhang, H.J. An integrated view of oxidative stress in aging: Basic mechanisms, functional effects, and pathological considerations. *Am. J. Physiol. Regul. Integr. Comp. Physiol.* **2007**, *292*, R18–R36. [CrossRef]

142. Kim, J.S.; Wilson, J.M.; Lee, S.R. Dietary implications on mechanisms of sarcopenia: Roles of protein, amino acids and antioxidants. *J. Nutr. Biochem.* **2010**, *21*, 1–13. [CrossRef]

143. Beaudart, C.; Dawson, A.; Shaw, S.C.; Harvey, N.C.; Kanis, J.A.; Binkley, N.; Reginster, J.Y.; Chapurlat, R.; Chan, D.C.; Bruyère, O.; et al. Nutrition and physical activity in the prevention and treatment of sarcopenia: Systematic review. *Osteoporos. Int.* **2017**, *28*, 1817–1833. [CrossRef]

144. Teixeira, V.H.; Valente, H.F.; Casal, S.I.; Marques, A.F.; Moreira, P.A. Antioxidants do not prevent postexercise peroxidation and may delay muscle recovery. *Med. Sci. Sports Exerc.* **2009**, *41*, 1752–1760. [CrossRef]

145. Peternelj, T.T.; Coombes, J.S. Antioxidant supplementation during exercise training: Beneficial or detrimental? *Sports Med.* **2011**, *41*, 1043–1069. [CrossRef]

146. Rowlands, D.S.; Pearce, E.; Aboud, A.; Gillen, J.B.; Gibala, M.J.; Donato, S.; Waddington, J.M.; Green, J.G.; Tarnopolsky, M.A. Oxidative stress, inflammation, and muscle soreness in an 894-km relay trail run. *Eur. J. Appl. Physiol.* **2012**, *112*, 1839–1848. [CrossRef]

147. Howard, A.C.; McNeil, A.K.; McNeil, P.L. Promotion of plasma membrane repair by vitamin E. *Nat. Commun.* **2011**, *2*, 597. [CrossRef]

148. Fusco, D.; Colloca, G.; Lo Monaco, M.R.; Cesari, M. Effects of antioxidant supplementation on the aging process. *Clin. Interv. Aging* **2007**, *2*, 377–387.

149. Mondola, P.; Annella, T.; Serù, R.; Santangelo, F.; Iossa, S.; Gioielli, A.; Santillo, M. Secretion and increase of intracellular CuZn superoxide dismutase content in human neuroblastoma SK-N-BE cells subjected to oxidative stress. *Brain Res. Bull.* **1998**, *45*, 517–520. [CrossRef]

150. Selman, C.; McLaren, J.S.; Meyer, C.; Duncan, J.S.; Redman, P.; Collins, A.R.; Duthie, G.G.; Speakman, J.R. Life-long vitamin C supplementation in combination with cold exposure does not affect oxidative damage or lifespan in mice, but decreases expression of antioxidant protection genes. *Mech. Ageing Dev.* **2006**, *127*, 897–904. [CrossRef]

151. Merry, T.L.; Ristow, M. Do antioxidant supplements interfere with skeletal muscle adaptation to exercise training? *J. Physiol.* **2016**, *594*, 5135–5147. [CrossRef]

International Journal of
Molecular Sciences

Review

Glutathione Metabolism in Renal Cell Carcinoma Progression and Implications for Therapies

Yi Xiao [1,2] and David Meierhofer [1,*]

1   Max Planck Institute for Molecular Genetics, Ihnestraße 63-73, 14195 Berlin, Germany
2   Freie Universität Berlin, Fachbereich Biologie, Chemie, Pharmazie, Takustraße 3, 14195 Berlin, Germany
*   Correspondence: meierhof@molgen.mpg.de; Tel.: +49-30-8413-1567; Fax: +49-30-8413-1960

Received: 27 June 2019; Accepted: 26 July 2019; Published: 26 July 2019

Abstract: A significantly increased level of the reactive oxygen species (ROS) scavenger glutathione (GSH) has been identified as a hallmark of renal cell carcinoma (RCC). The proposed mechanism for increased GSH levels is to counteract damaging ROS to sustain the viability and growth of the malignancy. Here, we review the current knowledge about the three main RCC subtypes, namely clear cell RCC (ccRCC), papillary RCC (pRCC), and chromophobe RCC (chRCC), at the genetic, transcript, protein, and metabolite level and highlight their mutual influence on GSH metabolism. A further discussion addresses the question of how the manipulation of GSH levels can be exploited as a potential treatment strategy for RCC.

Keywords: Renal cell carcinoma (RCC); reactive oxygen species (ROS); glutathione (GSH) metabolism; cancer therapy; clear cell RCC; papillary RCC; chromophobe RCC

---

1. Introduction

Increased reactive oxygen species (ROS) levels, including the superoxide anion, hydrogen peroxide, and hydroxyl radical, have been reported in many different cancer types. ROS can be either generated by genetic alterations and endogenous oxygen metabolism or by exogenous sources, such as UV light and radiation. ROS were long thought to be only damaging byproducts of the cellular metabolism that can negatively affect DNA, lipids, and proteins [1]. However, more recent studies have highlighted the important role of ROS in cell signaling, homeostasis, metabolism, and apoptosis [1]. One common characteristic of cancer is the ability to balance the increased level of oxidative stress with a high level of antioxidants. Glutathione (GSH), a tripeptide thiol antioxidant composed of the amino acids glutamic acid, cysteine, and glycine [2], is the main ROS scavenger in cells. GSH is highly reactive and exists in both a reduced (GSH) and oxidized disulfide (GSSG) form [3]. The predominant form is in the reduced state, which is the most abundant low molecular weight thiol in the cell, ranging from 0.5 to 10 mM in most cell types, whereas extracellular GSH exists in concentrations lower by magnitudes [4]. The de novo biosynthesis of GSH involves two ATP-dependent enzymatic reactions: The first step is catalyzed by glutamate cysteine ligase (GCL), which ligates the amino group of cysteine to the γ-carboxylate of glutamic acid to form the dipeptide γ-glutamyl cysteine. The second reaction involves GSH synthetase (GSS), which catalyzes a combination of the cysteinyl carboxylate of the dipeptide and the amino group of glycine to synthesize GSH [5] (outlined in Figure 1).

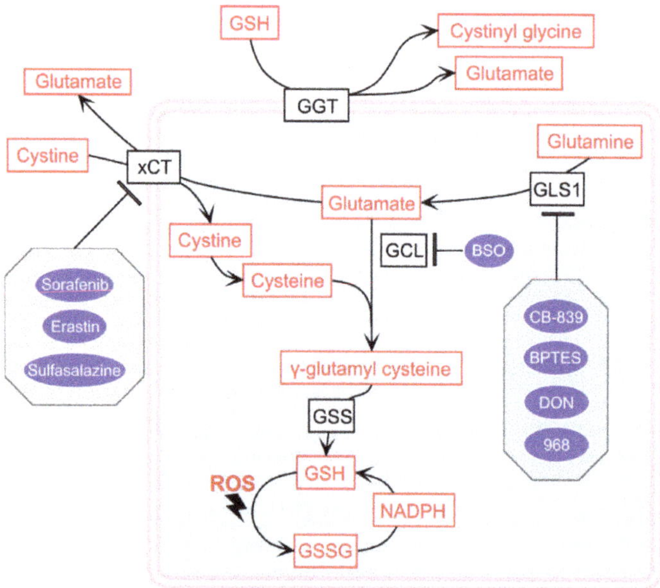

Figure 1. Schematic overview of glutathione (GSH) metabolism and the targeting sites of inhibitors. Color codes are defined as follows: black = enzymes or transporters; red = metabolites; blue = inhibitors. GGT: γ-glutamyl transferase; xCT: solute carrier family 7 member 11, a cystine-glutamate antiporter; GLS1: glutaminase 1; GCL: glutamate cysteine ligase; GSS: glutathione synthetase. GSSG: glutathione oxidized form; ROS: reactive oxygen species.

Besides the classical role of GSH acting as an ROS scavenger by being prey for radicals, GSH has several additional functions, including but not limited to providing a cysteine reservoir [6], being involved in the maturation of iron–sulfur proteins [7], detoxifying xenobiotics [8], regulating protein bioactivity by S-glutathionylation [9,10], and regulating redox signaling [11]. In cancer, GSH plays the role of a double-edged sword in its initiation and progression. Moderate ROS levels are widely recognized to trigger cancer initiation and progression by inducing mutations and promoting genome instability, eventually activating oncogenic signaling pathways that promote cell survival, proliferation, and stress resistance [12]. On the contrary, massive ROS accumulations can also limit cancer growth by causing severe oxidative damage of biomolecules, which finally can lead to cell death [13]. As a consequence, cancer cells are required to deliberately balance the levels of ROS and antioxidants (mainly GSH) to maintain redox homeostasis, which sustains viability and growth. For many years, one of the most obvious therapeutic strategies to overcome new balanced redox homeostasis in renal cell carcinoma (RCC) was to fight elevated ROS levels with the supplementation of antioxidants such as vitamins to actively force the tumor into apoptosis. Many clinical trials were initiated, and the outcomes showed mixed results, including worse survival rates upon supplementation with ROS inhibitors [14].

In this review, we will focus on the role of the ROS scavenger GSH in RCC and discuss possible strategies that can potentially exploit the manipulation of GSH levels for therapeutic strategies in RCC.

2. Renal Cell Carcinoma: An Overview

RCC represents approximately 4% of adult malignancies [15] and was ranked as the sixth deadliest cancer worldwide in 2018 [16]. The American Cancer Society estimated that about 73,820 new RCC cases would be diagnosed by the end of 2019 and more than 14,770 deaths would be caused by RCC this year in the USA alone [15]. RCC can be classified according to distinct morphologic and molecular genetic features and is composed of different subtypes, such as clear cell RCC (ccRCC), papillary RCC (pRCC), and chromophobe RCC (chRCC, Table 1). Many studies have been performed recently to characterize RCC to better understand its classification and subclassification and to elucidate pathway remodeling in these cancers [17–19]. A new classification concept based on molecular clustering of chromosoms, DNA, RNA, miRNA, and protein data was proposed [18,20], where the organ of origin does not fully determine the tumor type as the only factor [21]. Instead, the cancer classification should be based on the similarity of molecular features across different tissue types, which was considered to be more relevant for targeting the same mutations and oncogenic signaling pathways [21].

Here, we review the genetic foundation of the main RCC types and shed light on the sparse data of transcriptome, proteome, and metabolome profiles performed in these malignancies.

Table 1. Summary of the three renal cell carcinoma (RCC) subtypes, incidences, main mutations, and GSH regulation.

RCC Subtypes	Clear Cell	Papillary	Chromophobe
Incidence	75%	15%	5%
Main mutations	*VHL*	*MET, FH*	*TP53, PTEN*
Metabolites	GSH, GSSG increased	GSH, GSSG increased	GSH, GSSG increased
GSH regulation	1. GCL protein abundance increases; 2. Increased serum GGT as a marker for metastatic ccRCC; 3. GLS1, glutamine importers, and cysteine antiporter xCT enhance to favor GSH synthesis; 4. Increased PPP flux to produce NADPH for GSH conversion.	1. *FH* mutation causes HIF stabilization; 2. *FH* mutation activates NRF2–ARE pathway, leading to increased GSH synthesis and enhanced expression of antioxidant proteins.	Loss of GGT1 increases sensitivity to oxidative stress in chRCC cells.

ccRCC: clear cell RCC; chRCC: chromophobe RCC; pRCC: papillary RCC; *VHL*: von Hippel-Lindau; *MET*: proto-oncogene c-Met; *FH*: fumarate hydrotase; *TP53*: tumor antigen p53; *PTEN*: phosphatase and tensin homolog; PPP: pentose phosphate pathway; HIF: hypoxia-inducible factor; NRF2: nuclear factor erythroid 2-related factor 2; ARE: antioxidant response element; GGT1: γ-glutamyl transferase 1.

2.1. Clear Cell Renal Cell Carcinoma

Clear cell RCC is the most prevalent subtype and accounts for about 75% of all RCCs (Table 1) [22]. It is an aggressive cancer that originates from the proximal convoluted tubule, with a recurrence rate of up to 40% after the initial treatment of a localized tumor [23]. In its metastatic form, it is associated with a high mortality rate [24]. Clear cell RCC cells have, in general, a clear cytoplasm, (which helped coin the name "clear cell") that is circled by an easily distinguishable cell membrane and uniform round nuclei [25]. About 90% of all ccRCCs carry mutations in the von Hippel-Lindau (*VHL*) tumor suppressor gene [17,26], which was originally identified in a hereditary disease called VHL syndrome [27]. The VHL protein is a target recruitment subunit in an E3 ubiquitin ligase complex and recruits the hydroxylated hypoxia-inducible factor (HIF) under normoxic conditions for subsequent proteasomal degradation. Thereby, VHL can repress the transcription of more than 100 target genes through interaction with HIF1α and HIF1AN, which plays a vital role in forming the phenotype of ccRCC [28]. HIF1α is a master transcription factor that contributes substantially to the regulation of gene expression that is dependent on oxygen levels. Under normoxic conditions, VHL interacts with HIF1α and hydroxylates the proline residues in the oxygen-dependent degradation (ODD) domains of HIF1α by recruiting members of the Egl-nine homolog (EGLN) family [28–30]. With hypoxia or loss of function of VHL, these proline residues cannot be hydroxylated, which stabilizes HIF1α. HIF1α subsequently forms a HIF1α–HIF1β heterodimer, and this dimer translocates into the nucleus to enhance the transcription of HIF target genes, which are associated with crucial oncogenic pathways, including glucose uptake, glycolysis (e.g., glucose transporter type 1, *GLUT1*), cell proliferation (e.g., epidermal growth factor receptor, *EGFR*), and angiogenesis (vascular endothelial growth factor, *VEGF*) [30–33].

Furthermore, the gluconeogenic enzyme fructose 1,6-bisphosphatase 1 (FBP1) has been found to be decreased in over 600 ccRCCs and has been associated with poor disease prognosis. FBP1 has two distinct functions, antagonizing the glycolytic flux and inhibiting the nuclear function of HIFα [34], which can explain its ubiquitous loss in ccRCC [34]. Besides FBP1, the whole gluconeogenesis pathway has been shown to be severely diminished in ccRCC at the transcriptome [34] and proteome level [35]. This stimulates the metabolic switch by increasing glycolytic target genes [34], which is reflected by the metabolomic analysis of ccRCC, where metabolites in the glycolysis pathway show over two-fold increases in abundance compared to the normal kidney [36]. Furthermore, GSH metabolism-related metabolites, including cysteine, γ-glutamyl cysteine, and GSH, have all been shown to increase in late-stage ccRCC and are associated with worse survival outcomes in ccRCC patients [36].

2.2. Papillary Renal Cell Carcinoma

Papillary RCC represents about 15% of all RCCs (Table 1) and also derives from the proximal convoluted tubule, similarly to ccRCC [22]. It is a less aggressive subtype compared to ccRCC and has a high five-year survival rate of 80% to 85% [37]. The term "papillary" describes the papilla-like protuberances in most of the tumors. It can be further subdivided into type I and type II tumors based on morphological features. Type I pRCC is more common and shows small fibrovascular papillae that are covered by a single layer of small cuboidal cells with scant pale cytoplasm and usually grows slowly [38]. In contrast, type II pRCC consists of papillae, is lined by large columnar pseudostratified cells with an eosinophilic cytoplasm, and is often more aggressive [38,39]. Type I and type II pRCC have also been shown to be clinically and biologically distinct, as alterations in the MET pathway were associated with type I [18,40]. The proto-oncogene c-Met (MET) protein, a transmembrane receptor tyrosine kinase, can bind to its ligand hepatocyte growth factor (HGF) and activate several downstream intracellular pathways, including focal adhesion kinase (FAK), RAS/RAF/MEK/ERK, and PI3K/AKT [41]. The frequently activating mutations and amplification of *MET* in type I pRCC enable the activation of MET/HGF signaling and its above-mentioned downstream pathways to promote cancer cell proliferation, angiogenesis, and malignant transformation [41].

Frequent mutations in type II pRCC include *CDKN2A* silencing, *SETD2* mutations, and *TFE3* fusions. Type II tumors are characterized by increased expression of the nuclear factor erythroid 2-related factor 2 (*NRF2*)–antioxidant response element (ARE) pathway [18]. The NRF2–ARE pathway is a major regulator of cellular redox balance, and its activation under oxidative stress favors cell survival. Furthermore, fumarate hydratase (*FH*) mutations are also frequently found in type II pRCC [42,43]. The *FH* gene encodes a TCA cycle enzyme that catalyzes the hydration of fumarate to malate, and its deficiency causes fumarate and succinate accumulation [44,45]. Accumulated fumarate and succinate are believed to be able to suppress the hydroxylation of the proline residues in the ODD domain of HIFα, and thus *FH* mutations in type II pRCC also cause the stabilization of HIFα, similarly to ccRCC [44,45]. Some genes (such as *CDKN2A/B* and *TERT*) where mutations can be found in both types [40] play a pivotal role as tumor suppressors by regulating the cell cycle. Mutations of the above-mentioned genes and activation of the oncopathways are the main driver mutations in the progression of pRCC.

How do these genetic alterations in pRCC translate to the protein and metabolite level? Proteome profiles of pRCC versus matching healthy tissues have indicated a tremendous reprogramming of main metabolic pathways. Oxidative phosphorylation, the TCA cycle, branched-chain amino acids, cytochrome P450 drug metabolism, peroxisomes, fatty acid metabolism, and several amino acid metabolism pathways were significantly decreased in pRCC, whereas the spliceosome, the ribosome, and the cell cycle were significantly increased [46]. A striking anticorrelation between the proteome and the transcriptome data [18] was identified for oxidative phosphorylation. Transcripts of the respiratory chain were significantly increased in pRCC, whereas the entire pathway was significantly decreased on the proteome level. Most likely, the lower protein abundance of the respiratory chain was a consequence of the reduced mitochondrial DNA (mtDNA) content in pRCC, as a similar association was observed in ρ^0 cells [47]. ρ^0 cells entirely lacking mtDNA, which encodes for 13 core respiratory chain subunits and consequently miss all of the respiratory chain complexes [47]. Furthermore, the discrepancy between the transcripts and proteins of the respiratory chain in RCC can also originate from the regulation of post-translational modifications. The metabolome indicated a tremendous increase of reduced and oxidized GSH levels in pRCC tissues [46], as well as significantly increased rates of GSH de novo synthesis based on glutamine consumption in the pRCC-derived cell lines Caki-2 and ACHN [46]. All of these alterations can in principle serve as potential therapeutic targets. Specifically, a dysregulated respiratory chain can cause electron leakage [48,49]. This, in turn, leads frequently to an increase in ROS stress [50,51], which is subsequently compensated for by increased GSH levels in RCC and might serve as a main therapeutic site to eradicate RCC, as discussed later on.

2.3. Chromophobe Renal Cell Carcinoma

Chromophobe RCC accounts for approximately 5% of all RCCs (Table 1) [22], is thought to originate from the cortical collecting duct, and was first reported in 1985 [52]. Different morphological and ultrastructural features of the cytoplasm lead to the identification of the classical chromophobe and the eosinophilic variant. The cells of the classical type usually have abundant clear cytoplasm and a perinuclear halo caused by cytoplasmic organelles being pushed away from the center to form a rim along the cell membrane [53]. The eosinophilic type has, in general, smaller cells with an inconstant level of cytoplasmic organelles in the periphery. Both cell types frequently coexist in chRCC tumors, usually with one cell type predominating [53]. One of the most characteristic genetic features of chRCC is the monosomy of chromosomes 1, 2, 6, 10, 13, 17, and often 21 [54–57]. The most commonly mutated genes in chRCC are *TP53* (32%), *PTEN* (20%), and gene fusions involving the *TERT* promoter [19,54]. Mutations in these tumor suppressors combined with the deletion of one of their chromosomes leads to a complete loss of function. Further mutations with a lower frequency were observed in *MTOR*, *NRAS*, *TSC1*, and *TSC2*, indicating that the genomic targeting of the mTOR pathway occurred in 23% of all chRCC [19]. Hence, the anticancer functions of TP53 in apoptosis, genomic stability, and the inhibition of angiogenesis and the role of PTEN in the intracellular signaling pathway PI3K/AKT/mTOR are both disrupted and can thus be regarded as major driving events in chRCC tumorigenesis.

Proteome profiling has identified metabolic reprogramming in chRCC, including stalled gluconeogenesis, downregulated oxidative phosphorylation, and fatty acid and amino acid metabolism [57]. A similar anticorrelation between transcripts and proteins (as in pRCC) was also identified in chRCC. As chRCC has a significantly lower microvessel density and a lower glucose uptake rate compared to ccRCC and pRCC [58,59], it seems that chRCC cells prefer a different way to acquire nutrients to compensate for the nutrient-poor microenvironment. Chromophobe RCC cells can activate the endocytosis and downstream lysosomal pathways to gain extracellular macromolecules as a nutrition source for cell survival and proliferation, which is indicated by the abundance increase of proteins involved in these pathways and their enzymatic activities [57]. Metabolome profiling in chRCC [57,60] and the closely related but hardly distinguishable benign renal oncocytomas [61,62] has also elucidated a striking increase of GSH and GSSG levels in kidney tumors, thus a hallmark in all RCCs.

In the next chapters, we will investigate the role of GSH metabolism in RCC progression and how this adaption to increased ROS levels can be exploited therapeutically.

3. Rewired Glutathione Metabolism in RCC Is a Key Metabolic Alteration Involved in Tumor Progression

3.1. γ-Glutamyl Cycle and ccRCC Progression

The γ-glutamyl cycle was originally proposed by Meister in 1970 and involves the de novo biosynthesis and degradation of GSH [63]. Its main functions rely on the two enzymes GCL and GSS for biosynthesis and γ-glutamyl transferases (GGTs) for degradation (Figure 1).

GCL is involved in the first step of de novo GSH synthesis, which catalyzes the reaction of γ-glutamyl cysteine production. Other than substrate availability, GCL is the rate-limiting enzyme of GSH biosynthesis. It is composed of two different subunits: GCLC, the 73-kD catalytic subunit, which contains the active site for catalyzing the reaction; and GCLM, the 31-kD modulatory subunit, which interacts with GCLC to increase catalytic efficiency [64]. Under physiological conditions, the activity of the GCLC/GCLM heterodimer can be regulated by a negative feedback loop by its product GSH [5]. GCLC and GCLM were reported to have increased protein abundances, matching the significantly increased enzymatic activity in the tumor tissues of ccRCC patients [34,65,66]. Validation by siRNA-mediated silencing of GCLC led to a strong cell number reduction of ccRCC cell lines, which was accompanied by significantly decreased GSH levels [66]. The direct inhibition of GSH synthesis caused ferroptosis, a nonapoptotic form of cell death, in ccRCC cells [66]. These data support the substantial role of GSH metabolism in ccRCC progression.

GGTs are membrane-bound, N-terminal nucleophile hydrolases that catalyze the breakdown of extracellular GSH and transfer the γ-glutamyl group from GSH to produce the constituents glutamate and cysteine, which can be further used for intracellular GSH synthesis (Figure 1) [67]. Increased serum GGT was reported to be a sensitive marker for metastatic ccRCC [68], as GGT levels positively correlated with advanced stages, higher grades, and the presence of tumor necrosis, and it was further associated with worse survival rates in ccRCC patients [69].

3.2. Precursor Amino Acid Availability for GSH de novo Synthesis

Apart from the γ-glutamyl cycle, GSH de novo synthesis also relies on the availability of its three composing amino acids—glutamate, cysteine, and glycine—and the activity of their corresponding transporters, as summarized in Figure 2.

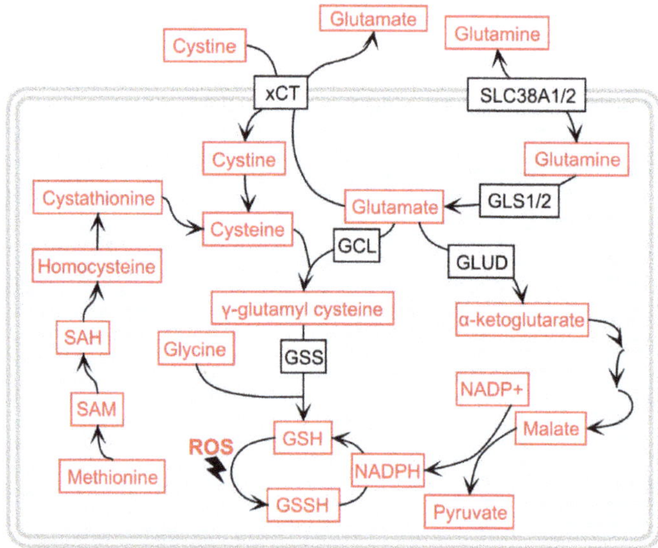

Figure 2. The availability of the precursor amino acids that influence GSH synthesis. Color codes are defined as follows: black = enzymes or transporters; red = metabolites. SLC38A1/2: solute carrier family 38 member 1 and 2, glutamine transporters; GLUD: glutamate dehydrogenase; SAM: S-adenosyl methionine; SAH: S-adenosyl homocysteine.

One of the metabolic hallmarks of ccRCC is the addiction to glutamine. The malignancies, therefore, require exogenous glutamine for growth and feature reprogrammed glutamine metabolism [70,71]. The availability of glutamine can directly or indirectly influence GSH de novo synthesis in three different ways. First, glutamine can be converted to glutamate by two isozymes, glutaminase 1 and

2 (GLS1 and GLS2). GLS1 is increased in many cancer types and is the main isoform within the kidney [72]. Second, glutamine-related transporter transcripts are consistently increased in ccRCC tumors, e.g., the glutamine importers *SLC38A1* and *SLC38A2* [73], to sustain glutaminolysis in ccRCC. Interestingly, SLC38A1 expression is also regulated by MYC in K562 and HeLa cells: Considering that MYC has been shown to be upregulated and that MYC pathway activation cooperates with VHL loss to induce ccRCC [73–75], SLC38A1 probably plays an important role in the progression of ccRCC. Third, glutamine can contribute to the de novo synthesis of GSH through the generation of NADPH via glutamate dehydrogenase (GLUD) or malate regulation [76]. Overall, glutamine is of vital importance to GSH de novo synthesis to modulate the oxidative stress level in ccRCC.

Cysteine, although a nonessential amino acid, plays an important role in protein synthesis by forming intraprotein disulfide bonds to stabilize proteins. There are multiple cellular pathways and transporters, which contribute to the availability of cysteine in cells. Aside from the γ-glutamyl cycle, the trans-sulfuration pathway serves as an important source for cysteine recruitment, which has been reported to be dysregulated in ccRCC [34,36]. Thereby, methionine is converted to S-adenosyl methionine (SAM), is hydrolyzed to homocysteine, and can then enter the trans-sulfuration pathway, where it gets converted into cystathionine, which forms cysteine in subsequent reactions [77]. Metabolome profiling of ccRCC tumors has identified elevated SAM, S-adenosyl homocysteine (SAH), and homocysteine in ccRCC compared to healthy kidney tissues [34,36], indicating a high demand for cysteine, which is synthesized through the trans-sulfuration pathway. Cystine, the oxidized dimer of cysteine, can also be absorbed from the tumor microenvironment by xCT (*SLC7A11*), which is a heterodimeric cystine–glutamate antiporter [78], where overexpression is associated with overall poor survival in ccRCC [26].

3.3. Increased Flux of the Pentose Phosphate Pathway in ccRCC to Support GSH Synthesis

The pentose phosphate pathway (PPP) is, in part, a metabolic pathway parallel to glycolysis. It generates NADPH and pentoses, including ribose 5-phosphate (R5P). Its primary role is considered to be anabolic rather than catabolic, as it provides R5P, a precursor for the synthesis of nucleotides. NADPH is an important cofactor for the enzyme GSH reductase (GR) to catalyze the reduction of GSSG to GSH and hence links PPP directly to GSH synthesis. Therefore, high GSH/GSSG ratios in ccRCC can also be explained by increased flux through PPP, which provides the necessary molecules of NADPH for GSH conversion [26,34,79]. Indeed, glucose-6-phosphate dehydrogenase (G6PD), which determines the production of NADPH and R5P within PPP, was found to be increased in ccRCC, and its elevation was associated with higher levels of NADPH and PPP-derived metabolites [79]. Furthermore, the inhibition of G6PD in chRCC cells decreased the NADPH level and increased ROS production to significantly impair cancer cell survival, suggesting that PPP plays a fundamental role in the regulation of redox homeostasis and progression in ccRCC [79]. Recently, fructose 1,6-bisphosphate (FBP), a glycolytic intermediate, was found to be accumulated in ccRCC, leading to the suppression of NADPH oxidase 4 (NOX4), which caused an increase in NADPH and a decrease in ROS, independent of PPP [80]. These changes were caused by the downregulation of aldolase B (ALDOB), which portended significantly worse survival in ccRCC patients [80].

3.4. Fumarate Hydratase Mutations and GSH in Type II pRCC

Fumarate hydratase (*FH*) is frequently mutated in type II pRCC, causing FH deficiency and affecting the normal flux of the TCA cycle [42,43]. The FH deficiency of this tumor leads to metabolic reprogramming, including impaired oxidative phosphorylation and aerobic glycolysis, known as the "Warburg effect", which is probably caused by the stabilization of HIF through fumarate accumulation [44,45]. A further study showed that the accumulation of fumarate was fueled by glutamine rather than glucose in type II pRCC cells [81]. Apart from causing HIF stabilization, fumarate accumulation has been reported to activate NRF2 and its downstream ARE pathway [82,83]. Moreover, somatic mutations of *NRF2* and its regulator Kelch-like ECH-associated protein 1 (*KEAP1*) have been reported to be highly correlated with poor prognosis in type II pRCC and pancreatic cancer [82,84,85]. NRF2 plays an important role in cellular redox balance as a transcription factor that regulates the expression of various genes to combat the harmful effects of extrinsic and intrinsic damage, such as xenobiotics and oxidative stress. NRF2 is primarily regulated by KEAP1, a substrate adapter protein of Cullin 3 (CUL3), which contains an E3 ubiquitin ligase activity. Under normal conditions, NRF2 builds a complex with KEAP1 via its Kelch domain for ubiquitination, and NRF2 is then targeted for subsequent proteasomal degradation [84,86,87]. However, in response to a diverse array of stimuli, such as oxidative stress, the cysteine residues within KEAP1, Cys151, Cys273, and Cys288 can be modified, which results in a conformational change along with the dissociation of NRF2 to avoid KEAP1-mediated degradation [84,87]. Stabilized NRF2 can then translocate to the nucleus and bind to ARE to activate the downstream effector genes of at least two pathways involved in cytoprotection [84]. First, NRF2 can activate genes involved in regulating GSH synthesis and metabolism by activating GCL [85], and second, NRF2 promotes the expression of genes coding for antioxidant proteins, such as GSH peroxidases (GPXes) and GSH S-transferases (GSTs). GPX is an enzyme family with peroxidase activity that protects cells from oxidative damage, and GSTs are comprised of a family of isozymes that catalyze the conjugation of GSH to xenobiotic substrates for detoxification [78,84]. Additionally, fumarate can directly bind to GPX1 through interaction with the Thr143 and Asp144 residues, and fumarate accumulation is thus able to activate GPX1 and decrease the ROS level in cells [88].

Apart from mutations in *FH*, *NRF2*, *CUL3*, or *KEAP1* in type II pRCC, several other mechanisms can also lead to increased NRF2 activity in other cancers, including epigenetic silencing, modifications of cysteine residues, metabolic alterations, and oncogene-dependent signaling [85].

3.5. Glutathione Salvage Pathway in chRCC

One member of the membrane transpeptidase family GGT is γ-glutamyl transferase 1 (GGT1), which can remove and transfer the γ-glutamyl moiety from extracellular GSH, GSSG, or even GSH conjugates to an amino acid acceptor, known as the GSH salvage pathway. This degradation of extracellular GSH species fuels the cytoplasm of cells to maintain intracellular GSH levels [67]. Recent metabolomic profiling studies have identified significantly increased amounts of GSH, GSSG, and its precursor γ-glutamyl cysteine in chRCC compared to normal kidney tissue [57,60]. Unlike in ccRCC, significantly lower expression of GGT1 has been reported in chRCC [57,60]. The specific loss of GGT1 in chRCC leads to an increased sensitivity to oxidative stress, mitochondrial damage, and reprogramming of glutamine and glucose metabolism [60]. Interestingly, renal oncocytomas, which are considered to be the benign counterpart of chRCC, were found to have a similar increase of GSH moieties and decreased levels of GGT1 relative to normal kidney tissue [61].

4. Therapeutic Strategies to Exploit Increased GSH Levels in RCC

Nonmetastatic primary RCC can be removed by partial or complete nephrectomy. Metastases that occur in about one-third of all RCC patients must be treated with various therapeutic agents [89]. One severe problem is that the malignancies gain a fast treatment resistance through the activation of alternative metabolic pathways, or parts of the cancer cells that are not responsive outgrow the responsive tumor cells. For example, angiopoietin 2, MET, or Interleukin (IL) can serve as alternative angiogenesis factors, or the AKT/PI3K/mTOR pathway can stimulate proliferation upon its activation [90,91]. Depletion of GSH alone has been shown to be insufficient to induce cell death in most cancer cell lines. The imposed selective pressure during cancer initiation and progression led to a robust adaption mechanism to tolerate these stress conditions [92]. To overcome these limitations, combinatory therapies targeting two independent mechanisms are promising. The following subchapters highlight current strategies that manipulate GSH metabolism and intend to eradicate RCC.

4.1. The Cystine–Glutamate Shuttle Inhibitor

xCT is a cystine–glutamate antiporter that is essential to the uptake of cystine (Figure 1). After the conversion of cystine into cysteine, it serves as a building block for the synthesis of intracellular GSH, as discussed before. xCT is upregulated in a variety of cancers, where the antiporter-assisted production of GSH reduces oxidative stress levels to protect cancer from apoptosis [93]. Pharmacological inhibition of xCT decreases cystine uptake and induces ferroptosis in cancer cells [94], which makes xCT inhibitors potential treatment agents for cancer.

Sorafenib, an FDA-approved kinase inhibitor drug used for almost 15 years for the treatment of RCC, has multiple kinase inhibition activities, including cell surface tyrosine kinases (e.g., vascular endothelial growth factor receptor, VEGFR; platelet-derived growth factor receptor, PDGFR; tyrosine-protein kinase kit, KIT; Fms-like tyrosine kinase 3, FLT3; RET proto-oncogene, RET) and downstream intracellular serine/threonine kinases (e.g., both wild-type and mutant BRAF and CRAF) [95]. As these kinases play important roles in cancer cell proliferation, angiogenesis, and apoptosis, sorafenib has been shown to inhibit the proliferation of cancer cells and induce apoptosis in vitro, as well as reduce angiogenesis and inhibit tumor growth in vivo [95]. Recently, sorafenib, but not other kinase inhibitors of the same class, has been reported by several studies to have novel inhibition activity versus xCT, leading to decreased cysteine uptake, GSH depletion, and ROS accumulation, finally causing endoplasmic reticulum stress and ferroptosis [94,96,97].

Besides sorafenib, there are two other xCT inhibitors worth discussing in more detail, erastin and sulfasalazine. Erastin is a small molecule that inhibits xCT activity through the mitochondrial voltage-dependent anion channel 2 and 3 (VDAC2 and VDAC3), causing abolition of the antioxidant defenses of the cell, and it furthermore has selectively lethal activity toward oncogenic *RAS* mutant cell lines [98,99]. A cystine addiction of VHL-deficient RCC cells was identified, and the deprivation thereof or treatment with erastin or sulfasalazine in RCC cells induced cell death [100]. Chromophobe RCC, but not pRCC, was found to have significantly increased abundances of the VDAC1, VDAC2, and VDAC3 proteins [46,57] and should have a good response to treatment with erastin. Sulfasalazine has been in use for over 50 years for the treatment of inflammatory conditions such as arthritis. It is a well-characterized specific inhibitor of xCT and shows anticancer effects on multiple types of cancers, including RCC [94,98,100]. The safety and side effects of sulfasalazine are well investigated and understood, and this old drug has the potential to be a novel, effective, and economical treatment option for RCC patients.

4.2. Glutaminase 1 Inhibitor

Glutaminase 1 (GLS1), a key mitochondrial enzyme that controls glutamine metabolism and contributes to de novo GSH synthesis (Figure 1), is very important for tumor proliferation and survival. The glutaminase inhibition of glutamine-addicted cancer cells leads to the disruption of metabolic pathways, such as macromolecule synthesis, ATP production, and the intracellular redox balance [101]. Thus, targeting glutaminase to disrupt vital metabolic pathways of tumors is considered to be a novel strategy to treat cancer.

CB-839, a potent, selective, and orally bioavailable GLS1 inhibitor, has been reported to exhibit significant antiproliferative activity in multiple cancer cell lines and has shown an antitumor effect in tumor xenografts and cancer patients [102]. CB-839 is currently being investigated in multiple phase 1 and 2 clinical trials for patients with locally advanced, metastatic, and/or refractory solid tumors, including ccRCC [102]. Emberley et al. [102] reported a cytotoxic effect of CB-839 in 18 out of 23 tested RCC cell lines and 0 out of 6 non-RCC cell lines. MacKinnon et al. [103] found that the abundance of pyruvate carboxylase (PC), which catalyzes the conversion of pyruvate to oxaloacetate to fuel the TCA cycle, strongly correlated with resistance, and knockdown of PC reduced TCA cycle activity and sensitized cells to CB-839 treatment, suggesting that PC expression may be a biomarker of resistance to CB-839. Chromophobe RCC and pRCC have been reported recently to have low PC expression [46,57]: They may lack this mechanism and therefore would probably be sensitive to

CB-839. In addition, decreases in mTOR signaling were also observed in RCC cell lines that were sensitive to CB-839, indicating that CB-839-induced glutamate deprivation has a direct influence on the mTOR pathway [102]. These observations suggest that receptor tyrosine kinase (RTK) signaling or mTOR inhibitors would have synergistic effects with CB-839 to increase cytotoxicity in RCC cell lines. The combined CB-839 and cabozantinib (RTK inhibitor) therapy, which reached phase 2 clinical evaluation (CANTATA: NCT03428217), showed pronounced reductions in TCA cycle activity and in signaling via AKT and ERK compared to single-agent treatments. When applying CB-839 in combination with everolimus (mTOR inhibitor) to RCC cell lines in vitro and to a Caki-1 RCC xenograft model in vivo, synergistic antitumor activity and inhibition of both glucose and glutamine utilization were observed [104]. Furthermore, the combined therapy of CB-839 with everolimus in a phase 1 clinical trial showed a 100% disease control rate (DCR) in ccRCC and 67% in pRCC [104], and it is currently in a phase 2 investigation in patients with advanced ccRCC (ENTRATA: NCT03163667). Moreover, a phase 1/2 study of CB-839 in combination with nivolumab (anti-PD-1 antibody) is currently ongoing (NCT02771626) and has shown a 74% DCR in ccRCC patients. Furthermore, CB-839 was found to have a synergistic effect in selectively suppressing the growth of ccRCC cells in vitro and in vivo when combined with poly(ADP-ribose) polymerase (PARP) inhibitors [105]. Currently, a phase 1b/2 clinical trial of CB-839 in combination with talazoparib (PARP inhibitor) is under investigation (NCT03875313).

At present, CB-839 is the only small-molecule GLS1 inhibitor being studied in a clinical setting, but there are other GLS1 inhibitors in preclinical investigations, including bis-2-(5-phenylacetamido-1,2,4-thiadiazol-2-yl)ethyl sulfide (BPTES), 6-diazo-5-oxo-L-norleucine (DON), and 5-[3-bromo-4-(dimethylamino)phenyl]-2,3,5,6-tetrahydro-2,2-dimethyl-benzo[a]phenanth-ridin-4(1H)-one (968). Two of these GLS1 inhibitors have not been further applied and investigated in clinical studies due to the low solubility and potency of BPTES and the high toxicity and poor binding selectivity of DON [106]. However, 968 is known to be a noncompetitive inhibitor of GLS1 and is currently still in the preclinical stage [106].

4.3. The Glutamate–Cysteine Ligase Inhibitor Buthionine Sulfoximine

GCL, the enzyme catalyzing the first reaction in GSH de novo synthesis (Figure 1), plays an important function in maintaining intracellular GSH levels to combat oxidative stress in RCC [65,66]. Thus, targeting GCL for the treatment of RCC remains a potentially effective strategy to benefit patients.

Buthionine sulfoximine (BSO) is a specific and competitive inhibitor of GCL [107]. Developed in 1979 by Griffith and Meister [108], BSO is able to decrease the intracellular GSH level and sensitize different types of cancers both in vitro and in vivo to various chemotherapies and other cytotoxic therapies, e.g., irradiation and hyperthermia [107,109]. BSO has been reported to enhance the activity of melphalan, doxorubicin, daunorubicin, and other cytotoxic agents in myeloma, breast cancer, and lung cancer [107,109]. Two clinical trials of BSO in combination with melphalan were conducted for the treatment of neuroblastoma. In a pilot study and a phase I clinical trial, BSO in combination with melphalan was well tolerated and had therapeutic activity toward recurrent and refractory high-risk neuroblastoma (NCT00002730, NCT00005835) [110,111]. Sorafenib, as discussed earlier, has xCT and multiple kinase inhibition activity. However, it has been reported that some RCC patients were initially resistant or acquired resistance to sorafenib within a median of 5–9 months [112]. Mechanistic studies have shown that the resistance to sorafenib in RCC was mediated by enhanced expression of *HIF* and numerous *HIF*-regulated genes, such as *VEGF* [112,113]. As the redox state could regulate *HIF* expression and downstream substrates to cause drug resistance [29,114], BSO, which can regulate the redox environment through GSH, was shown to be able to decrease the expression of *HIF* [114–116]. These studies indicate that combination therapies with BSO and sorafenib might overcome the drug resistance of sorafenib in resistant RCC patients.

4.4. Inhibition of Deubiquitinating Enzymes Initiates Proteotoxicity

A recent report outlined a combinatory treatment strategy applying the before-mentioned GCL inhibitor BSO together with deubiquitinating enzyme (DUB) inhibitors [92]. Individual DUB inhibitors, such as MI-2, PR-619, and EERI, were not effective in inducing cell death alone but led to an induction of proteotoxic stress and cell death in combination with BSO in many different cancer cell lines. Though the exact molecular role of how DUBs can protect cells from oxidative stress is still elusive, a dependency on DUB activity to maintain protein homeostasis by eliminating the accumulation of damaged and potentially cytotoxic polyubiquitinated proteins and cell viability have been proposed [92]. This hypothesis was supported by a study profiling the ubiquitination status between ρ^0 cells, which entirely lack mitochondrial DNA, and their parent cell line 143B.TK$^-$. The significant decrease in the global ubiquitination pattern in ρ^0 cells can be explained by the lack of main ROS generators localized within the oxidative phosphorylation system, namely complex I and III, which reduce the oxidative damage of proteins to a minimal level [47].

This combination of DUB and GSH inhibitors has not been applied to RCC cell lines yet but would present a valuable new strategy to trigger proteotoxic stress as a potential beneficial treatment in RCC cell lines and animal models.

4.5. The Role of GSH Metabolism in the Immune Microenvironment of the Tumor

Traditional immunotherapies using interleukin-2 or interferon-alfa on metastatic RCC have presented limited efficacy and highly toxic side effects [117,118]. In recent years, a new generation of immunotherapy utilizing a novel strategy to block immune checkpoints has shown promising efficacy and manageable toxicity and has emerged as a new milestone for RCC treatment [119]. Currently, the approved immune checkpoint inhibitors for RCC are ipilimumab (a cytotoxic T lymphocyte-associated protein 4 [CTLA-4] inhibitor) [120], the programmed cell death 1 (PD-1)-specific antibodies nivolumab [121] and pembrolizumab, and the programmed death-ligand 1 (PD-L1) antibody avelumab.

A main metabolic feature of RCC is the reprogramming of the main metabolic pathways, which helps cancer cells adapt to and simultaneously shape the tumor microenvironment. Immune cells utilize different metabolic programs for their differentiation and effective functions. These immune–metabolic pathways can be modified or "highjacked" in the tumor microenvironment and thus affect the normal functions of immune cells, e.g., by infiltrating tumor tissues and presenting tumor-associated antigens to T-cells [122]. Therefore, a huge effort has been put into the development of immunotherapies to take advantage of the complex crosstalk between immune cells and the tumor [123].

How does GSH metabolism influence this crosstalk in RCC? Though there is sparse literature on this topic, it has been shown that GSH metabolism can influence the immune microenvironment in cancer at least in the following two aspects. First, glutamine is a crucial nutrient for the effector function of T-cells. Glutamine deprivation or its transporter deficiency blocks the differentiation of T-helper 1 and 17 cells [124,125]. Similarly, the proliferation and differentiation of B-cells also requires glutamine [126]. As glutamine addiction is one of the main features of RCC, glutamine may be a limiting nutrient factor in the tumor microenvironment, and thus a lack of glutamine can induce immunosuppression. Second, T-cell stimulation activates the cystine–glutamate antiporter xCT and leads to increased uptake of cystine and subsequent GSH synthesis [122]. Reduced GSH levels in antigen-presenting cells have been shown to influence antigen processing and presentation as well as T-cell differentiation into T-helper 1 or 2 phenotypes [127]. Furthermore, GSH can bind to anticancer drugs, and these conjugates can be effluxed out of the cell via multiple resistance-associated protein transporters, which are the underlying reasons for therapeutic resistance in some cancers [128]. All the points discussed above impressively show how tumor cells compete with immune cells for GSH-related nutrients in the tumor microenvironment, but more research on the immune response in RCC is needed to exploit these mechanisms for new therapy development.

5. Conclusions

Altered GSH metabolism contributes significantly to the development and progression of all renal malignancies but could, at the same time, be the key to potential therapies. All RCCs have a reduced oxidative phosphorylation capacity in common. The dysregulated respiratory chain is the main source of electron leakage, resulting in excessive ROS. Raised oxidative stress levels in RCC are counteracted by tremendously increased GSH levels and thereby potentially prevent immune reactions, apoptosis, or other forms of cell death as a strategy to foster the survival of the malignancy. Many applied chemotherapeutics initiate an additional production of ROS as one potent mechanism to eradicate RCC, frequently accompanied by the supplementation of antioxidants in the past. Not surprisingly, targeting ROS by antioxidants and the simultaneous generation of ROS by chemotherapeutics has led to mixed results in the treatability of RCC [14]. New therapeutic ways exploit sensitivity toward inhibitors of the GSH metabolism, such as xCT, glutaminase, and GCL. However, the inhibition of just one altered pathway to cure RCC turned out to be not successful either, as cells are fitted by a very flexible system to compensate for the impairment of one pathway or mechanism. New studies have shown that a combinatory therapy targeting two independent pathways and one involved in ROS metabolism is key to improving the survival rate and eventually curing RCC. Although the main role of GSH and other antioxidants is to scavenge intracellular ROS to maintain an overall healthy pro- and antioxidant exposure status in cells, GSH can also function as a signaling molecule or as a donor of the post-translational modification S-glutathionylation to regulate protein bioactivity, which was not discussed in this review. These diverse functions of GSH, a molecule that was identified more than 100 years ago, still need to be further investigated for better understanding of the underlying disease mechanisms in cancer. This might facilitate the development of GSH-related modulators with improved therapeutic efficiencies in the future.

Author Contributions: Both authors contributed equally.

Funding: This research was funded by the Max Planck Society and the China Scholarship Council and the APC was funded by the Max Planck Society.

Conflicts of Interest: The authors declare no conflicts of interest.

Abbreviations

968	5-[3-bromo-4-(dimethylamino)phenyl]-2,3,5,6-tetrahydro-2,2-dimethyl-benzo[A]phenanthridin-4(1H)-one
ALDOB	Aldolase B
ARE	Antioxidant response element
BPTES	Bis-2-(5-phenylacetamido-1,2,4-thiadiazol-2-Yl)ethyl sulfide
BRAF	Proto-oncogene B-RAF
BSO	Buthionine sulfoximine
ccRCC	Clear cell renal cell carcinoma
CDKN2A	Cyclin-dependent kinase inhibitor 2A
CDKN2B	Cyclin-dependent kinase inhibitor 2B
chRCC	Chromophobe renal cell carcinoma
CRAF	Proto-oncogene C-RAF
CTLA-4	Cytotoxic T lymphocyte-associated protein 4
CUL3	Cullin 3
DCR	Disease control rate
DON	6-diazo-5-oxo-L-norleucine
DUB	Deubiquitinating enzyme
EGFR	Epidermal growth factor receptor
EGLN	Egl-nine homolog
FAK	Focal adhesion kinase
FBP	Fructose 1,6-bisphosphate
FBP1	Fructose 1,6-bisphosphatase 1

FH	Fumarate hydratase
FLT3	Fms-like tyrosine kinase 3
G6PD	Glucose-6-phosphate dehydrogenase
GCL	Glutamate cysteine ligase
GCLC	Glutamate cysteine ligase catalytic subunit
GCLM	Glutamate cysteine ligase modulatory subunit
GGT	γ-glutamyl transferase
GGT1	γ-glutamyl transferase 1
GLS1	Glutaminase 1
GLS2	Glutaminase 2
GLUD	Glutamate dehydrogenase
GLUT1	Glucose transporter type 1
GPX	Glutathione peroxidase
GR	Glutathione reductase
GSH	Glutathione reduced form
GSS	Glutathione synthetase
GSSG	Glutathione oxidized form
GST	Glutathione S-transferase
HGF	Hepatocyte growth factor
HIF	Hypoxia inducible factor
KEAP1	Kelch-like ECH-associated protein 1
KIT	Tyrosine-protein kinase kit
MET	Proto-oncogene c-Met
MYC	MYC proto-oncogene
NADPH	Nicotinamide adenine dinucleotide phosphate
NOX4	NADPH oxidase 4
NRAS	NRAS proto-oncogene, GTPase
NRF2	Nuclear factor erythroid 2-related factor 2
ODD	Oxygen-dependent degradation
PC	Pyruvate carboxylase
PD-1	Programmed cell death 1
PD-L1	Programmed cell death-ligand 1
PDGFR	Platelet-derived growth factor receptor
PPP	Pentose phosphate pathway
pRCC	Papillary renal cell carcinoma
R5P	Ribose 5-phosphate
RCC	Renal cell carcinoma
RET	RET proto-oncogene
ROS	Reactive oxygen species
RTK	Receptor tyrosine kinase
SAH	S-adenosyl homocysteine
SAM	S-adenosyl methionine
SETD2	Set domain-containing 2, histone lysine methyltransferase
SLC38A1	Solute carrier family 38 member 1
SLC7A11	Solute carrier family 7 member 11
TERT	Telomerase reverse transcriptase
TFE3	Transcription factor E3
TSC1	Tuberous sclerosis 1 protein
TSC2	Tuberous sclerosis 2 protein
VDAC1	Voltage-dependent anion channel 1
VDAC2	Voltage-dependent anion channel 2
VDAC3	Voltage-dependent anion channel 3
VEGF	Vascular endothelial growth factor
VEGFR	Vascular endothelial growth factor receptor
VHL	von Hippel-Lindau

References

1. Kim, J.; Kim, J.; Bae, J.S. ROS homeostasis and metabolism: A critical liaison for cancer therapy. *Exp. Mol. Med.* **2016**, *48*, e269. [CrossRef] [PubMed]
2. Meister, A. On the discovery of glutathione. *Trends. Biochem. Sci.* **1988**, *13*, 185–188. [CrossRef]
3. Kaplowitz, N.; Aw, T.Y.; Ookhtens, M. The regulation of hepatic glutathione. *Annu. Rev. Pharmacol. Toxicol.* **1985**, *25*, 715–744. [CrossRef] [PubMed]
4. Meister, A.; Anderson, M.E. Glutathione. *Annu. Rev. Biochem.* **1983**, *52*, 711–760. [CrossRef] [PubMed]
5. Lu, S.C. Regulation of glutathione synthesis. *Mol. Asp. Med.* **2009**, *30*, 42–59. [CrossRef] [PubMed]
6. Cho, E.S.; Johnson, N.; Snider, B.C. Tissue glutathione as a cyst(e)ine reservoir during cystine depletion in growing rats. *J. Nutr.* **1984**, *114*, 1853–1862. [CrossRef]
7. Sipos, K.; Lange, H.; Fekete, Z.; Ullmann, P.; Lill, R.; Kispal, G. Maturation of cytosolic iron-sulfur proteins requires glutathione. *J. Biol. Chem.* **2002**, *277*, 26944–26949. [CrossRef]
8. Awasthi, Y.C.; Misra, G.; Rassin, D.K.; Srivastava, S.K. Detoxification of xenobiotics by glutathione S-transferases in erythrocytes: The transport of the conjugate of glutathione and 1-chloro-2,4-dinitrobenzene. *Br. J. Haematol.* **1983**, *55*, 419–425. [CrossRef]
9. Duan, J.; Kodali, V.K.; Gaffrey, M.J.; Guo, J.; Chu, R.K.; Camp, D.G.; Smith, R.D.; Thrall, B.D.; Qian, W.J. Quantitative Profiling of Protein S-Glutathionylation Reveals Redox-Dependent Regulation of Macrophage Function during Nanoparticle-Induced Oxidative Stress. *ACS Nano* **2016**, *10*, 524–538. [CrossRef]
10. Zhang, X.; Liu, P.; Zhang, C.; Chiewchengchol, D.; Zhao, F.; Yu, H.; Li, J.; Kambara, H.; Luo, K.Y.; Venkataraman, A.; et al. Positive Regulation of Interleukin-1beta Bioactivity by Physiological ROS-Mediated Cysteine S-Glutathionylation. *Cell. Rep.* **2017**, *20*, 224–235. [CrossRef]
11. Ren, X.; Zou, L.; Zhang, X.; Branco, V.; Wang, J.; Carvalho, C.; Holmgren, A.; Lu, J. Redox Signaling Mediated by Thioredoxin and Glutathione Systems in the Central Nervous System. *Antioxid. Redox Signal.* **2017**, *27*, 989–1010. [CrossRef] [PubMed]
12. Hussain, S.P.; Hofseth, L.J.; Harris, C.C. Radical causes of cancer. *Nat. Rev. Cancer* **2003**, *3*, 276–285. [CrossRef] [PubMed]
13. Hatem, E.; El Banna, N.; Huang, M.E. Multifaceted Roles of Glutathione and Glutathione-Based Systems in Carcinogenesis and Anticancer Drug Resistance. *Antioxid. Redox Signal.* **2017**, *27*, 1217–1234. [CrossRef] [PubMed]
14. Thyagarajan, A.; Sahu, R.P. Potential Contributions of Antioxidants to Cancer Therapy: Immunomodulation and Radiosensitization. *Integr. Cancer* **2018**, *17*, 210–216. [CrossRef] [PubMed]
15. Siegel, R.L.; Miller, K.D.; Jemal, A. Cancer statistics, 2019. *CA. Cancer J. Clin.* **2019**, *69*, 7–34. [CrossRef] [PubMed]
16. Bray, F.; Ferlay, J.; Soerjomataram, I.; Siegel, R.L.; Torre, L.A.; Jemal, A. Global cancer statistics 2018: GLOBOCAN estimates of incidence and mortality worldwide for 36 cancers in 185 countries. *CA Cancer J. Clin.* **2018**, *68*, 394–424. [CrossRef] [PubMed]
17. Sato, Y.; Yoshizato, T.; Shiraishi, Y.; Maekawa, S.; Okuno, Y.; Kamura, T.; Shimamura, T.; Sato-Otsubo, A.; Nagae, G.; Suzuki, H.; et al. Integrated molecular analysis of clear-cell renal cell carcinoma. *Nat. Genet.* **2013**, *45*, 860–867. [CrossRef]
18. Cancer Genome Atlas Research Network; Linehan, W.M.; Spellman, P.T.; Ricketts, C.J.; Creighton, C.J.; Fei, S.S.; Davis, C.; Wheeler, D.A.; Murray, B.A.; Schmidt, L.; et al. Comprehensive Molecular Characterization of Papillary Renal-Cell Carcinoma. *N. Engl. J. Med.* **2016**, *374*, 135–145. [CrossRef]
19. Davis, C.F.; Ricketts, C.J.; Wang, M.; Yang, L.; Cherniack, A.D.; Shen, H.; Buhay, C.; Kang, H.; Kim, S.C.; Fahey, C.C.; et al. The somatic genomic landscape of chromophobe renal cell carcinoma. *Cancer Cell* **2014**, *26*, 319–330. [CrossRef]
20. Ricketts, C.J.; De Cubas, A.A.; Fan, H.; Smith, C.C.; Lang, M.; Reznik, E.; Bowlby, R.; Gibb, E.A.; Akbani, R.; Beroukhim, R.; et al. The Cancer Genome Atlas Comprehensive Molecular Characterization of Renal Cell Carcinoma. *Cell. Rep.* **2018**, *23*, 313–326 e5. [CrossRef]
21. Hoadley, K.A.; Yau, C.; Hinoue, T.; Wolf, D.M.; Lazar, A.J.; Drill, E.; Shen, R.; Taylor, A.M.; Cherniack, A.D.; Thorsson, V.; et al. Cell-of-Origin Patterns Dominate the Molecular Classification of 10,000 Tumors from 33 Types of Cancer. *Cell* **2018**, *173*, 291–304 e6. [CrossRef] [PubMed]

22. Moch, H.; Cubilla, A.L.; Humphrey, P.A.; Reuter, V.E.; Ulbright, T.M. The 2016 WHO Classification of Tumours of the Urinary System and Male Genital Organs-Part A: Renal, Penile, and Testicular Tumours. *Eur. Urol.* **2016**, *70*, 93–105. [CrossRef] [PubMed]

23. Chin, A.I.; Lam, J.S.; Figlin, R.A.; Belldegrun, A.S. Surveillance strategies for renal cell carcinoma patients following nephrectomy. *Rev. Urol.* **2006**, *8*, 1–7. [PubMed]

24. Motzer, R.J.; Bacik, J.; Mazumdar, M. Prognostic factors for survival of patients with stage IV renal cell carcinoma: Memorial sloan-kettering cancer center experience. *Clin. Cancer Res.* **2004**, *10*, 6302S–6303S. [CrossRef] [PubMed]

25. Delahunt, B.; Srigley, J.R.; Montironi, R.; Egevad, L. Advances in renal neoplasia: Recommendations from the 2012 International Society of Urological Pathology Consensus Conference. *Urology* **2014**, *83*, 969–974. [CrossRef] [PubMed]

26. Cancer Genome Atlas Research Network. Comprehensive molecular characterization of clear cell renal cell carcinoma. *Nature* **2013**, *499*, 43–49. [CrossRef] [PubMed]

27. Latif, F.; Tory, K.; Gnarra, J.; Yao, M.; Duh, F.M.; Orcutt, M.L.; Stackhouse, T.; Kuzmin, I.; Modi, W.; Geil, L.; et al. Identification of the von Hippel-Lindau disease tumor suppressor gene. *Science* **1993**, *260*, 1317–1320. [CrossRef]

28. Jaakkola, P.; Mole, D.R.; Tian, Y.M.; Wilson, M.I.; Gielbert, J.; Gaskell, S.J.; von Kriegsheim, A.; Hebestreit, H.F.; Mukherji, M.; Schofield, C.J.; et al. Targeting of HIF-alpha to the von Hippel-Lindau ubiquitylation complex by O2-regulated prolyl hydroxylation. *Science* **2001**, *292*, 468–472. [CrossRef]

29. Ivan, M.; Kondo, K.; Yang, H.; Kim, W.; Valiando, J.; Ohh, M.; Salic, A.; Asara, J.M.; Lane, W.S.; Kaelin, W.G., Jr. HIFalpha targeted for VHL-mediated destruction by proline hydroxylation: Implications for O_2 sensing. *Science* **2001**, *292*, 464–468. [CrossRef]

30. Gossage, L.; Eisen, T.; Maher, E.R. VHL, the story of a tumour suppressor gene. *Nat. Rev. Cancer* **2015**, *15*, 55–64. [CrossRef]

31. Riazalhosseini, Y.; Lathrop, M. Precision medicine from the renal cancer genome. *Nat. Rev. Nephrol.* **2016**, *12*, 655–666. [CrossRef] [PubMed]

32. Wettersten, H.I.; Aboud, O.A.; Lara, P.N., Jr.; Weiss, R.H. Metabolic reprogramming in clear cell renal cell carcinoma. *Nat. Rev. Nephrol.* **2017**, *13*, 410–419. [CrossRef] [PubMed]

33. Posadas, E.M.; Limvorasak, S.; Figlin, R.A. Targeted therapies for renal cell carcinoma. *Nat. Rev. Nephrol.* **2017**, *13*, 496–511. [CrossRef] [PubMed]

34. Li, B.; Qiu, B.; Lee, D.S.; Walton, Z.E.; Ochocki, J.D.; Mathew, L.K.; Mancuso, A.; Gade, T.P.; Keith, B.; Nissim, I.; et al. Fructose-1,6-bisphosphatase opposes renal carcinoma progression. *Nature* **2014**, *513*, 251–255. [CrossRef] [PubMed]

35. Guo, T.; Kouvonen, P.; Koh, C.C.; Gillet, L.C.; Wolski, W.E.; Rost, H.L.; Rosenberger, G.; Collins, B.C.; Blum, L.C.; Gillessen, S.; et al. Rapid mass spectrometric conversion of tissue biopsy samples into permanent quantitative digital proteome maps. *Nat. Med.* **2015**, *21*, 407–413. [CrossRef]

36. Hakimi, A.A.; Reznik, E.; Lee, C.H.; Creighton, C.J.; Brannon, A.R.; Luna, A.; Aksoy, B.A.; Liu, E.M.; Shen, R.; Lee, W.; et al. An Integrated Metabolic Atlas of Clear Cell Renal Cell Carcinoma. *Cancer Cell* **2016**, *29*, 104–116. [CrossRef]

37. Steffens, S.; Janssen, M.; Roos, F.C.; Becker, F.; Schumacher, S.; Seidel, C.; Wegener, G.; Thüroff, J.W.; Hofmann, R.; Stöckle, M.; et al. Incidence and long-term prognosis of papillary compared to clear cell renal cell carcinoma—A multicentre study. *Eur. J. Cancer* **2012**, *48*, 2347–2352. [CrossRef]

38. Delahunt, B.; Eble, J.N. Papillary renal cell carcinoma: A clinicopathologic and immunohistochemical study of 105 tumors. *Mod. Pathol.* **1997**, *10*, 537–544.

39. Pignot, G.; Elie, C.; Conquy, S.; Vieillefond, A.; Flam, T.; Zerbib, M.; Debre, B.; Amsellem-Ouazana, D. Survival analysis of 130 patients with papillary renal cell carcinoma: Prognostic utility of type 1 and type 2 subclassification. *Urology* **2007**, *69*, 230–235. [CrossRef]

40. Pal, S.K.; Ali, S.M.; Yakirevich, E.; Geynisman, D.M.; Karam, J.A.; Elvin, J.A.; Frampton, G.M.; Huang, X.; Lin, D.I.; Rosenzweig, M.; et al. Characterization of Clinical Cases of Advanced Papillary Renal Cell Carcinoma via Comprehensive Genomic Profiling. *Eur. Urol.* **2018**, *73*, 71–78. [CrossRef]

41. Fay, A.P.; Signoretti, S.; Choueiri, T.K. MET as a target in papillary renal cell carcinoma. *Clin. Cancer Res.* **2014**, *20*, 3361–3363. [CrossRef]

42. Li, S.; Shuch, B.M.; Gerstein, M.B. Whole-genome analysis of papillary kidney cancer finds significant noncoding alterations. *PLoS Genet.* **2017**, *13*, e1006685. [CrossRef]

43. Tomlinson, I.; Alam, N.; Rowan, A.; Barclay, E.; Jaeger, E.; Kelsell, D.; Leigh, I.; Gorman, P.; Lamlum, H.; Rahman, S. Multiple Leiomyoma Consortium: Germline mutations in FH predispose to dominantly inherited uterine fibroids, skin leiomyomata and papillary renal cell cancer. *Nat. Genet.* **2002**, *30*, 406–410.

44. Pollard, P.J.; Briere, J.J.; Alam, N.A.; Barwell, J.; Barclay, E.; Wortham, N.C.; Hunt, T.; Mitchell, M.; Olpin, S.; Moat, S.J.; et al. Accumulation of Krebs cycle intermediates and over-expression of HIF1alpha in tumours which result from germline FH and SDH mutations. *Hum. Mol. Genet.* **2005**, *14*, 2231–2239. [CrossRef]

45. Sullivan, L.B.; Martinez-Garcia, E.; Nguyen, H.; Mullen, A.R.; Dufour, E.; Sudarshan, S.; Licht, J.D.; Deberardinis, R.J.; Chandel, N.S. The proto-oncometabolite fumarate binds glutathione to amplify ROS-dependent signaling. *Mol. Cell* **2013**, *51*, 236–248. [CrossRef]

46. Alahmad, A.; Paffrath, V.; Clima, R.; Busch, J.F.; Rabien, A.; Kilic, E.; Villegas, S.; Timmermann, B.; Attimonelli, M.; Jung, K.; et al. Papillary renal cell carcinomas rewire glutathione metabolism and are deficient in anabolic glucose synthesis. *bioRxiv* **2019**, 651265.

47. Aretz, I.; Hardt, C.; Wittig, I.; Meierhofer, D. An Impaired Respiratory Electron Chain Triggers Down-regulation of the Energy Metabolism and De-ubiquitination of Solute Carrier Amino Acid Transporters. *Mol. Cell. Proteom. MCP* **2016**, *15*, 1526–1538. [CrossRef]

48. Hu, H.; Nan, J.; Sun, Y.; Zhu, D.; Xiao, C.; Wang, Y.; Zhu, L.; Wu, Y.; Zhao, J.; Wu, R.; et al. Electron leak from NDUFA13 within mitochondrial complex I attenuates ischemia-reperfusion injury via dimerized STAT3. *Proc. Natl. Acad. Sci. USA* **2017**, *114*, 11908–11913. [CrossRef]

49. Jastroch, M.; Divakaruni, A.S.; Mookerjee, S.; Treberg, J.R.; Brand, M.D. Mitochondrial proton and electron leaks. *Essays Biochem.* **2010**, *47*, 53–67. [CrossRef]

50. Hirst, J.; Roessler, M.M. Energy conversion, redox catalysis and generation of reactive oxygen species by respiratory complex I. *Biochim. Biophys. Acta* **2016**, *1857*, 872–883. [CrossRef]

51. Reichart, G.; Mayer, J.; Zehm, C.; Kirschstein, T.; Tokay, T.; Lange, F.; Baltrusch, S.; Tiedge, M.; Fuellen, G.; Ibrahim, S.; et al. Mitochondrial complex IV mutation increases reactive oxygen species production and reduces lifespan in aged mice. *Acta Physiol. (Oxf.)* **2019**, *225*, e13214. [CrossRef]

52. Thoenes, W.; Storkel, S.; Rumpelt, H.J. Human chromophobe cell renal carcinoma. *Virchows Arch. B Cell Pathol. Incl. Mol. Pathol.* **1985**, *48*, 207–217. [CrossRef]

53. Yusenko, M.V. Molecular pathology of chromophobe renal cell carcinoma: A review. *Int. J. Urol.* **2010**, *17*, 592–600. [CrossRef]

54. Haake, S.M.; Weyandt, J.D.; Rathmell, W.K. Insights into the Genetic Basis of the Renal Cell Carcinomas from The Cancer Genome Atlas. *Mol. Cancer Res.* **2016**, *14*, 589–598. [CrossRef]

55. Brunelli, M.; Eble, J.N.; Zhang, S.; Martignoni, G.; Delahunt, B.; Cheng, L. Eosinophilic and classic chromophobe renal cell carcinomas have similar frequent losses of multiple chromosomes from among chromosomes 1, 2, 6, 10, and 17, and this pattern of genetic abnormality is not present in renal oncocytoma. *Mod. Pathol. Off. J. United States Can. Acad. Pathol. Inc.* **2005**, *18*, 161–169. [CrossRef]

56. Casuscelli, J.; Weinhold, N.; Gundem, G.; Wang, L.; Zabor, E.C.; Drill, E.; Wang, P.I.; Nanjangud, G.J.; Redzematovic, A.; Nargund, A.M.; et al. Genomic landscape and evolution of metastatic chromophobe renal cell carcinoma. *JCI Insight* **2017**, *2*. [CrossRef]

57. Xiao, Y.; Clima, R.; Busch, J.F.; Rabien, A.; Kilic, E.; Villegas, S.; Türkmen, S.; Timmermann, B.; Attimonelli, M.; Jung, K.; et al. Metabolic reprogramming and elevation of glutathione in chromophobe renal cell carcinomas. *bioRxiv.* **2019**, 649046.

58. Jinzaki, M.; Tanimoto, A.; Mukai, M.; Ikeda, E.; Kobayashi, S.; Yuasa, Y.; Narimatsu, Y.; Murai, M. Double-phase helical CT of small renal parenchymal neoplasms: Correlation with pathologic findings and tumor angiogenesis. *J. Comput. Assist. Tomogr.* **2000**, *24*, 835–842. [CrossRef]

59. Nakajima, R.; Nozaki, S.; Kondo, T.; Nagashima, Y.; Abe, K.; Sakai, S. Evaluation of renal cell carcinoma histological subtype and fuhrman grade using (18)F-fluorodeoxyglucose-positron emission tomography/computed tomography. *Eur. Radiol.* **2017**, *27*, 4866–4873. [CrossRef]

60. Priolo, C.; Khabibullin, D.; Reznik, E.; Filippakis, H.; Ogorek, B.; Kavanagh, T.R.; Nijmeh, J.; Herbert, Z.T.; Asara, J.M.; Kwiatkowski, D.J.; et al. Impairment of gamma-glutamyl transferase 1 activity in the metabolic pathogenesis of chromophobe renal cell carcinoma. *Proc. Natl. Acad. Sci. USA* **2018**, *115*, E6274–E6282. [CrossRef]

61. Kurschner, G.; Zhang, Q.; Clima, R.; Xiao, Y.; Busch, J.F.; Kilic, E.; Jung, K.; Berndt, N.; Bulik, S.; Holzhutter, H.G.; et al. Renal oncocytoma characterized by the defective complex I of the respiratory chain boosts the synthesis of the ROS scavenger glutathione. *Oncotarget* **2017**, *8*, 105882–105904. [CrossRef]

62. Gopal, R.K.; Calvo, S.E.; Shih, A.R.; Chaves, F.L.; McGuone, D.; Mick, E.; Pierce, K.A.; Li, Y.; Garofalo, A.; Van Allen, E.M.; et al. Early loss of mitochondrial complex I and rewiring of glutathione metabolism in renal oncocytoma. *Proc. Natl. Acad. Sci. USA* **2018**, *115*, E6283–E6290. [CrossRef]

63. Orlowski, M.; Meister, A. The gamma-glutamyl cycle: A possible transport system for amino acids. *Proc. Natl. Acad. Sci. USA* **1970**, *67*, 1248–1255. [CrossRef]

64. Franklin, C.C.; Backos, D.S.; Mohar, I.; White, C.C.; Forman, H.J.; Kavanagh, T.J. Structure, function, and post-translational regulation of the catalytic and modifier subunits of glutamate cysteine ligase. *Mol. Asp. Med.* **2009**, *30*, 86–98. [CrossRef]

65. Li, M.; Zhang, Z.; Yuan, J.; Zhang, Y.; Jin, X. Altered glutamate cysteine ligase expression and activity in renal cell carcinoma. *Biomed. Rep.* **2014**, *2*, 831–834. [CrossRef]

66. Miess, H.; Dankworth, B.; Gouw, A.M.; Rosenfeldt, M.; Schmitz, W.; Jiang, M.; Saunders, B.; Howell, M.; Downward, J.; Felsher, D.W.; et al. The glutathione redox system is essential to prevent ferroptosis caused by impaired lipid metabolism in clear cell renal cell carcinoma. *Oncogene* **2018**, *37*, 5435–5450. [CrossRef]

67. Terzyan, S.S.; Burgett, A.W.; Heroux, A.; Smith, C.A.; Mooers, B.H.; Hanigan, M.H. Human Gamma-Glutamyl Transpeptidase 1: Structures Of The Free Enzyme, Inhibitor-Bound Tetrahedral Transition States, And Glutamate-Bound Enzyme Reveal Novel Movement Within The Active Site During Catalysis. *J. Biol. Chem.* **2015**, *290*, 17576–17586. [CrossRef]

68. Simic, T.; Dragicevic, D.; Savic-Radojevic, A.; Cimbaljevic, S.; Tulic, C.; Mimic-Oka, J. Serum gamma glutamyl-transferase is a sensitive but unspecific marker of metastatic renal cell carcinoma. *Int. J. Urol.* **2007**, *14*, 289–293. [CrossRef]

69. Hofbauer, S.L.; Stangl, K.I.; de Martino, M.; Lucca, I.; Haitel, A.; Shariat, S.F.; Klatte, T. Pretherapeutic gamma-glutamyltransferase is an independent prognostic factor for patients with renal cell carcinoma. *Br. J. Cancer* **2014**, *111*, 1526–1531. [CrossRef]

70. Fu, Q.; Xu, L.; Wang, Y.; Jiang, Q.; Liu, Z.; Zhang, J.; Zhou, Q.; Zeng, H.; Tong, S.; Wang, T.; et al. Tumor-associated Macrophage-derived Interleukin-23 Interlinks Kidney Cancer Glutamine Addiction with Immune Evasion. *Eur. Urol.* **2019**, *75*, 752–763. [CrossRef]

71. Abu Aboud, O.; Habib, S.L.; Trott, J.; Stewart, B.; Liang, S.; Chaudhari, A.J.; Sutcliffe, J.; Weiss, R.H. Glutamine Addiction in Kidney Cancer Suppresses Oxidative Stress and Can Be Exploited for Real-Time Imaging. *Cancer Res.* **2017**, *77*, 6746–6758. [CrossRef]

72. Hoerner, C.R.; Chen, V.J.; Fan, A.C. The 'Achilles Heel' of Metabolism in Renal Cell Carcinoma: Glutaminase Inhibition as a Rational Treatment Strategy. *Kidney Cancer* **2019**, *3*, 15–29. [CrossRef]

73. Broer, A.; Rahimi, F.; Broer, S. Deletion of Amino Acid Transporter ASCT2 (SLC1A5) Reveals an Essential Role for Transporters SNAT1 (SLC38A1) and SNAT2 (SLC38A2) to Sustain Glutaminolysis in Cancer Cells. *J. Biol. Chem.* **2016**, *291*, 13194–131205. [CrossRef]

74. Bailey, S.T.; Smith, A.M.; Kardos, J.; Wobker, S.E.; Wilson, H.L.; Krishnan, B.; Saito, R.; Lee, H.J.; Zhang, J.; Eaton, S.C.; et al. MYC activation cooperates with Vhl and Ink4a/Arf loss to induce clear cell renal cell carcinoma. *Nat. Commun.* **2017**, *8*, 15770. [CrossRef]

75. Tang, S.W.; Chang, W.H.; Su, Y.C.; Chen, Y.C.; Lai, Y.H.; Wu, P.T.; Hsu, C.I.; Lin, W.C.; Lai, M.K.; Lin, J.Y. MYC pathway is activated in clear cell renal cell carcinoma and essential for proliferation of clear cell renal cell carcinoma cells. *Cancer Lett.* **2009**, *273*, 35–43. [CrossRef]

76. Son, J.; Lyssiotis, C.A.; Ying, H.; Wang, X.; Hua, S.; Ligorio, M.; Perera, R.M.; Ferrone, C.R.; Mullarky, E.; Shyh-Chang, N.; et al. Glutamine supports pancreatic cancer growth through a KRAS-regulated metabolic pathway. *Nature* **2013**, *496*, 101. [CrossRef]

77. Kredich, N.M. Biosynthesis of Cysteine. *EcoSal Plus* **2008**, *3*. [CrossRef]

78. Bansal, A.; Simon, M.C. Glutathione metabolism in cancer progression and treatment resistance. *J. Cell Biol.* **2018**, *217*, 2291–2298. [CrossRef]

79. Lucarelli, G.; Galleggiante, V.; Rutigliano, M.; Sanguedolce, F.; Cagiano, S.; Bufo, P.; Lastilla, G.; Maiorano, E.; Ribatti, D.; Giglio, A.; et al. Metabolomic profile of glycolysis and the pentose phosphate pathway identifies the central role of glucose-6-phosphate dehydrogenase in clear cell-renal cell carcinoma. *Oncotarget* **2015**, *6*, 13371–13386. [CrossRef]

80. Wang, J.; Wu, Q.; Qiu, J. Accumulation of fructose 1,6-bisphosphate protects clear cell renal cell carcinoma from oxidative stress. *Lab. Investig.* **2019**. [CrossRef]

81. Mullen, A.R.; Wheaton, W.W.; Jin, E.S.; Chen, P.H.; Sullivan, L.B.; Cheng, T.; Yang, Y.; Linehan, W.M.; Chandel, N.S.; DeBerardinis, R.J. Reductive carboxylation supports growth in tumour cells with defective mitochondria. *Nature* **2011**, *481*, 385–388. [CrossRef]

82. Ooi, A.; Wong, J.C.; Petillo, D.; Roossien, D.; Perrier-Trudova, V.; Whitten, D.; Min, B.W.; Tan, M.H.; Zhang, Z.; Yang, X.J.; et al. An antioxidant response phenotype shared between hereditary and sporadic type 2 papillary renal cell carcinoma. *Cancer Cell* **2011**, *20*, 511–523. [CrossRef]

83. Adam, J.; Hatipoglu, E.; O'Flaherty, L.; Ternette, N.; Sahgal, N.; Lockstone, H.; Baban, D.; Nye, E.; Stamp, G.W.; Wolhuter, K.; et al. Renal cyst formation in Fh1-deficient mice is independent of the Hif/Phd pathway: Roles for fumarate in KEAP1 succination and Nrf2 signaling. *Cancer Cell* **2011**, *20*, 524–537. [CrossRef]

84. Jaramillo, M.C.; Zhang, D.D. The emerging role of the Nrf2-Keap1 signaling pathway in cancer. *Genes Dev.* **2013**, *27*, 2179–2191. [CrossRef]

85. Kansanen, E.; Kuosmanen, S.M.; Leinonen, H.; Levonen, A.L. The Keap1-Nrf2 pathway: Mechanisms of activation and dysregulation in cancer. *Redox Biol.* **2013**, *1*, 45–49. [CrossRef]

86. Tong, K.I.; Katoh, Y.; Kusunoki, H.; Itoh, K.; Tanaka, T.; Yamamoto, M. Keap1 recruits Neh2 through binding to ETGE and DLG motifs: Characterization of the two-site molecular recognition model. *Mol. Cell Biol.* **2006**, *26*, 2887–2900. [CrossRef]

87. Taguchi, K.; Yamamoto, M. The KEAP1-NRF2 System in Cancer. *Front. Oncol.* **2017**, *7*, 85. [CrossRef]

88. Jin, L.; Li, D.; Alesi, G.N.; Fan, J.; Kang, H.B.; Lu, Z.; Boggon, T.J.; Jin, P.; Yi, H.; Wright, E.R.; et al. Glutamate dehydrogenase 1 signals through antioxidant glutathione peroxidase 1 to regulate redox homeostasis and tumor growth. *Cancer Cell* **2015**, *27*, 257–270. [CrossRef]

89. Cairns, P. Renal cell carcinoma. *Cancer Biomark* **2010**, *9*, 461–473. [CrossRef]

90. Shojaei, F.; Lee, J.H.; Simmons, B.H.; Wong, A.; Esparza, C.O.; Plumlee, P.A.; Feng, J.; Stewart, A.E.; Hu-Lowe, D.D.; Christensen, J.G. HGF/c-Met acts as an alternative angiogenic pathway in sunitinib-resistant tumors. *Cancer Res.* **2010**, *70*, 10090–10100. [CrossRef]

91. Bielecka, Z.F.; Czarnecka, A.M.; Solarek, W.; Kornakiewicz, A.; Szczylik, C. Mechanisms of Acquired Resistance to Tyrosine Kinase Inhibitors in Clear - Cell Renal Cell Carcinoma (ccRCC). *Curr. Signal. Transduct.* **2014**, *8*, 218–228. [CrossRef]

92. Harris, I.S.; Endress, J.E.; Coloff, J.L.; Selfors, L.M.; McBrayer, S.K.; Rosenbluth, J.M.; Takahashi, N.; Dhakal, S.; Koduri, V.; Oser, M.G.; et al. Deubiquitinases Maintain Protein Homeostasis and Survival of Cancer Cells upon Glutathione Depletion. *Cell Metab.* **2019**, *29*, 1166–1181. [CrossRef]

93. Lewerenz, J.; Hewett, S.J.; Huang, Y.; Lambros, M.; Gout, P.W.; Kalivas, P.W.; Massie, A.; Smolders, I.; Methner, A.; Pergande, M. The cystine/glutamate antiporter system xc– in health and disease: From molecular mechanisms to novel therapeutic opportunities. *Antioxid. Redox Signal.* **2013**, *18*, 522–555. [CrossRef]

94. Dixon, S.J.; Patel, D.N.; Welsch, M.; Skouta, R.; Lee, E.D.; Hayano, M.; Thomas, A.G.; Gleason, C.E.; Tatonetti, N.P.; Slusher, B.S.; et al. Pharmacological inhibition of cystine-glutamate exchange induces endoplasmic reticulum stress and ferroptosis. *eLife* **2014**, *3*, e02523. [CrossRef]

95. Keating, G.M. Sorafenib: A Review in Hepatocellular Carcinoma. *Target. Oncol.* **2017**, *12*, 243–253. [CrossRef]

96. Sehm, T.; Rauh, M.; Wiendieck, K.; Buchfelder, M.; Eyupoglu, I.Y.; Savaskan, N.E. Temozolomide toxicity operates in a xCT/SLC7a11 dependent manner and is fostered by ferroptosis. *Oncotarget* **2016**, *7*, 74630–74647. [CrossRef]

97. Roh, J.L.; Kim, E.H.; Jang, H.; Shin, D. Aspirin plus sorafenib potentiates cisplatin cytotoxicity in resistant head and neck cancer cells through xCT inhibition. *Free Radic. Biol. Med.* **2017**, *104*, 1–9. [CrossRef]

98. Dixon, S.J.; Lemberg, K.M.; Lamprecht, M.R.; Skouta, R.; Zaitsev, E.M.; Gleason, C.E.; Patel, D.N.; Bauer, A.J.; Cantley, A.M.; Yang, W.S.; et al. Ferroptosis: An iron-dependent form of nonapoptotic cell death. *Cell* **2012**, *149*, 1060–1072. [CrossRef]

99. Yagoda, N.; von Rechenberg, M.; Zaganjor, E.; Bauer, A.J.; Yang, W.S.; Fridman, D.J.; Wolpaw, A.J.; Smukste, I.; Peltier, J.M.; Boniface, J.J.; et al. RAS–RAF–MEK-dependent oxidative cell death involving voltage-dependent anion channels. *Nature* **2007**, *447*, 865. [CrossRef]

100. Tang, X.; Wu, J.; Ding, C.K.; Lu, M.; Keenan, M.M.; Lin, C.C.; Lin, C.A.; Wang, C.C.; George, D.; Hsu, D.S.; et al. Cystine Deprivation Triggers Programmed Necrosis in VHL-Deficient Renal Cell Carcinomas. *Cancer Res.* **2016**, *76*, 1892–1903. [CrossRef]

101. Momcilovic, M.; Bailey, S.T.; Lee, J.T.; Fishbein, M.C.; Magyar, C.; Braas, D.; Graeber, T.; Jackson, N.J.; Czernin, J.; Emberley, E.; et al. Targeted Inhibition of EGFR and Glutaminase Induces Metabolic Crisis in EGFR Mutant Lung Cancer. *Cell Rep.* **2017**, *18*, 601–610. [CrossRef]

102. Emberley, E.; Bennett, M.; Chen, J.; Gross, M.; Huang, T.; Li, W.; Mackinnon, A.; Pan, A.; Rodriguez, M.; Steggerda, S.; et al. CB-839, a selective glutaminase inhibitor, has anti-tumor activity in renal cell carcinoma and synergizes with everolimus and receptor tyrosine kinase inhibitors. *Eur. J. Cancer* **2016**, *69*, S124. [CrossRef]

103. MacKinnon, A.L.; Bennett, M.; Gross, M.; Janes, J.; Li, W.Q.; Rodriquez, M.; Wang, T.; Zhang, W.; Parlati, F. Metabolomic, Proteomic and Genomic Profiling Identifies Biomarakers of Sensitivity to Glutaminase Inhibitor CB-839 in Multiple Myeloma. *Blood* **2015**, *126*, 1802.

104. Meric-Bernstam, F.; Tannir, N.M.; Mier, J.W.; DeMichele, A.; Telli, M.L.; Fan, A.C.; Munster, P.N.; Carvajal, R.D.; Orford, K.W.; Bennett, M.K.; et al. Phase 1 study of CB-839, a small molecule inhibitor of glutaminase (GLS), alone and in combination with everolimus (E) in patients (pts) with renal cell cancer (RCC). *J. Clin. Oncol.* **2016**, *34*, 4568. [CrossRef]

105. Okazaki, A.; Gameiro, P.A.; Christodoulou, D.; Laviollette, L.; Schneider, M.; Chaves, F.; Stemmer-Rachamimov, A.; Yazinski, S.A.; Lee, R.; Stephanopoulos, G.; et al. Glutaminase and poly(ADP-ribose) polymerase inhibitors suppress pyrimidine synthesis and VHL-deficient renal cancers. *J. Clin. Investig.* **2017**, *127*, 1631–1645. [CrossRef]

106. Song, M.; Kim, S.H.; Im, C.Y.; Hwang, H.J. Recent Development of Small Molecule Glutaminase Inhibitors. *Curr. Top. Med. Chem.* **2018**, *18*, 432–443. [CrossRef]

107. Bailey, H.H. l-S,R-buthionine sulfoximine: Historical development and clinical issues. *Chem. Biol. Interact.* **1998**, *111–112*, 239–254. [CrossRef]

108. Griffith, O.W.; Meister, A. Potent and specific inhibition of glutathione synthesis by buthionine sulfoximine (Sn-butyl homocysteine sulfoximine). *J. Biol. Chem.* **1979**, *254*, 7558–7560.

109. Tagde, A.; Singh, H.; Kang, M.H.; Reynolds, C.P. The glutathione synthesis inhibitor buthionine sulfoximine synergistically enhanced melphalan activity against preclinical models of multiple myeloma. *Blood Cancer J.* **2014**, *4*, e229. [CrossRef]

110. Anderson, C.P.; Matthay, K.K.; Perentesis, J.P.; Neglia, J.P.; Bailey, H.H.; Villablanca, J.G.; Groshen, S.; Hasenauer, B.; Maris, J.M.; Seeger, R.C.; et al. Pilot study of intravenous melphalan combined with continuous infusion l-S,R-buthionine sulfoximine for children with recurrent neuroblastoma. *Pediatr. Blood Cancer* **2015**, *62*, 1739–1746. [CrossRef]

111. Villablanca, J.G.; Volchenboum, S.L.; Cho, H.; Kang, M.H.; Cohn, S.L.; Anderson, C.P.; Marachelian, A.; Groshen, S.; Tsao-Wei, D.; Matthay, K.K.; et al. A Phase I New Approaches to Neuroblastoma Therapy Study of Buthionine Sulfoximine and Melphalan With Autologous Stem Cells for Recurrent/Refractory High-Risk Neuroblastoma. *Pediatr. Blood Cancer* **2016**, *63*, 1349–1356. [CrossRef]

112. Zhang, L.; Bhasin, M.; Schor-Bardach, R.; Wang, X.; Collins, M.P.; Panka, D.; Putheti, P.; Signoretti, S.; Alsop, D.C.; Libermann, T.; et al. Resistance of renal cell carcinoma to sorafenib is mediated by potentially reversible gene expression. *PLoS ONE* **2011**, *6*, e19144. [CrossRef]

113. Panka, D.; Kumar, M.; Schor-Bardach, R.; Zhang, L.; Atkins, M.; Libermann, T.; Goldberg, N.; Bhatt, R.; Mier, J. Mechanism of acquired resistance to sorafenib in RCC. *Cancer Res.* **2008**, *68*, 2500.

114. Nikinmaa, M.; Pursiheimo, S.; Soitamo, A.J. Redox state regulates HIF-1alpha and its DNA binding and phosphorylation in salmonid cells. *J. Cell Sci.* **2004**, *117*, 3201–3206. [CrossRef]

115. Chen, H.; Shi, H. A reducing environment stabilizes HIF-2alpha in SH-SY5Y cells under hypoxic conditions. *FEBS Lett.* **2008**, *582*, 3899–3902. [CrossRef]

116. Jin, W.-s.; Kong, Z.-l.; Shen, Z.-f.; Jin, Y.-z.; Zhang, W.-k.; Chen, G.-f. Regulation of hypoxia inducible factor-1α expression by the alteration of redox status in HepG2 cells. *J. Exp. Clin. Cancer Res. CR* **2011**, *30*, 61. [CrossRef]

117. Fyfe, G.; Fisher, R.I.; Rosenberg, S.A.; Sznol, M.; Parkinson, D.R.; Louie, A.C. Results of treatment of 255 patients with metastatic renal cell carcinoma who received high-dose recombinant interleukin-2 therapy. *J. Clin. Oncol. Off. J. Am. Soc. Clin. Oncol.* **1995**, *13*, 688–696. [CrossRef]

118. Negrier, S.; Escudier, B.; Lasset, C.; Douillard, J.Y.; Savary, J.; Chevreau, C.; Ravaud, A.; Mercatello, A.; Peny, J.; Mousseau, M.; et al. Recombinant human interleukin-2, recombinant human interferon alfa-2a, or both in metastatic renal-cell carcinoma. Groupe Francais d'Immunotherapie. *N. Engl. J. Med.* **1998**, *338*, 1272–1278. [CrossRef]
119. Alsharedi, M.; Katz, H. Check point inhibitors a new era in renal cell carcinoma treatment. *Med. Oncol.* **2018**, *35*, 85. [CrossRef]
120. Motzer, R.J.; Escudier, B.; McDermott, D.F.; George, S.; Hammers, H.J.; Srinivas, S.; Tykodi, S.S.; Sosman, J.A.; Procopio, G.; Plimack, E.R.; et al. Nivolumab versus Everolimus in Advanced Renal-Cell Carcinoma. *N. Engl. J. Med.* **2015**, *373*, 1803–1813. [CrossRef]
121. McDermott, D.F.; Atkins, M.B. PD-1 as a potential target in cancer therapy. *Cancer Med.* **2013**, *2*, 662–673. [CrossRef]
122. Siska, P.J.; Kim, B.; Ji, X.; Hoeksema, M.D.; Massion, P.P.; Beckermann, K.E.; Wu, J.; Chi, J.T.; Hong, J.; Rathmell, J.C. Fluorescence-based measurement of cystine uptake through xCT shows requirement for ROS detoxification in activated lymphocytes. *J. Immunol. Methods* **2016**, *438*, 51–58. [CrossRef]
123. Li, Y.; Zhu, B. Editorial: Metabolism of Cancer Cells and Immune Cells in the Tumor Microenvironment. *Front. Immunol.* **2018**, *9*, 3080. [CrossRef]
124. Nakaya, M.; Xiao, Y.; Zhou, X.; Chang, J.H.; Chang, M.; Cheng, X.; Blonska, M.; Lin, X.; Sun, S.C. Inflammatory T cell responses rely on amino acid transporter ASCT2 facilitation of glutamine uptake and mTORC1 kinase activation. *Immunity* **2014**, *40*, 692–705. [CrossRef]
125. Klysz, D.; Tai, X.; Robert, P.A.; Craveiro, M.; Cretenet, G.; Oburoglu, L.; Mongellaz, C.; Floess, S.; Fritz, V.; Matias, M.I. Glutamine-dependent α-ketoglutarate production regulates the balance between T helper 1 cell and regulatory T cell generation. *Sci. Signal.* **2015**, *8*, ra97. [CrossRef]
126. Crawford, J.; Cohen, H.J. The essential role of L-glutamine in lymphocyte differentiation in vitro. *J. Cell. Physiol.* **1985**, *124*, 275–282. [CrossRef]
127. Fraternale, A.; Brundu, S.; Magnani, M. Glutathione and glutathione derivatives in immunotherapy. *Biol. Chem.* **2017**, *398*, 261–275. [CrossRef]
128. Gamcsik, M.P.; Kasibhatla, M.S.; Teeter, S.D.; Colvin, O.M. Glutathione levels in human tumors. *Biomarkers* **2012**, *17*, 671–691. [CrossRef]

International Journal of
Molecular Sciences

MDPI

Review

Evolution Shapes the Gene Expression Response to Oxidative Stress

Rima Siauciunaite [1], Nicholas S. Foulkes [1], Viola Calabrò [2] and Daniela Vallone [1,*

[1] Institute of Toxicology and Genetics, Karlsruhe Institute of Technology, Hermann-von-Helmholtz-Platz 1, 76344 Eggenstein-Leopoldshafen, Germany; rima.siauciunaite@kit.edu (R.S.); nicholas.foulkes@kit.edu (N.S.F.)

[2] Department of Biology, Monte Sant'Angelo Campus, University of Naples Federico II, Via Cinthia 4, 80126 Naples, Italy; vcalabro@unina.it

* Correspondence: daniela.vallone@kit.edu; Tel.: +49-721-6082-8728

Received: 26 May 2019; Accepted: 18 June 2019; Published: 21 June 2019

Abstract: Reactive oxygen species (ROS) play a key role in cell physiology and function. ROS represents a potential source of damage for many macromolecules including DNA. It is thought that daily changes in oxidative stress levels were an important early factor driving evolution of the circadian clock which enables organisms to predict changes in ROS levels before they actually occur and thereby optimally coordinate survival strategies. It is clear that ROS, at relatively low levels, can serve as an important signaling molecule and also serves as a key regulator of gene expression. Therefore, the mechanisms that have evolved to survive or harness these effects of ROS are ancient evolutionary adaptations that are tightly interconnected with most aspects of cellular physiology. Our understanding of these mechanisms has been mainly based on studies using a relatively small group of genetic models. However, we know comparatively little about how these mechanisms are conserved or have adapted during evolution under different environmental conditions. In this review, we describe recent work that has revealed significant species-specific differences in the gene expression response to ROS by exploring diverse organisms. This evidence supports the notion that during evolution, rather than being highly conserved, there is inherent plasticity in the molecular mechanisms responding to oxidative stress.

Keywords: ROS; light; DNA damage; evolution; D-box; cavefish; *Spalax*

1. Background

Since the origin of life on earth, oxidative stress has posed a major challenge for living systems. From the evolution of the first plants and photosynthesis to the development of aerobic oxidative respiration, living systems have faced the challenge of exposure to elevated oxygen levels and consequently Reactive Oxygen Species (ROS) (including peroxides (e.g., H_2O_2), superoxide ($O_2^{\bullet-}$), hydroxyl radicals ($^{\bullet}OH$) and singlet oxygen ($^1O_2$) [1]). Oxidative stress accompanies exposure to environmental stressors such as hypoxia, UV radiation, as well as visible light and so frequently changes across the day-night cycle. In more recent evolutionary time, in relation to the impact of human activities on the environment, the toxic effects of many man-made compounds also induce oxidative stress.

ROS levels are modulated by a balance between pro-oxidant and antioxidant elements. When increased levels of ROS are not countered by increases of antioxidant activity or reducing equivalents, a cell undergoes an oxidative stress state. Higher concentrations of ROS represent a potential source of damage for many macromolecules due to the induction of single- and double-stranded DNA breaks, oxidative decarboxylation of α-ketoacids such as pyruvate, and irreversible denaturation of proteins through oxidation and carbonylation of arginine, proline, lysine, and threonine residues [2]. Many cellular mechanisms have evolved to counteract these effects which are based on enzymatic as well as non-enzymatic processes. One of the best described is the role of glutathione in its reduced state (GSH). GSH serves as an antioxidant in plants, animals, fungi and also in some bacteria. It neutralizes ROS by directly donating a reduced equivalent ($H^+ + e^-$). Furthermore, GSH activates oxidative stress response cascades by promoting transcription and post-translational modifications of proteins that affect their functionality. Another conserved antioxidant mechanism is based on the rapid rerouting of carbohydrate flux from glycolysis to the pentose phosphate pathway (PPP) via inhibiting the activity of glycolytic enzymes such as GAPDH (glyceraldehyde-3-phosphate dehydrogenase). The PPP pathway then has the role of regulating cellular NADPH levels that serve as the fuel for antioxidant systems [3–5].

Important antioxidant enzymes used in nearly all cells exposed to oxygen are the superoxide dismutases (SODs) and the catalases, as well as peroxiredoxins and glutathione peroxidases. SODs are a group of metalloproteins catalyzing the dismutation of superoxide $O_2^{\bullet-}$ radicals into two less damaging species, O_2 and H_2O_2. There are three major families of superoxide dismutase depending on the metal cofactor used. The SOD Cu/Zn family which binds copper and zinc is mainly used in eukaryotes including humans; the SOD families which either bind iron and manganese or nickel are used by prokaryotic and protozoa. The physiological importance of SODs is illustrated by the severe pathologies observed in genetically engineered model organisms lacking these enzymes spanning from mouse and *Drosophila* to yeast [6,7]. Hydrogen peroxide is subsequently degraded by catalase activity, usually localized in peroxisomes. This highly active enzyme that catalyzes the decomposition of millions of H_2O_2 molecules to water and oxygen each second [8], also plays a central role in aging and degenerative disorders in humans [9]. Other enzymes involved in scavenging H_2O_2 outside of the peroxisomes are the Peroxiredoxins [10] and the Glutathione peroxidases [11] that detoxify a broad range of peroxides to the corresponding alcohols or water.

Together with strategies aimed at reducing the levels of ROS, a key adaptation for surviving the harmful effects is the evolution of DNA repair mechanisms such as base excision repair, (BER) [12,13], which targets ROS-induced covalent modifications of bases. Several regulatory systems, including cell cycle control and apoptosis, also protect organisms from the negative effects of ROS and are themselves activated by oxidative stress. Cell fate decisions involving cell cycle arrest and apoptosis represent key cellular responses to the damaging effects of ROS. Furthermore, it is increasingly clear that ROS, at relatively low levels, can reversibly oxidize redox-sensitive cysteine and methionine residues, acting as a second messenger via the targeted inactivation of enzymes bearing active site cysteines, for example, phosphotyrosine phosphatases [14].

The gene expression control mechanisms that enable organisms to survive elevated ROS levels or harness their effects are frequently ancient evolutionary adaptations that are tightly interconnected with most aspects of cellular physiology. These mechanisms have received significant attention in studies involving a small number of genetically accessible model organisms. However, comparatively little is known about how these mechanisms are conserved or have adapted during evolution under different environmental conditions.

Even the simplest unicellular organisms possess mechanisms to counter the damaging effects of oxidative stress. Plants and animals frequently excrete hydrogen peroxide or superoxide-generating redox-cycling compounds as a strategy to inhibit microbial growth [15,16] and so bacteria frequently inhabit oxidizing environments. They protect themselves by activating regulons controlled by the OxyR, PerR and, SoxR transcription factors [16–18]. In *Escherichia coli*, the SoxR transcription factor induces the expression of the SoxS protein that in turn activates the transcription of several other genes including the antioxidant enzyme superoxide dismutase [19]. In *Rhodobacter*, the redox signal is detected by the membrane-bound sensor kinase, RegB via a redox-active cysteine located in its cytosolic domain that in turn regulates autophosphorylation. The active form of RegB is able to phosphorylate its regulatory partner RegA, capable of activating or repressing a variety of genes [20]. The conservation of both RegB and RegA homologs in a broad range of bacteria points to a central role for the RegB/RegA two component system in the transcriptional regulation of redox-regulated gene expression [21].

In the yeast *Saccharomyces cerevisiae*, peroxiredoxins are ubiquitous, thiol-containing antioxidant proteins that serve to reduce hydroperoxides. They also regulate hydrogen peroxide-mediated signal transduction via activation of transcription factors such as NF-κB [22,23]. Particularly relevant for the antioxidant function of the peroxiredoxins is a conserved protein called Sulphiredoxin. Sulphiredoxin can reduce cysteine-sulphinic acid of the antioxidant peroxiredoxins via activation by phosphorylation followed by a thiol-mediated reduction step. It has been speculated that Sulphiredoxin is also involved in the repair of proteins containing cysteine-sulphinic acid modifications [24].

In higher organisms, there is evidence for diversity in the response to oxidative stress. Particularly in plants, the production of O_2 by photosynthesis represents a major potential source of oxidative stress. Furthermore, the generation of H_2O_2 in response to various pathogens can elicit localized cell death to limit pathogen spread [25] and thereby serves as part of a systemic response involving the induction of defense genes regulating plant immunity [26]. Therefore, in plants, regulatory systems are required to minimize ROS production without decreasing photosynthetic activity or inhibiting the light-driven production of ROS that is an important signaling molecule controlling plant growth and development [27]. The most abundant ROS scavengers in plants, ascorbate (AsA) and GSH, are typically concentrated in chloroplasts. These metabolites function together in the AsA-GSH cycle to metabolize H_2O_2 and thereby to dissipate excess excitation energy in chloroplasts. Moreover, the AsA-GSH detoxification pathway cross-talks with other detoxification pathways including the peroxiredoxin (PRX) and glutathione peroxidase (GPX) pathways which are also important for the detoxification of lipid peroxides. It has been speculated that all these detoxification pathways are tailored to suit specific stressors and that their relative importance probably varies according to the prevailing environmental conditions [27].

In animals, there is evidence for species-specific differences in the exploitation of the effects of ROS. For example, in echinoderms fertilization triggers a burst of extracellular production of H_2O_2 by a plasma membrane NADPH oxidase with a simultaneous release of ovoperoxidase. This elevated H_2O_2 is the oxidant responsible for the extracellular cross-linking reaction involving in the formation of a protective envelope around the freshly fertilized oocyte [28].

In insects as in plants, reactive oxygen species can also function as immune effector molecules which exert microbicidal activity. In *Drosophila*, a burst of ROS is generated by DUOX (dual oxidase) upon gut microbe infection and regulates the production of antimicrobial peptides (AMPs) by the fat body, a major immune organ in the fly [29]. As in mammals, the ROS dependent mechanism which activates antimicrobial peptide production involves the activation of the Toll and the NF-κB pathways which both play essential roles in antibacterial and antifungal responses [29,30]. At the same time, the pathogen-induced ROS levels activate the JAK-STAT (Janus kinase–signal transducers and activators of transcription) and JNK (c-Jun NH_2 terminal kinase) pathways to induce stem cell proliferation counteracting the cellular damage generated by the burst of ROS [31].

Another mechanism underlying oxidative stress tolerance in *Drosophila* that influences life-span and xenobiotic response is the conserved Keap1 (Kelch-like ECH-associated protein (1)/Nrf2 (NF-E2-related factor (2) signaling pathway [32]. The Keap1/Nrf2 dimer activated by ROS plays a crucial role in reducing oxidative stress in the germline stem cells from *Drosophila* testis. This pathway regulates the expression of antioxidant and detoxification genes [32,33] and has also been shown to play a critical role in ROS detoxification in mammalian systems.

1.1. ROS Regulation of Gene Expression in Vertebrate Systems

Many studies have focused on the regulation of gene expression by ROS in mammalian systems. It is well known that a moderate level of ROS synthesis is physiologically normal and acts as a specific signal in the control of cell proliferation, blood circulation, myoblast differentiation and the regulation of immune and endocrine processes [34]. However, external factors such as xenobiotics and UV-radiation as well as hormones, cytokines, and other physiological stimuli can enhance the production of cellular ROS up to toxic levels [34–36]. To protect against the potentially damaging effects of ROS, mammalian cells can enhance the level of endogenous non-enzymatic antioxidant metabolites such as lipoic acid, glutathione, L-arginine and coenzyme Q10 [37] as well as inducing the synthesis of several antioxidant enzymes including superoxide dismutase, catalase and glutathione peroxidase. The mechanism underlying this transcriptional response to ROS involves the activation of MAP protein kinases and redox-regulated transcription factors such as c-Jun, ATF2, ATF4, NFκB, Nrf2 and p53.

Mitogen-activated protein kinases (MAPKs) have been reported to orchestrate ROS-responsive signaling pathways by activating redox-responsive transcription factors such as AP-1, NFκB/IκB and the Nrf2/Keap1 systems. The prevention of ROS accumulation by antioxidants blocks MAPK activation [38,39]. MAPK activation consists of a kinase cascade initiated by a MAPK kinase kinase (MAPKKK), that phosphorylates and thereby activates a MAPK kinase (MEK, or MKK), which in turn phosphorylates and activates one or more MAPKs. In vertebrates, three subtypes of MAPKs have been described. The extracellular signal-regulated kinases (ERKs), the c-Jun N-terminal kinases (JNKs), and the p38 MAPKs. MAPKs can be activated by a wide variety of different stimuli, but in general, ERK-1 and ERK-2 are preferentially activated in response to growth factors, while the JNKs and p38 MAPKs are more responsive to stress stimuli including elevated levels of ROS. The p38 MAPKs represent stress-activated protein kinases activated by extracellular stress and cytokines, (e.g., tumor necrosis factor-a (TNF-α) and interleukin-1b (IL-1β)), and are thereby involved in the inflammation response [40].

The stress-activated protein kinases JNKs, were originally identified by their ability to activate the transcription factor c-Jun via phosphorylation of its transactivation domain [41]. However, it is now clear that they also phosphorylate other target proteins. A key question concerns precisely how ROS activates the JNK and p38 MAPK pathways. Redox-sensitive proteins, such as Trx and glutaredoxin (Grx) have been implicated in this regulatory mechanism [42]. For example, the oxidation of Trx by ROS results in dissociation of the Trx/ASK-1 complex leading to the activation of the two stress-responsive MAPK pathways [42]. ASK1 has been extensively characterized as a ROS-responsive kinase [43,44].

The protooncoprotein c-Jun, as well as ATF2 (Activating transcription factor 2) belong to the activating protein-1 (AP1) transcription factor family. These factors regulate gene expression in the context of homo- or hetero-dimeric complexes with other AP1 members (e.g., CREB, Fos, Maf, Jun-B and Jun-D) [45–48]. AP1 dimers containing ATF2 and c-Jun have been reported to bind to the promoter consensus sequence $T^G/_T$ACNTCA that is encountered in the promoters of many genes involved in DNA repair and apoptosis [49].

ATF2 was originally identified from a human brain cDNA library screen as a CRE-binding protein [50]. It is activated by various stimuli including oxidative stress, growth factors, ultraviolet (UV) radiation, and cytokines. ATF2 can shuttle between the nucleus and cytoplasm under basal conditions and following stress stimuli via an autoinhibition mechanism. Specifically, in the inactive state, the ATF2 N-terminal transcriptional activation domain (TAD) interacts with its C-terminal basic leucine zipper (bZIP) DNA-binding domain, inhibiting the ability of ATF2 to activate transcription. Following stress stimuli, ATF2 undergoes phosphorylation at threonine (T69, T71) mediated by stress-activated protein kinases (e.g., p38 and JNK) and is able to translocate to the nucleus as homo- or hetero-dimers with other AP1 transcription factors to modulate the expression of hundreds of genes [47,51]. The proto-oncogene c-Jun is the cellular homolog of the viral oncoprotein v-jun discovered in the avian sarcoma virus 17 [48,52,53]. The c-Jun protein was originally described as a driver of malignant transformation. One characteristic of c-Jun is that it can activate its own expression [54] and therefore can drive a positive autoregulatory loop. It has been shown that in response to oxidative damage, c-Jun is phosphorylated at N-terminal serine residues (S63, S73) by JNK [41,55,56] and that c-Jun play an important role in cell cycle re-entry after DNA damage induced by UV exposure since immortalized fibroblasts lacking c-Jun undergo a prolonged UV-induced growth arrest [57].

The nuclear factor NF-κB is a widely investigated dimeric transcription factor involved in the regulation of genes that control various aspects of the immune and inflammatory response and is responsive to ROS. Redox signaling plays a critical role in NF-κB activation by various stimuli via thioredoxin peroxidases [23]. In unstimulated cells, NF-κB dimers are sequestered in the cytosol through noncovalent interactions with inhibitory proteins termed IκBs [58]. The nuclear translocation and activation of NF-κB as a transcription factor by cytokines, microbial agents, oxidative challenge (ROS) and irradiation occurs through the signal-induced phosphorylation of IκB and its proteolytic degradation by thioredoxin peroxidases. IκB degradation exposes the nuclear localization signal on NF-κB, which allows its nuclear translocation and activation of the transcription of its target genes. Interestingly, the DNA binding function of NF-κB is also regulated by the intracellular redox status. Specifically, the p50 subunit is targeted by S-glutathionylation which reversibly inhibits its DNA binding activity [59].

One of those NF-κB target genes which includes in its promoter the GGGDNWTTCC enhancer element, is the redox-regulating thioredoxin gene (Trx). Trx is an oxidoreductase that works together with the glutathione system to establish and maintain a reduced intracellular redox state. Other NF-κB target genes that play a protective antioxidant role include the peroxiredoxin, heme oxygenase-1, the cystine transporter xc2 and manganese SOD (mnSOD) genes [60].

The redox stress-sensitive transcription factor Nrf2 (nuclear erythroid-derived 2-like) regulates the expression of several antioxidant and detoxification genes. In the absence of ROS, Nrf2 is retained in the cytoplasmic compartment in complex with another protein, KEAP1, ensuring effective Nrf2 repression. Upon oxidative stress increase, NRF2 protein is rapidly released from KEAP1 and thereby translocated into the nucleus of affected cells. Nuclear Nrf2 heterodimerizes with Maf and binds to the antioxidant response element ARE sequence (ARE; $5'-^A/_GTGA^C/_GNNNGC^A/_G-3'$) located in the promoter regions of antioxidant and detoxification enzymes and by activating expression of these target genes, counteracts oxidative stress [61–63]. Specific Nrf2-regulatory targets include elements of the glutathione and thioredoxin antioxidant systems, as well as enzymes catalyzing the detoxification of exogenous and endogenous products, NADPH regeneration and heme metabolism. Consistently, loss of Nrf2 function is associated with increased susceptibility to many environmental stressors. Recently, it has been demonstrated that Nrf2 can cooperate in a ROS detoxification program with the cAMP responsive transcription factor ATF4 [64]. Moreover, Nrf2 is involved in other cellular processes such as autophagy, metabolism, stem cell quiescence and unfolding protein responses [65]. More recently, De Nicola and colleagues [66] have pointed out a potential link between the "reduced" cellular environment and tumor initiation. They have shown that oncogene-mediated induction of Nrf2 in mice promotes ROS detoxification required for tumor initiation [66]. Furthermore, Nrf2 mutations have been isolated from patients with lung, gall bladder, head and neck cancers supporting a pro-tumorigenic role for Nrf2 [62].

The cytokine responsive transcription factor STAT3 has also been shown to be regulated by the redox sensor peroxiredoxin 2 (Prx2) [67]. Specifically, Prx2 act as a sensitive receptor for H_2O_2 and transmits oxidative equivalents to STAT3. This, in turn, induces the formation of STAT3 oligomers with reduced transcriptional activity. This observation is consistent with reports that ROS plays a key role in the regulation of tissue regeneration and development [68].

The tumor-suppressor p53 is potently induced by oxidative stress and mediates all the antiproliferative cellular responses to oxidative signals, including transient cell-cycle arrest, cellular senescence and apoptosis [69,70]. Specifically, after oxidative stress the p53 transcriptional response is dependent on the p66Shc protein, the redox enzyme implicated in ROS generation and the translation of oxidative signals into apoptosis [71,72]. The tumor suppressor p53 is critically involved in oxidative stress-dependent apoptosis and is upregulated upon treatment with H_2O_2 and UV [73]. Strikingly, p53−/− MEFs show resistance to UV and H_2O_2-induced apoptosis [70]. p53 activation leads to a significant increase in ROS levels and apoptosome assembly via the release of cytochrome C from the mitochondria. This release of cytochrome C upon oxidative stress is p66Shc-dependent since p66Shc−/− cells fail to increase ROS levels. Therefore, p66Shc serves as a downstream effector of p53 [72]. However, p66Shc is not involved in p53 functions such as cell cycle arrest but it does regulate p53-dependent apoptosis pointing to a crucial role for p66Shc and p53 crosstalk in the regulation of intracellular ROS levels [72].

1.2. Circadian Clocks and Timing of the Response to ROS

Close links exist between oxidative stress and the circadian clock. It has been speculated that during the origin of life on earth, one of the first driving forces for the evolution of an internal timing mechanism, was the great oxidation event that occurred following the evolution of green plants and photosynthesis [74]. Hydrogen peroxide production and scavenging are strongly time-of-day dependent, therefore, the evolution of an endogenous 24 h clock mechanism was a fundamental step enabling a temporally coordinated homeostatic response to ROS.

The circadian clock is a highly conserved mechanism, from cyanobacteria to humans, that regulates rhythms with a period of 24 h in almost every aspect of biology and behavior. The circadian clock is regularly "reset" or "entrained" by external time cues including light, food and temperature (so-called zeitgebers, time givers) to remain synchronized with the external environment. However, its timing function also persists in the absence of zeitgebers with a rhythm of circa 24 h and based on this property is termed "circadian" (for Latin: circa-diem, around one day). At the molecular level, the vertebrate circadian timing system can be subdivided into three parts: an input pathway that detects and processes zeitgeber information; a core oscillator that is entrained by the input pathway and generates endogenous and self-sustained rhythms; and an output pathway that relays this integrated timing information to various aspects of behavior and physiology [75]. Genetic screens for circadian clock mutants in several model organisms (*Drosophila, Neurospora, Arabidopsis* and mouse) led to the identification of many genes involved in this mechanism [76–81]. Although, the circadian clock genes identified were not conserved between the different groups of organisms, they all share a functional property, namely that they generate circadian rhythmicity by serving as elements of transcriptional-translational feedback loops (TTFL) [81]. In vertebrates, the BMAL and CLOCK basic helix–loop–helix (bHLH), Per-Arnt-Single minded (PAS) transcription factors serve as positive elements of the core TTFL clock mechanism. CLOCK-BMAL hetero-dimers bind to E-box enhancer elements (5′-CACGTG-3′) located in the promoter regions of the negative elements of the TTFL (the period (Per) and cryptochrome (Cry) genes) as well as in the promoters of other clock-controlled genes [82,83], and thereby activate their transcription. The induced PER and CRY proteins in turn form a hetero-dimeric complex and translocate to the nucleus where they inhibit their own transcription by interfering with CLOCK-BMAL driven transcriptional activation [81]. The stability of this core regulatory TTFL is enhanced by additional feedback loops [84]. For their discovery of the first elements of the molecular mechanisms generating circadian rhythms, Michael Rosbash, Michael W. Young and Jeffrey C. Hall obtained the Nobel prize in Physiology and Medicine in 2017.

In mammals, cell autonomous clocks located in most cell types and tissues (so-called peripheral clocks) are light-synchronized via systemic signals provided by a specialized clock located in the SCN (Suprachiasmatic nucleus) of the hypothalamus, a paired neuronal structure located in the anteroventral hypothalamus above the optic chiasm that is synchronized by light via light-dependent input from the retina-hypothalamic tract [85,86]. In lower vertebrates, including fish and insects such as *Drosophila*, peripheral clocks are reset directly by light without the presence of a light entrainable central pacemaker [87].

Many lines of evidence point to extensive links between the circadian clock and the redox state of the cell. Thus, model organisms with genetically disrupted circadian clocks show many features of abnormal metabolism, such as obesity and diabetes [88–91]. Furthermore, the early aging phenotype observed in the clock protein BMAL1 knockout mouse has been attributed to an accumulation of ROS due to mitochondrial uncoupling [92,93].

The core clock mechanism has been shown to respond to redox state in a range of model organisms [94,95]. This enables the circadian clock to respond to changes in metabolic activity [96,97]. In vertebrates, a feedback loop links redox homeostasis and clock function. The circadian clock controls the NAD pathway via regulation of the enzyme NAMPT, crucial for the synthesis of NAD [98]. This regulation governs the cellular NAD^+:NADH ratio [99]. Conversely, a NAD^+-dependent deacetylase, the protein SIRT1, directly regulates the expression of clock and clock-controlled genes via deacetylation of clock proteins and histones [100,101]. The mechanism of protein deacetylation by SIRT1 is dependent upon the availability of NAD^+. Both NADH and NADPH enhance the binding of CLOCK:BMAL1 and NPAS2:BMAL1 hetero-dimers to the E-box enhancer element, whereas NAD^+ and $NADP^+$ inhibit this activity [102].

It has been shown that circadian clock regulation of the redox stress-sensitive transcription factor NRF2 occurs via the E-box enhancer element. NRF2 regulates the circadian rhythmic expression of antioxidant genes, and genes involved in NADPH production [103]. Consistent with a clock control of ROS homeostasis, time of day dependent differences in the levels of DNA damage, lipid peroxidation and protein oxidation have been documented [104,105].

The clock also regulates the production of the hormone melatonin, a potent antioxidant molecule [106]. Melatonin can act as a direct free radical scavenger or as an indirect antioxidant by stimulating antioxidant enzymes including SOD, GPX, and glutathione reductase [107]. Moreover, melatonin is able to regulate the mitochondrial electron transport chain, thereby reducing electron leakage and free radical generation [107]. In addition, it has also been hypothesized that intracellular melatonin can directly neutralize H_2O_2 by generating N'-acetyl-5-methoxikynuramine (AFMK) [107].

1.3. Stress Response and Cytoplasmic Granules

Following exposure to environmental stress, cells can undergo two alternative fates, either inducing apoptosis or cell cycle arrest to repair the stress-induced damage. The activation of one or the other mechanism is strictly dependent on the extent of cellular damage. These processes minimize cell loss and prevent the survival of cells with genetic and protein alterations. An additional evolutionary conserved strategy for dealing with an imbalanced redox state is Stress Granule (SG) formation. Stress Granules are cytoplasmic non-membrane bound aggregates of RNA and proteins whose formation is associated with inhibition of translation initiation and the disassembly of polysomes [108].

In mammalian cells, stress granule formation is orchestrated by PI3K and p38MAPK that act in a hierarchical manner to drive mTORC1 activity thus facilitating stress granule assembly [109]. SGs are composed of only 10% of bulk mRNA molecules as well as some non-coding RNA (ncRNAs). Recent transcriptomic analyses reveal that SGs are enriched with long mRNAs that exhibit poor translation efficiency [110]. In addition to RNAs, SGs contain various proteins, including G3BP1, T-cell restricted intracellular antigen-related protein (TIAR), PABP1, RACK1, HDAC6 and Y box binding protein 1 (YB-1). Although the role of each protein in SG assembly is not fully elucidated, each appears to play an essential role in SG-associated functions. For example, YB-1 directly binds to and translationally activates the 5′ untranslated region (UTR) of G3BP1mRNAs, thereby controlling the availability of the G3BP1 nucleator for SG assembly.

In mammals, irradiation and genotoxic drugs are prone to trigger apoptosis. Instead, arsenite, hydrogen peroxide and heat shock treatment induce stress granule formation, that suppresses ROS elevation and thereby inhibits apoptosis. This antioxidant property results from the function of two key SG components, namely the GTPase-activating protein SH3 domain binding protein 1 (G3BP1) and ubiquitin-specific protease 10 (USP10). Under normal conditions, G3BP1 elevates ROS by inhibiting the antioxidant activity of USP10. However, under oxidative stress or heat shock, G3BP1 and USP10 trigger SG formation, which then activates the antioxidant activity of USP 10 [111].

Interestingly, it has been shown that upon oxidative insults, SG assembly is associated with an enhancement of YB-1 protein secretion. Furthermore, an enriched fraction of extracellular YB-1 (exYB-1) caused a G2/M cell cycle arrest in receiving cells thus inhibiting cell proliferation [112].

These observations suggest that the release of extracellular YB-1 can function as a stress-dependent paracrine/autocrine signal that controls cell cycle progression. Despite the considerable literature on SGs in mammals, little is known about how much these membrane-less organelles and their functions are conserved during evolution. Interestingly, it has been shown that the nuclear localization of the SG marker YB-1 is robustly regulated in fish by the circadian clock [113]. Specifically, a daily nuclear entry of YB-1 at the beginning of the light phase appears to be mediated by clock-controlled changes in YB-1 SUMOylation. Moreover, in zebrafish, the YB-1 nuclear protein is able to downregulate cyclin A2 transcript levels thus providing a direct link between the circadian clock, YB-1 and the control of cell proliferation. Thus, in response to oxidative stress, YB-1 seems to play distinct roles depending on its subcellular localization. In the nucleus, YB-1 restrains cell cycle progression while in the cytoplasm it participates in SG assembly and inhibition of translation. Overall, it appears that YB-1 acts to prevent and then eventually repair genotoxic damage.

Thus, oxidative stress response pathways have been extensively studied in mammalian systems. However, given the diversity of antioxidative stress pathways in other species, are these pathways conserved in all vertebrates or may there be species-specific differences in their function depending on the ecological niche occupied? In the next section, we will explore recent progress made in studying the oxidative stress response in fish.

2. Fish as Models for Studying ROS Responses

Recent work has compared the response to oxidative stress in selected species of fish with mammalian models such as human and mouse. Fish represent the most diverse and largest vertebrate group which occupy an important ecological position and have great commercial value. Furthermore, zebrafish and medaka are also powerful genetic models that provide a wealth of molecular and genetic tools for studying various biological processes at the molecular level. Another unique and attractive advantage of using fish for studying ROS pathways, is that their peripheral tissues are directly light responsive [87] and central to these light-driven responses is an increase in intracellular ROS levels [114, 115]. Specifically, most tissues and cell types in fish express photoreceptors, directly light-regulated clocks, light enhanced DNA repair capacity and importantly the expression of many genes is induced in fish cells upon direct exposure to visible as well as UV light and ROS. This cell-autonomous property is even observed in cell lines derived from various fish tissues [116–118]. Therefore, fish cell lines represent powerful in vitro models for exploring the capacity of light via ROS to regulate physiological mechanisms in vertebrates. The utility of fish also extends to studying the effects of evolution in response to a changing environment. Notable examples are species of blind cavefish that inhabit perpetually dark subterranean environments and exhibit a set of striking anatomical adaptations including eye loss, enhancement of non-visual senses and loss of body pigment, so called troglomorphisms. One particular species, the Somalian cavefish *Phreatichthys andruzzii*, represents one of the most extreme examples of cavefish since it has been completely isolated from surface water and surface-dwelling forms for at least 2–3 million years. Interestingly, this species also exhibits complete loss of light, UV and ROS induced gene expression. Therefore, comparing the responses to ROS in different fish species subjected to different evolutionary pressures, represents a unique approach to gain insight into how the ROS response is shaped by the environment during evolution.

2.1. Sunlight, ROS and Regulation of the Circadian Clock in Fish

A remarkable feature of the organization of the circadian timing system in vertebrates is a fundamental difference in the mechanisms whereby light entrains the multiple peripheral tissue clocks. Thus, while peripheral clocks are directly light-regulated in lower vertebrates such as fish, in modern mammals the photic entrainment of peripheral clocks is dependent upon a centralized, retina-based photoreception system [119]. These observations predict the existence of fundamental changes in the regulatory networks of peripheral clocks over the course of vertebrate evolution. In zebrafish, functional genomic analysis has identified more than 40 opsins of which 32 are non-visual opsins expressed in peripheral organs [85,120]. It has also been shown that exposure to visible light can stimulate H_2O_2 production in zebrafish embryonic cell lines and that this production is a key signal for mediating light-dependent circadian gene expression [114,115]. It is also well documented that exposure of cultured mouse, monkey and human cells to violet-blue visible light, as well as UVA, also stimulates H_2O_2 production via photoreduction of flavoproteins [121]. Hirayama et al., in 2007 [115], demonstrated that ROS species act as a second messenger coupling photoreception to circadian clock entrainment and consequentially regulate the circadian expression and activity of clock genes as well as antioxidant enzymes such as catalase that is responsible for H_2O_2 degradation. A more recent study [114], revealed a key role for NADPH-flavin-containing oxidases (NOXes) in the regulation of light-inducible clock gene expression in zebrafish. Using a pharmacological approach and by studying zebrafish cell lines, visible light was shown to trigger increases in ROS levels via NADPH oxidase activity, which in turn activates the expression of the light regulated clock genes *zfcry1a* and *zfper2*. This induction of *zfcry1a* and *zfper2* gene expression is also dependent on the activation of the JNK and p38MAPK stress pathways. Surprisingly, the promoter enhancer elements targeted by these ROS pathways was not one of the classical enhancers regulated by ROS. Instead, exposure to ROS species, as well as visible and UV light, activated the expression of circadian clock and DNA repair genes via D-box enhancer elements located in their promoters [114,122]. Interestingly, in mammalian cells, neither blue light, UV nor H_2O_2 exposure activates gene expression via D-box enhancer elements [114,122].

2.2. The D-Box and the Transcriptional Response to ROS

The D-box element was first described in mouse liver as one of the transcriptional enhancers present in the albumin gene promoter [123]. In 1990, Uli Schibler's group cloned the first transcription factor that binds with high specificity to the D-box element of the rat liver albumin promoter, thereby called DBP (D-binding protein) [124]. Subsequently, two other transcription factors were shown to bind to the D-box element and to serve as transcriptional activators, namely the thyrotrophic embryonic factor (TEF) [125] and the hepatocyte leukemia factor (HLF) [126]. DBP, TEF and HLF belong to the so-called PAR bZip (proline and acidic amino acid-rich basic leucine zipper) transcription factor family which share a proline-and acidic amino acid-rich domain positioned proximal to a C-terminal bZip domain as well as an extended basic region at their N-termini and which are highly conserved throughout evolution [127,128]. These transcription factors constitute a subfamily of the basic leucine zipper proteins (bZip), characterized by a C-terminal α-helical region that allows homodimerization or heterodimerization between other bZip proteins [129]. All PAR bZip family members activate transcription of downstream genes by binding as homo- or hetero-dimers to D box elements matching the consensus sequences RTTAYGTAAY [130]. These transcriptional activators compete for DNA binding with another bZip protein which shares a similar DNA-binding profile but which lacks the PAR domain and functions as a repressor, namely E4BP4 (E 4 binding protein 4) [129,131].

A striking observation is that the D-box enhancer element operates in a completely different fashion in mammalian cells compared with fish cells. In mammalian species, the D-box is not involved in the light input pathway of the clock. The expression of genes which contain D-boxes in their promoters do not respond to visible light, UV or ROS exposure (Figure 1). Instead, the mammalian D-box is clock regulated and thereby plays a role in the activation of genes belonging to the output pathway of the clock machinery [132]. The expression of both DBP [133] and HLF [134] is strongly clock regulated in the mammalian liver and the D-box enhancer has been shown to mediate the clock-driven regulation of genes encoding enzymes such as CAT and SOD [115,135] as well as the production of low molecular weight antioxidants such as glutathione (GSH) [136–138] with important functions in the defense against xenobiotic and oxidative stress [139].

In contrast, D-box regulation in fish is tightly linked with coordinating the gene expression response to oxidative stress as well as sunlight. Furthermore, as a result of genome duplication events in early teleost ancestors, fish species possess extra D-box binding factors (TEF1, TEF2, HLF1, HLF2,

DBP1, DBP2 and 6 E4BP4 homologs) [140]. These factors exhibit significant differences in their tissue- and cell type-specific expression patterns and also can bind to D-boxes as hetero- or homo-dimers. Therefore, it is clear that in fish, transcriptional regulation at the D-box is inherently complex. A clear challenge for the future is to identify precisely how elevated ROS levels lead to the regulation of this complex combination of transcription factors via activated JNK and p38 MAPK activity.

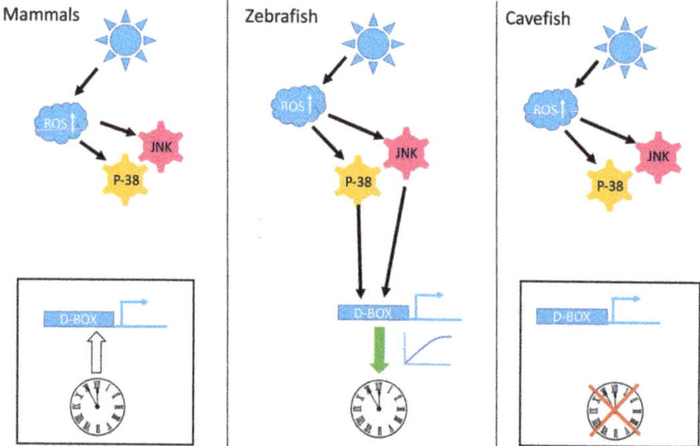

Figure 1. Light, reactive oxygen species (ROS) and the D-box enhancer. Schematic representation of how D-box enhancer-driven gene expression is differentially influenced by blue light exposure in mammalian, zebrafish and cavefish cells. In all three cell types light triggers an increase in intracellular ROS levels that in turn activates the p-38 mitogen-activated protein kinase (MAPK) and c-Jun NH_2 terminal kinase (JNK) stress pathways. In zebrafish cells (central panel), this signaling results in the activation of D-box-driven gene expression, ultimately leading to circadian clock entrainment (indicated by the green arrow). In mammalian cells (left panel) and in cavefish cells (right panel) this signaling fails to activate gene expression via the D-box enhancer element and does not entrain the circadian clock. The white arrow starting from the clock and pointing on the D-box element indicates the circadian clock regulation of this enhancer in mammals [132]. The red cross over the clock in the cavefish cells indicates the blind circadian clock observed in these cells [141].

2.3. Adaptation of Mechanisms Responding to Oxidative Stress during Evolution

Insight into how the light-ROS-D-box signaling pathway has adapted during evolution under extreme environmental conditions has been gained from a set of comparative studies focusing on a species of blind cavefish (*Phreatichthys andruzzii*). This species has evolved over 2–3 million years completely isolated from sunlight, in layers of water locked beneath the Somalian desert [114,122]. Like other species inhabiting perpetually dark cave environments, these cavefish exhibit a set of striking anatomical adaptations including complete eye loss and absence of body pigmentation, so-called troglomorphisms. By comparing cavefish- and zebrafish-derived cell lines, it was revealed that light entrainment of the clock as well as photoreactivation DNA repair has also been lost during evolution of this cavefish [122,141,142]. Specifically, this originates from a loss of light-, UV- and ROS-induced gene expression. Similar to the situation in zebrafish, blue or UV light triggers an increase in cellular ROS levels as well as an activation of the MAP kinase stress pathways in the cavefish cell lines. However, these events do not result in the transcriptional activation of D-box enhancer-regulated clock genes such as *per2* and *cry1a* or the *CPD*, *DASH* and *6-4 photolyase* DNA-repair genes (Figures 1 and 2). Precisely how the D-box regulatory factors have been modified during cavefish evolution remains unclear. Furthermore, it will be fascinating to compare these cavefish regulatory mechanisms with those of mammals which also fail to respond to ROS.

Interestingly, the cavefish *P. andruzzii* appears to be only species described to date, apart from placental mammals, that lacks the highly evolutionary conserved photoreactivation DNA repair function. It has been speculated that in the DNA repair systems of *P. andruzzii*, we are witnessing the first stages of a process that occurred previously in the ancestors of placental mammals during the Mesozoic era. This speculation is based on the "nocturnal bottleneck" theory [143–145]. This theory predicts that the ancestors of modern mammals became exclusively nocturnal in order to avoid predation by diurnal carnivorous dinosaurs. This adaptation to a dark ecological niche may also explain many features of present-day mammals including a general loss of extraretinal photoreception, as well as adaptations in the eye and retina to facilitate vision under low-lighting conditions [143,145]. Adaptation to a nocturnal lifestyle is also predicted to have entailed a general loss of light-dependent repair mechanisms that target UV-induced DNA damage, namely the photolyase genes and photoreactivation function [143]. It is also interesting to note that similar to cavefish cells, in mammalian cell lines light and UV exposure all trigger an increase in cellular ROS levels followed by activation of the MAP kinase stress pathways but this does not result in activation of D-box enhancer-mediated gene expression [114,122].

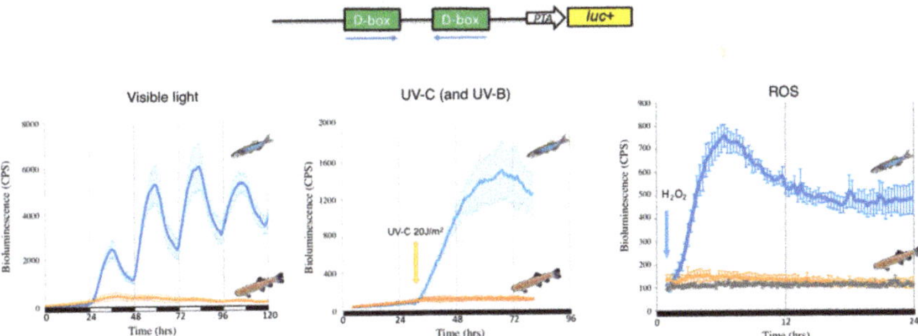

Figure 2. Loss of D-box function in cavefish. Representative in vivo bioluminescence assays performed in zebrafish (blue traces) and cavefish (orange traces) of cells transfected with the D-box enhancer luciferase reporter derived from the zebrafish *6-4 phr* promoter [122] (upper part of the figure). Cells were exposed to three different stressors: light-dark cycles (left panel), a UVC pulse (central panel) or 300 μM H_2O_2 (right panel). The grey (right panel) trace indicates luciferase expression of cavefish cells not exposed to H_2O_2 (negative controls). Black and white bars below each panel indicate the lighting conditions.

Another example where changes in the response to ROS are coupled with evolution in an extreme environment is the naked blind mole rat (*Spalax*). These are small subterranean rodents common in the Middle East that are distinguished by their adaptation to life in a hypoxic underground environment. Specifically, these animals undergo cycles of digging closer to the surface and then burrowing in their subterranean tunnel systems. As a consequence, at the cellular level, these animals experience cycles of hypoxia followed by reoxygenation. In a normal animal, these alternating shifts in oxygen levels can serve as a source of genomic instability, which underlies both aging and cancer. Severe hypoxia results in S-phase arrest, dNTPs depletion and replication stress as well as a general repression of DNA repair activity. Then, upon reoxygenation, S-phase restarts during an initial period when DNA repair has still not recovered to normal levels. Therefore, DNA replication occurs in the presence of ROS-induced DNA damage, leading to the accumulation of mutations. [146,147]. However, *Spalax* exhibits remarkable longevity and a striking resistance to cancer [148]. It is therefore evident that a key aspect of the adaptations *Spalax* has made for life in its hypoxic environment are fundamental changes in the molecular mechanisms responding to cycles of hypoxia and oxidative stress. It has been speculated that the cancer resistance in *Spalax* results from an amino acid substitution in the p53 gene leading to a R174K substitution. p53 serves as a master regulator of the DNA damage response and has been termed the "guardian of the genome." This amino acid substitution in *Spalax* impairs the ability of the p53 protein to trigger apoptosis, although the protein is still able to induce cell cycle arrest [149].

3. Conclusions

Our current understanding of the molecular mechanisms responding to oxidative stress reveals extensive conservation of many of the main regulatory pathways between organisms as diverse as mammals, yeast and *Drosophila*. However, there are also examples of fundamental changes in these mechanisms during evolution. For example, the switch in the function of the D-box enhancer from a ROS regulatory target in fish, to a clock regulatory target in mammals as well as the loss of ROS responsiveness of the D-box enhancer during cavefish evolution. Furthermore, fundamental changes in the functionality of the p53 protein have accompanied the adaptation of the naked blind mole rat to its life in a hypoxic environment. These findings reveal that the functions of ROS responsive mechanisms are also potentially plastic and are shaped by the selective pressure of the environment. By identifying precisely which elements of these regulatory mechanisms are "targeted" during the process of evolution, we will gain considerable insight into how organisms adapt to oxidative stress and damage over evolutionary time scales.

Author Contributions: Writing—original draft preparation, D.V., V.C., N.S.F. and R.S.; writing—review and editing, D.V., V.C. and N.S.F.; literature research, D.V., V.C., N.S.F. and R.S.; funding acquisition, D.V., V.C., N.S.F. and R.S.

Funding: The research of D. Vallone and N.S Foulkes was funded through the Helmholtz funding program BIFTM and through the Fazit-Stiftung for R. Siauciunaite. V. Calabrò acknowledges research grants from the University of Naples "Federico II", MIUR 2017–2018. Exchanges between the N.S Foulkes and V. Calabro' laboratories are supported by the MIUR-DAAD Joint Mobility Program 2018.

Acknowledgments: We thank C. Pagano, H. Zhao, A.M. Guarino and N. Geyer for their work and we acknowledge support by the Deutsche Forschungsgemeinschaft and the open access publishing fund of Karlsruhe Institute of Technology.

Conflicts of Interest: The authors declare no conflict of interest.

References

1. Hayyan, M.; Hashim, M.A.; AlNashef, I.M. Superoxide Ion: Generation and Chemical Implications. *Chem. Rev.* **2016**, *116*, 3029–3085. [CrossRef] [PubMed]

2. Nathan, C.; Ding, A. SnapShot: Reactive Oxygen Intermediates (ROI). *Cell* **2010**, *140*, 951. [CrossRef] [PubMed]

3. Ralser, M.; Wamelink, M.M.; Kowald, A.; Gerisch, B.; Heeren, G.; Struys, E.A.; Klipp, E.; Jakobs, C.; Breitenbach, M.; Lehrach, H.; et al. Dynamic rerouting of the carbohydrate flux is key to counteracting oxidative stress. *J. Biol.* **2007**, *6*, 10. [CrossRef] [PubMed]

4. Janero, D.R.; Hreniuk, D.; Sharif, H.M. Hydroperoxide-induced oxidative stress impairs heart muscle cell carbohydrate metabolism. *Am. J. Physiol. Cell Physiol.* **1994**, *266*, C179–C188. [CrossRef] [PubMed]

5. Le Goffe, C.; Vallette, G.; Charrier, L.; Candelon, T.; Bou-Hanna, C.; Bouhours, J.F.; Laboisse, C.L. Metabolic control of resistance of human epithelial cells to H2O2 and NO stresses. *Biochem. J.* **2002**, *364*, 349–359. [CrossRef] [PubMed]

6. Elchuri, S.; Oberley, T.D.; Qi, W.; Eisenstein, R.S.; Jackson Roberts, L.; Van Remmen, H.; Epstein, C.J.; Huang, T.T. CuZnSOD deficiency leads to persistent and widespread oxidative damage and hepatocarcinogenesis later in life. *Oncogene* **2005**, *24*, 367–380. [CrossRef] [PubMed]

7. Muid, K.A.; Karakaya, H.C.; Koc, A. Absence of superoxide dismutase activity causes nuclear DNA fragmentation during the aging process. *Biochem. Biophys. Res. Commun.* **2014**, *444*, 260–263. [CrossRef]

8. Chelikani, P.; Fita, I.; Loewen, P.C. Diversity of structures and properties among catalases. *Cell. Mol. Life Sci.* **2004**, *61*, 192–208. [CrossRef]

9. Sinitsyna, O.; Krysanova, Z.; Ishchenko, A.; Dikalova, A.E.; Stolyarov, S.; Kolosova, N.; Vasunina, E.; Nevinsky, G. Age-associated changes in oxidative damage and the activity of antioxidant enzymes in rats with inherited overgeneration of free radicals. *J. Cell. Mol. Med.* **2006**, *10*, 206–215. [CrossRef]

10. Perkins, A.; Nelson, K.J.; Parsonage, D.; Poole, L.B.; Karplus, P.A. Peroxiredoxins: Guardians against oxidative stress and modulators of peroxide signaling. *Trends Biochem. Sci.* **2015**, *40*, 435–445. [CrossRef]

11. Margis, R.; Dunand, C.; Teixeira, F.K.; Margis-Pinheiro, M. Glutathione peroxidase family—An evolutionary overview. *FEBS J.* **2008**, *275*, 3959–3970. [CrossRef] [PubMed]

12. Hitomi, K.; Iwai, S.; Tainer, J.A. The intricate structural chemistry of base excision repair machinery: Implications for DNA damage recognition, removal, and repair. *DNA Repair* **2007**, *6*, 410–428. [CrossRef] [PubMed]

13. Robertson, A.B.; Klungland, A.; Rognes, T.; Leiros, I. DNA repair in mammalian cells: Base excision repair: The long and short of it. *Cell. Mol. Life Sci.* **2009**, *66*, 981–993. [CrossRef] [PubMed]

14. Karisch, R.; Fernandez, M.; Taylor, P.; Virtanen, C.; St-Germain, J.R.; Jin, L.L.; Harris, I.S.; Mori, J.; Mak, T.W.; Senis, Y.A.; et al. Global proteomic assessment of the classical protein-tyrosine phosphatome and "Redoxome". *Cell* **2011**, *146*, 826–840. [CrossRef] [PubMed]

15. Mehdy, M.C. Active Oxygen Species in Plant Defense against Pathogens. *Plant Physiol.* **1994**, *105*, 467–472. [CrossRef] [PubMed]

16. Imlay, J.A. Transcription Factors That Defend Bacteria against Reactive Oxygen Species. *Annu. Rev. Microbiol.* **2015**, *69*, 93–108. [CrossRef] [PubMed]

17. Storz, G.; Tartaglia, L.A.; Ames, B.N. Transcriptional regulator of oxidative stress-inducible genes: Direct activation by oxidation. *Science* **1990**, *248*, 189–194. [CrossRef]

18. Christman, M.F.; Morgan, R.W.; Jacobson, F.S.; Ames, B.N. Positive control of a regulon for defenses against oxidative stress and some heat-shock proteins in Salmonella typhimurium. *Cell* **1985**, *41*, 753–762. [CrossRef]

19. Demple, B.; Amabile-Cuevas, C.F. Redox redux: The control of oxidative stress responses. *Cell* **1991**, *67*, 837–839. [CrossRef]

20. Elsen, S.; Swem, L.R.; Swem, D.L.; Bauer, C.E. RegB/RegA, a highly conserved redox-responding global two-component regulatory system. *Microbiol. Mol. Biol. Rev.* **2004**, *68*, 263–279. [CrossRef]

21. Wu, J.; Bauer, C.E. RegB/RegA, a global redox-responding two-component system. *Adv. Exp. Med. Biol.* **2008**, *631*, 131–148. [CrossRef] [PubMed]

22. Wood, Z.A.; Poole, L.B.; Karplus, P.A. Peroxiredoxin evolution and the regulation of hydrogen peroxide signaling. *Science* **2003**, *300*, 650–653. [CrossRef] [PubMed]

23. Jin, D.Y.; Chae, H.Z.; Rhee, S.G.; Jeang, K.T. Regulatory role for a novel human thioredoxin peroxidase in NF-kappaB activation. *J. Biol. Chem.* **1997**, *272*, 30952–30961. [CrossRef] [PubMed]

24. Biteau, B.; Labarre, J.; Toledano, M.B. ATP-dependent reduction of cysteine-sulphinic acid by S. cerevisiae sulphiredoxin. *Nature* **2003**, *425*, 980–984. [CrossRef] [PubMed]

25. Levine, A.; Tenhaken, R.; Dixon, R.; Lamb, C. H_2O_2 from the oxidative burst orchestrates the plant hypersensitive disease resistance response. *Cell* **1994**, *79*, 583–593. [CrossRef]

26. Alvarez, M.E.; Pennell, R.I.; Meijer, P.J.; Ishikawa, A.; Dixon, R.A.; Lamb, C. Reactive oxygen intermediates mediate a systemic signal network in the establishment of plant immunity. *Cell* **1998**, *92*, 773–784. [CrossRef]

27. Foyer, C.H.; Shigeoka, S. Understanding oxidative stress and antioxidant functions to enhance photosynthesis. *Plant Physiol.* **2011**, *155*, 93–100. [CrossRef]

28. Shapiro, B.M. The control of oxidant stress at fertilization. *Science* **1991**, *252*, 533–536. [CrossRef]

29. Wu, S.C.; Liao, C.W.; Pan, R.L.; Juang, J.L. Infection-induced intestinal oxidative stress triggers organ-to-organ immunological communication in Drosophila. *Cell Host Microbe* **2012**, *11*, 410–417. [CrossRef]

30. Ferrandon, D.; Imler, J.L.; Hetru, C.; Hoffmann, J.A. The Drosophila systemic immune response: Sensing and signalling during bacterial and fungal infections. *Nat. Rev. Immunol.* **2007**, *7*, 862–874. [CrossRef]

31. Buchon, N.; Broderick, N.A.; Chakrabarti, S.; Lemaitre, B. Invasive and indigenous microbiota impact intestinal stem cell activity through multiple pathways in Drosophila. *Genes Dev.* **2009**, *23*, 2333–2344. [CrossRef] [PubMed]

32. Tan, S.W.S.; Lee, Q.Y.; Wong, B.S.E.; Cai, Y.; Baeg, G.H. Redox Homeostasis Plays Important Roles in the Maintenance of the Drosophila Testis Germline Stem Cells. *Stem Cell Rep.* **2017**, *9*, 342–354. [CrossRef] [PubMed]

33. Nguyen, T.; Nioi, P.; Pickett, C.B. The Nrf2-antioxidant response element signaling pathway and its activation by oxidative stress. *J. Biol. Chem.* **2009**, *284*, 13291–13295. [CrossRef] [PubMed]

34. Thannickal, V.J.; Fanburg, B.L. Reactive oxygen species in cell signaling. *Am. J. Physiol. Lung Cell. Mol. Physiol.* **2000**, *279*, L1005–L1028. [CrossRef] [PubMed]

35. Mohania, D.; Chandel, S.; Kumar, P.; Verma, V.; Digvijay, K.; Tripathi, D.; Choudhury, K.; Mitten, S.K.; Shah, D. Ultraviolet Radiations: Skin Defense-Damage Mechanism. *Adv. Exp. Med. Biol.* **2017**, *996*, 71–87. [CrossRef]

36. Klotz, L.O.; Steinbrenner, H. Cellular adaptation to xenobiotics: Interplay between xenosensors, reactive oxygen species and FOXO transcription factors. *Redox Biol.* **2017**, *13*, 646–654. [CrossRef]

37. Pizzino, G.; Irrera, N.; Cucinotta, M.; Pallio, G.; Mannino, F.; Arcoraci, V.; Squadrito, F.; Altavilla, D.; Bitto, A. Oxidative Stress: Harms and Benefits for Human Health. *Oxid. Med. Cell. Longev.* **2017**, *2017*, 8416763. [CrossRef]

38. Aggeli, I.K.; Gaitanaki, C.; Beis, I. Involvement of JNKs and p38-MAPK/MSK1 pathways in H_2O_2-induced upregulation of heme oxygenase-1 mRNA in H9c2 cells. *Cell. Signal.* **2006**, *18*, 1801–1812. [CrossRef]

39. McCubrey, J.A.; Lahair, M.M.; Franklin, R.A. Reactive oxygen species-induced activation of the MAP kinase signaling pathways. *Antioxid. Redox Signal.* **2006**, *8*, 1775–1789. [CrossRef]

40. Yong, H.Y.; Koh, M.S.; Moon, A. The p38 MAPK inhibitors for the treatment of inflammatory diseases and cancer. *Expert Opin. Investig. Drugs* **2009**, *18*, 1893–1905. [CrossRef]

41. Davies, C.; Tournier, C. Exploring the function of the JNK (c-Jun N-terminal kinase) signalling pathway in physiological and pathological processes to design novel therapeutic strategies. *Biochem. Soc. Trans.* **2012**, *40*, 85–89. [CrossRef] [PubMed]

42. Katagiri, K.; Matsuzawa, A.; Ichijo, H. Regulation of apoptosis signal-regulating kinase 1 in redox signaling. *Methods Enzymol.* **2010**, *474*, 277–288. [CrossRef] [PubMed]

43. Fujisawa, T.; Takeda, K.; Ichijo, H. ASK family proteins in stress response and disease. *Mol. Biotechnol.* **2007**, *37*, 13–18. [CrossRef] [PubMed]

44. Matsuzawa, A.; Ichijo, H. Redox control of cell fate by MAP kinase: Physiological roles of ASK1-MAP kinase pathway in stress signaling. *Biochim. Biophys. Acta* **2008**, *1780*, 1325–1336. [CrossRef] [PubMed]

45. Hai, T.; Curran, T. Cross-family dimerization of transcription factors Fos/Jun and ATF/CREB alters DNA binding specificity. *Proc. Natl. Acad. Sci. USA* **1991**, *88*, 3720–3724. [CrossRef] [PubMed]

46. Shaulian, E.; Karin, M. AP-1 as a regulator of cell life and death. *Nat. Cell Biol.* **2002**, *4*, E131–E136. [CrossRef] [PubMed]

47. Watson, G.; Ronai, Z.A.; Lau, E. ATF2, a paradigm of the multifaceted regulation of transcription factors in biology and disease. *Pharmacol. Res.* **2017**, *119*, 347–357. [CrossRef]

48. Bohmann, D.; Bos, T.J.; Admon, A.; Nishimura, T.; Vogt, P.K.; Tjian, R. Human proto-oncogene c-jun encodes a DNA binding protein with structural and functional properties of transcription factor AP-1. *Science* **1987**, *238*, 1386–1392. [CrossRef]

49. Hayakawa, J.; Mittal, S.; Wang, Y.; Korkmaz, K.S.; Adamson, E.; English, C.; Ohmichi, M.; McClelland, M.; Mercola, D. Identification of promoters bound by c-Jun/ATF2 during rapid large-scale gene activation following genotoxic stress. *Mol. Cell* **2004**, *16*, 521–535. [CrossRef]

50. Maekawa, T.; Sakura, H.; Kanei-Ishii, C.; Sudo, T.; Yoshimura, T.; Fujisawa, J.; Yoshida, M.; Ishii, S. Leucine zipper structure of the protein CRE-BP1 binding to the cyclic AMP response element in brain. *EMBO J.* **1989**, *8*, 2023–2028. [CrossRef]

51. Li, X.Y.; Green, M.R. Intramolecular inhibition of activating transcription factor-2 function by its DNA-binding domain. *Genes Dev.* **1996**, *10*, 517–527. [CrossRef] [PubMed]

52. Vogt, P.K. Fortuitous convergences: The beginnings of JUN. *Nat. Rev. Cancer* **2002**, *2*, 465–469. [CrossRef] [PubMed]

53. Angel, P.; Allegretto, E.A.; Okino, S.T.; Hattori, K.; Boyle, W.J.; Hunter, T.; Karin, M. Oncogene jun encodes a sequence-specific trans-activator similar to AP-1. *Nature* **1988**, *332*, 166–171. [CrossRef] [PubMed]

54. Eferl, R.; Wagner, E.F. AP-1: A double-edged sword in tumorigenesis. *Nat. Rev. Cancer* **2003**, *3*, 859–868. [CrossRef] [PubMed]

55. Binetruy, B.; Smeal, T.; Karin, M. Ha-Ras augments c-Jun activity and stimulates phosphorylation of its activation domain. *Nature* **1991**, *351*, 122–127. [CrossRef] [PubMed]

56. Smeal, T.; Binetruy, B.; Mercola, D.A.; Birrer, M.; Karin, M. Oncogenic and transcriptional cooperation with Ha-Ras requires phosphorylation of c-Jun on serines 63 and 73. *Nature* **1991**, *354*, 494–496. [CrossRef] [PubMed]

57. Shaulian, E.; Schreiber, M.; Piu, F.; Beeche, M.; Wagner, E.F.; Karin, M. The mammalian UV response: C-Jun induction is required for exit from p53-imposed growth arrest. *Cell* **2000**, *103*, 897–907. [CrossRef]

58. Liu, T.; Zhang, L.; Joo, D.; Sun, S.C. NF-kappaB signaling in inflammation. *Signal Transduct. Target. Ther.* **2017**, *2*. [CrossRef]

59. Pineda-Molina, E.; Klatt, P.; Vazquez, J.; Marina, A.; Garcia de Lacoba, M.; Perez-Sala, D.; Lamas, S. Glutathionylation of the p50 subunit of NF-kappaB: A mechanism for redox-induced inhibition of DNA binding. *Biochemistry* **2001**, *40*, 14134–14142. [CrossRef]

60. Droge, W. Redox regulation in anabolic and catabolic processes. *Curr. Opin. Clin. Nutr. Metab. Care* **2006**, *9*, 190–195. [CrossRef]

61. Nioi, P.; McMahon, M.; Itoh, K.; Yamamoto, M.; Hayes, J.D. Identification of a novel Nrf2-regulated antioxidant response element (ARE) in the mouse NAD(P)H:quinone oxidoreductase 1 gene: Reassessment of the ARE consensus sequence. *Biochem. J.* **2003**, *374*, 337–348. [CrossRef]

62. Hayes, J.D.; McMahon, M. NRF2 and KEAP1 mutations: Permanent activation of an adaptive response in cancer. *Trends Biochem. Sci.* **2009**, *34*, 176–188. [CrossRef]

63. Motohashi, H.; Yamamoto, M. Nrf2-Keap1 defines a physiologically important stress response mechanism. *Trends Mol. Med.* **2004**, *10*, 549–557. [CrossRef]

64. Mimura, J.; Inose-Maruyama, A.; Taniuchi, S.; Kosaka, K.; Yoshida, H.; Yamazaki, H.; Kasai, S.; Harada, N.; Kaufman, R.J.; Oyadomari, S.; et al. Concomitant Nrf2- and ATF4-activation by Carnosic Acid Cooperatively Induces Expression of Cytoprotective Genes. *Int. J. Mol. Sci.* **2019**, *20*, 1706. [CrossRef]

65. Tonelli, C.; Chio, I.I.C.; Tuveson, D.A. Transcriptional Regulation by Nrf2. *Antioxid. Redox Signal.* **2018**, *29*, 1727–1745. [CrossRef]

66. De Nicola, G.M.; Karreth, F.A.; Humpton, T.J.; Gopinathan, A.; Wei, C.; Frese, K.; Mangal, D.; Yu, K.H.; Yeo, C.J.; Calhoun, E.S.; et al. Oncogene-induced Nrf2 transcription promotes ROS detoxification and tumorigenesis. *Nature* **2011**, *475*, 106–109. [CrossRef]

67. Sobotta, M.C.; Liou, W.; Stocker, S.; Talwar, D.; Oehler, M.; Ruppert, T.; Scharf, A.N.; Dick, T.P. Peroxiredoxin-2 and STAT3 form a redox relay for H_2O_2 signaling. *Nat. Chem. Biol.* **2015**, *11*, 64–70. [CrossRef]

68. Rampon, C.; Volovitch, M.; Joliot, A.; Vriz, S. Hydrogen Peroxide and Redox Regulation of Developments. *Antioxidants* **2018**, *7*, 159. [CrossRef]

69. Johnson, T.M.; Yu, Z.X.; Ferrans, V.J.; Lowenstein, R.A.; Finkel, T. Reactive oxygen species are downstream mediators of p53-dependent apoptosis. *Proc. Natl. Acad. Sci. USA* **1996**, *93*, 11848–11852. [CrossRef]

70. Gambino, V.; De Michele, G.; Venezia, O.; Migliaccio, P.; Dall'Olio, V.; Bernard, L.; Minardi, S.P.; Della Fazia, M.A.; Bartoli, D.; Servillo, G.; et al. Oxidative stress activates a specific p53 transcriptional response that regulates cellular senescence and aging. *Aging Cell* **2013**, *12*, 435–445. [CrossRef]

71. Giorgio, M.; Migliaccio, E.; Orsini, F.; Paolucci, D.; Moroni, M.; Contursi, C.; Pelliccia, G.; Luzi, L.; Minucci, S.; Marcaccio, M.; et al. Electron transfer between cytochrome c and p66Shc generates reactive oxygen species that trigger mitochondrial apoptosis. *Cell* **2005**, *122*, 221–233. [CrossRef]

72. Trinei, M.; Giorgio, M.; Cicalese, A.; Baruzzi, S.; Ventura, A.; Migliaccio, E.; Milia, E.; Padura, I.M.; Raker, V.A.; Maccarana, M.; et al. A p53-p66Shc signalling pathway controls intracellular redox status, levels of oxidation-damaged DNA and oxidative stress-induced apoptosis. *Oncogene* **2002**, *21*, 3872–3878. [CrossRef]

73. Bhat, S.S.; Anand, D.; Khanday, F.A. p66Shc as a switch in bringing about contrasting responses in cell growth: Implications on cell proliferation and apoptosis. *Mol. Cancer* **2015**, *14*, 76. [CrossRef]

74. Loudon, A.S. Circadian biology: A 2.5 billion year old clock. *Curr. Biol.* **2012**, *22*, R570–R571. [CrossRef]

75. Reppert, S.M.; Weaver, D.R. Molecular analysis of mammalian circadian rhythms. *Annu. Rev. Physiol.* **2001**, *63*, 647–676. [CrossRef]

76. Dushay, M.S.; Rosbash, M.; Hall, J.C. Mapping the clock rhythm mutation to the period locus of Drosophila melanogaster by germline transformation. *J. Neurogenet.* **1992**, *8*, 173–179. [CrossRef]

77. Reddy, P.; Jacquier, A.C.; Abovich, N.; Petersen, G.; Rosbash, M. The period clock locus of D. melanogaster codes for a proteoglycan. *Cell* **1986**, *46*, 53–61. [CrossRef]

78. Zehring, W.A.; Wheeler, D.A.; Reddy, P.; Konopka, R.J.; Kyriacou, C.P.; Rosbash, M.; Hall, J.C. P-element transformation with period locus DNA restores rhythmicity to mutant, arrhythmic Drosophila melanogaster. *Cell* **1984**, *39*, 369–376. [CrossRef]

79. McClung, C.R.; Fox, B.A.; Dunlap, J.C. The Neurospora clock gene frequency shares a sequence element with the Drosophila clock gene period. *Nature* **1989**, *339*, 558–562. [CrossRef]

80. Bargiello, T.A.; Jackson, F.R.; Young, M.W. Restoration of circadian behavioural rhythms by gene transfer in Drosophila. *Nature* **1984**, *312*, 752–754. [CrossRef]

81. Dunlap, J.C. Molecular bases for circadian clocks. *Cell* **1999**, *96*, 271–290. [CrossRef]

82. Ko, C.H.; Takahashi, J.S. Molecular components of the mammalian circadian clock. *Hum. Mol. Genet* **2006**, *15* (Suppl. 2), R271–R277. [CrossRef]

83. Bellet, M.M.; Sassone-Corsi, P. Mammalian circadian clock and metabolism—The epigenetic link. *J. Cell Sci.* **2010**, *123*, 3837–3848. [CrossRef]

84. Buhr, E.D.; Takahashi, J.S. Molecular components of the Mammalian circadian clock. *Handb. Exp. Pharmacol.* **2013**, 3–27. [CrossRef]

85. Peirson, S.N.; Halford, S.; Foster, R.G. The evolution of irradiance detection: Melanopsin and the non-visual opsins. *Philos. Trans. R. Soc. B Biol. Sci.* **2009**, *364*, 2849–2865. [CrossRef]

86. Roenneberg, T.; Foster, R.G. Twilight times: Light and the circadian system. *Photochem. Photobiol.* **1997**, *66*, 549–561. [CrossRef]

87. Whitmore, D.; Foulkes, N.S.; Strahle, U.; Sassone-Corsi, P. Zebrafish Clock rhythmic expression reveals independent peripheral circadian oscillators. *Nat. Neurosci.* **1998**, *1*, 701–707. [CrossRef]

88. Turek, F.W.; Joshu, C.; Kohsaka, A.; Lin, E.; Ivanova, G.; McDearmon, E.; Laposky, A.; Losee-Olson, S.; Easton, A.; Jensen, D.R.; et al. Obesity and metabolic syndrome in circadian Clock mutant mice. *Science* **2005**, *308*, 1043–1045. [CrossRef]

89. Asher, G.; Sassone-Corsi, P. Time for food: The intimate interplay between nutrition, metabolism, and the circadian clock. *Cell* **2015**, *161*, 84–92. [CrossRef]

90. Zarrinpar, A.; Chaix, A.; Panda, S. Daily Eating Patterns and Their Impact on Health and Disease. *Trends Endocrinol. Metab.* **2016**, *27*, 69–83. [CrossRef]

91. DiAngelo, J.R.; Erion, R.; Crocker, A.; Sehgal, A. The central clock neurons regulate lipid storage in Drosophila. *PLoS ONE* **2011**, *6*, e19921. [CrossRef] [PubMed]

92. Lee, J.; Moulik, M.; Fang, Z.; Saha, P.; Zou, F.; Xu, Y.; Nelson, D.L.; Ma, K.; Moore, D.D.; Yechoor, V.K. Bmal1 and beta-cell clock are required for adaptation to circadian disruption, and their loss of function leads to oxidative stress-induced beta-cell failure in mice. *Mol. Cell. Biol.* **2013**, *33*, 2327–2338. [CrossRef]

93. Kondratov, R.V.; Kondratova, A.A.; Gorbacheva, V.Y.; Vykhovanets, O.V.; Antoch, M.P. Early aging and age-related pathologies in mice deficient in BMAL1, the core componentof the circadian clock. *Genes Dev.* **2006**, *20*, 1868–1873. [CrossRef] [PubMed]

94. Tamaru, T.; Hattori, M.; Ninomiya, Y.; Kawamura, G.; Vares, G.; Honda, K.; Mishra, D.P.; Wang, B.; Benjamin, I.; Sassone-Corsi, P.; et al. ROS stress resets circadian clocks to coordinate pro-survival signals. *PLoS ONE* **2013**, *8*, e82006. [CrossRef] [PubMed]

95. Gyongyosi, N.; Kaldi, K. Interconnections of reactive oxygen species homeostasis and circadian rhythm in Neurospora crassa. *Antioxid. Redox Signal.* **2014**, *20*, 3007–3023. [CrossRef] [PubMed]

96. Brown, S.A. Circadian Metabolism: From Mechanisms to Metabolomics and Medicine. *Trends Endocrinol. Metab.* **2016**, *27*, 415–426. [CrossRef] [PubMed]

97. Brown, S.A.; Gaspar, L. Circadian Metabolomics: Insights for Biology and Medicine. In *A Time for Metabolism and Hormones*; Sassone-Corsi, P., Christen, Y., Eds.; Springer: Cham, Switzerland, 2016; pp. 79–85. [CrossRef]

98. Nakahata, Y.; Sahar, S.; Astarita, G.; Kaluzova, M.; Sassone-Corsi, P. Circadian control of the NAD$^+$ salvage pathway by CLOCK-SIRT1. *Science* **2009**, *324*, 654–657. [CrossRef]

99. Peek, C.B.; Affinati, A.H.; Ramsey, K.M.; Kuo, H.Y.; Yu, W.; Sena, L.A.; Ilkayeva, O.; Marcheva, B.; Kobayashi, Y.; Omura, C.; et al. Circadian clock NAD$^+$ cycle drives mitochondrial oxidative metabolism in mice. *Science* **2013**, *342*, 1243417. [CrossRef]

100. Nakahata, Y.; Kaluzova, M.; Grimaldi, B.; Sahar, S.; Hirayama, J.; Chen, D.; Guarente, L.P.; Sassone-Corsi, P. The NAD$^+$-dependent deacetylase SIRT1 modulates CLOCK-mediated chromatin remodeling and circadian control. *Cell* **2008**, *134*, 329–340. [CrossRef]

101. Asher, G.; Gatfield, D.; Stratmann, M.; Reinke, H.; Dibner, C.; Kreppel, F.; Mostoslavsky, R.; Alt, F.W.; Schibler, U. SIRT1 regulates circadian clock gene expression through PER2 deacetylation. *Cell* **2008**, *134*, 317–328. [CrossRef]

102. Wilking, M.; Ndiaye, M.; Mukhtar, H.; Ahmad, N. Circadian rhythm connections to oxidative stress: Implications for human health. *Antioxid. Redox Signal.* **2013**, *19*, 192–208. [CrossRef] [PubMed]

103. Putker, M.; O'Neill, J.S. Reciprocal Control of the Circadian Clock and Cellular Redox State—A Critical Appraisal. *Mol. Cells* **2016**, *39*, 6–19. [CrossRef] [PubMed]

104. Krishnan, N.; Davis, A.J.; Giebultowicz, J.M. Circadian regulation of response to oxidative stress in Drosophila melanogaster. *Biochem. Biophys. Res. Commun.* **2008**, *374*, 299–303. [CrossRef] [PubMed]

105. Tomas-Zapico, C.; Coto-Montes, A.; Martinez-Fraga, J.; Rodriguez-Colunga, M.J.; Tolivia, D. Effects of continuous light exposure on antioxidant enzymes, porphyric enzymes and cellular damage in the Harderian gland of the Syrian hamster. *J. Pineal Res.* **2003**, *34*, 60–68. [CrossRef] [PubMed]

106. Fanjul-Moles, M.L.; Lopez-Riquelme, G.O. Relationship between Oxidative Stress, Circadian Rhythms, and AMD. *Oxid. Med. Cell. Longev.* **2016**, *2016*, 7420637. [CrossRef] [PubMed]

107. Reiter, R.J.; Sainz, R.M.; Lopez-Burillo, S.; Mayo, J.C.; Manchester, L.C.; Tan, D.X. Melatonin ameliorates neurologic damage and neurophysiologic deficits in experimental models of stroke. *Ann. N.Y. Acad. Sci.* **2003**, *993*, 35–47, discussion 48–53. [CrossRef]

108. Anderson, P.; Kedersha, N. RNA granules: Post-transcriptional and epigenetic modulators of gene expression. *Nat. Rev. Mol. Cell Biol.* **2009**, *10*, 430–436. [CrossRef]

109. Heberle, A.M.; Razquin Navas, P.; Langelaar-Makkinje, M.; Kasack, K.; Sadik, A.; Faessler, E.; Hahn, U.; Marx-Stoelting, P.; Opitz, C.A.; Sers, C.; et al. The PI3K and MAPK/p38 pathways control stress granule assembly in a hierarchical manner. *Life Sci. Alliance* **2019**, *2*, e201800257. [CrossRef]

110. Khong, A.; Matheny, T.; Jain, S.; Mitchell, S.F.; Wheeler, J.R.; Parker, R. The Stress Granule Transcriptome Reveals Principles of mRNA Accumulation in Stress Granules. *Mol. Cell* **2017**, *68*, 808–820. [CrossRef]

111. Takahashi, M.; Higuchi, M.; Matsuki, H.; Yoshita, M.; Ohsawa, T.; Oie, M.; Fujii, M. Stress granules inhibit apoptosis by reducing reactive oxygen species production. *Mol. Cell. Biol.* **2013**, *33*, 815–829. [CrossRef]

112. Guarino, A.M.; Troiano, A.; Pizzo, E.; Bosso, A.; Vivo, M.; Pinto, G.; Amoresano, A.; Pollice, A.; La Mantia, G.; Calabro, V. Oxidative Stress Causes Enhanced Secretion of YB-1 Protein that Restrains Proliferation of Receiving Cells. *Genes* **2018**, *9*, 513. [CrossRef] [PubMed]

113. Pagano, C.; di Martino, O.; Ruggiero, G.; Maria Guarino, A.; Mueller, N.; Siauciunaite, R.; Reischl, M.; Simon Foulkes, N.; Vallone, D.; Calabro, V. The tumor-associated YB-1 protein: New player in the circadian control of cell proliferation. *Oncotarget* **2017**, *8*, 6193–6205. [CrossRef] [PubMed]

114. Pagano, C.; Siauciunaite, R.; Idda, M.L.; Ruggiero, G.; Ceinos, R.M.; Pagano, M.; Frigato, E.; Bertolucci, C.; Foulkes, N.S.; Vallone, D. Evolution shapes the responsiveness of the D-box enhancer element to light and reactive oxygen species in vertebrates. *Sci. Rep.* **2018**, *8*, 13180. [CrossRef] [PubMed]

115. Hirayama, J.; Cho, S.; Sassone-Corsi, P. Circadian control by the reduction/oxidation pathway: Catalase represses light-dependent clock gene expression in the zebrafish. *Proc. Natl. Acad. Sci. USA* **2007**, *104*, 15747–15752. [CrossRef] [PubMed]

116. Vallone, D.; Gondi, S.B.; Whitmore, D.; Foulkes, N.S. E-box function in a period gene repressed by light. *Proc. Natl. Acad. Sci. USA* **2004**, *101*, 4106–4111. [CrossRef] [PubMed]

117. Vallone, D.; Santoriello, C.; Gondi, S.B.; Foulkes, N.S. Basic protocols for zebrafish cell lines: Maintenance and transfection. *Methods Mol. Biol.* **2007**, *362*, 429–441. [CrossRef] [PubMed]

118. Vallone, D.; Lahiri, K.; Dickmeis, T.; Foulkes, N.S. Zebrafish cell clocks feel the heat and see the light! *Zebrafish* **2005**, *2*, 171–187. [CrossRef]

119. Menaker, M.; Takahashi, J.S.; Eskin, A. The physiology of circadian pacemakers. *Annu. Rev. Physiol.* **1978**, *40*, 501–526. [CrossRef]

120. Davies, W.I.; Tamai, T.K.; Zheng, L.; Fu, J.K.; Rihel, J.; Foster, R.G.; Whitmore, D.; Hankins, M.W. An extended family of novel vertebrate photopigments is widely expressed and displays a diversity of function. *Genome Res.* **2015**, *25*, 1666–1679. [CrossRef]

121. Hockberger, P.E.; Skimina, T.A.; Centonze, V.E.; Lavin, C.; Chu, S.; Dadras, S.; Reddy, J.K.; White, J.G. Activation of flavin-containing oxidases underlies light-induced production of H_2O_2 in mammalian cells. *Proc. Natl. Acad. Sci. USA* **1999**, *96*, 6255–6260. [CrossRef]

122. Zhao, H.; Di Mauro, G.; Lungu-Mitea, S.; Negrini, P.; Guarino, A.M.; Frigato, E.; Braunbeck, T.; Ma, H.; Lamparter, T.; Vallone, D.; et al. Modulation of DNA Repair Systems in Blind Cavefish during Evolution in Constant Darkness. *Curr. Biol.* **2018**, *28*, 3229–3243. [CrossRef] [PubMed]

123. Lichtsteiner, S.; Wuarin, J.; Schibler, U. The interplay of DNA-binding proteins on the promoter of the mouse albumin gene. *Cell* **1987**, *51*, 963–973. [CrossRef]

124. Mueller, C.R.; Maire, P.; Schibler, U. DBP, a liver-enriched transcriptional activator, is expressed late in ontogeny and its tissue specificity is determined posttranscriptionally. *Cell* **1990**, *61*, 279–291. [CrossRef]

125. Drolet, D.W.; Scully, K.M.; Simmons, D.M.; Wegner, M.; Chu, K.T.; Swanson, L.W.; Rosenfeld, M.G. TEF, a transcription factor expressed specifically in the anterior pituitary during embryogenesis, defines a new class of leucine zipper proteins. *Genes Dev.* **1991**, *5*, 1739–1753. [CrossRef] [PubMed]

126. Hunger, S.P.; Ohyashiki, K.; Toyama, K.; Cleary, M.L. Hlf, a novel hepatic bZIP protein, shows altered DNA-binding properties following fusion to E2A in t(17;19) acute lymphoblastic leukemia. *Genes Dev.* **1992**, *6*, 1608–1620. [CrossRef]

127. Lin, S.C.; Lin, M.H.; Horvath, P.; Reddy, K.L.; Storti, R.V. PDP1, a novel Drosophila PAR domain bZIP transcription factor expressed in developing mesoderm, endoderm and ectoderm, is a transcriptional regulator of somatic muscle genes. *Development* **1997**, *124*, 4685–4696.

128. Xu, X.; Liu, L.; Wong, K.C.; Ge, R. Cloning and characterization of two isoforms of the zebrafish thyrotroph embryonic factor (tef alpha and tefbeta). *Biochim. Biophys. Acta* **1998**, *1395*, 13–20. [CrossRef]

129. Cowell, I.G. E4BP4/NFIL3, a PAR-related bZIP factor with many roles. *Bioessays* **2002**, *24*, 1023–1029. [CrossRef] [PubMed]

130. Falvey, E.; Marcacci, L.; Schibler, U. DNA-binding specificity of PAR and C/EBP leucine zipper proteins: A single amino acid substitution in the C/EBP DNA-binding domain confers PAR-like specificity to C/EBP. *Biol. Chem.* **1996**, *377*, 797–809.

131. Li, S.; Hunger, S.P. The DBP transcriptional activation domain is highly homologous to that of HLF and TEF and is not responsible for the tissue type-specific transcriptional activity of DBP. *Gene* **2001**, *263*, 239–245. [CrossRef]

132. Gachon, F. Physiological function of PARbZip circadian clock-controlled transcription factors. *Ann. Med.* **2007**, *39*, 562–571. [CrossRef] [PubMed]

133. Wuarin, J.; Schibler, U. Expression of the liver-enriched transcriptional activator protein DBP follows a stringent circadian rhythm. *Cell* **1990**, *63*, 1257–1266. [CrossRef]

134. Falvey, E.; Fleury-Olela, F.; Schibler, U. The rat hepatic leukemia factor (HLF) gene encodes two transcriptional activators with distinct circadian rhythms, tissue distributions and target preferences. *EMBO J.* **1995**, *14*, 4307–4317. [CrossRef] [PubMed]

135. Jang, Y.S.; Lee, M.H.; Lee, S.H.; Bae, K. Cu/Zn superoxide dismutase is differentially regulated in period gene-mutant mice. *Biochem. Biophys. Res. Commun.* **2011**, *409*, 22–27. [CrossRef] [PubMed]

136. Hardeland, R.; Coto-Montes, A.; Poeggeler, B. Circadian rhythms, oxidative stress, and antioxidative defense mechanisms. *Chronobiol. Int.* **2003**, *20*, 921–962. [CrossRef] [PubMed]

137. Beaver, L.M.; Klichko, V.I.; Chow, E.S.; Kotwica-Rolinska, J.; Williamson, M.; Orr, W.C.; Radyuk, S.N.; Giebultowicz, J.M. Circadian regulation of glutathione levels and biosynthesis in Drosophila melanogaster. *PLoS ONE* **2012**, *7*, e50454. [CrossRef]

138. Blanco, R.A.; Ziegler, T.R.; Carlson, B.A.; Cheng, P.Y.; Park, Y.; Cotsonis, G.A.; Accardi, C.J.; Jones, D.P. Diurnal variation in glutathione and cysteine redox states in human plasma. *Am. J. Clin. Nutr.* **2007**, *86*, 1016–1023. [CrossRef] [PubMed]

139. Gachon, F.; Olela, F.F.; Schaad, O.; Descombes, P.; Schibler, U. The circadian PAR-domain basic leucine zipper transcription factors DBP, TEF, and HLF modulate basal and inducible xenobiotic detoxification. *Cell Metab.* **2006**, *4*, 25–36. [CrossRef]

140. Ben-Moshe, Z.; Vatine, G.; Alon, S.; Tovin, A.; Mracek, P.; Foulkes, N.S.; Gothilf, Y. Multiple PAR and E4BP4 bZIP transcription factors in zebrafish: Diverse spatial and temporal expression patterns. *Chronobiol. Int.* **2010**, *27*, 1509–1531. [CrossRef]

141. Cavallari, N.; Frigato, E.; Vallone, D.; Frohlich, N.; Lopez-Olmeda, J.F.; Foa, A.; Berti, R.; Sanchez-Vazquez, F.J.; Bertolucci, C.; Foulkes, N.S. A blind circadian clock in cavefish reveals that opsins mediate peripheral clock photoreception. *PLoS Biol.* **2011**, *9*, e1001142. [CrossRef]

142. Ceinos, R.M.; Frigato, E.; Pagano, C.; Frohlich, N.; Negrini, P.; Cavallari, N.; Vallone, D.; Fuselli, S.; Bertolucci, C.; Foulkes, N.S. Mutations in blind cavefish target the light-regulated circadian clock gene, period 2. *Sci. Rep.* **2018**, *8*, 8754. [CrossRef] [PubMed]

143. Gerkema, M.P.; Davies, W.I.; Foster, R.G.; Menaker, M.; Hut, R.A. The nocturnal bottleneck and the evolution of activity patterns in mammals. *Proc Biol Sci.* **2013**, *280*, 20130508. [CrossRef] [PubMed]

144. Maor, R.; Dayan, T.; Ferguson-Gow, H.; Jones, K.E. Temporal niche expansion in mammals from a nocturnal ancestor after dinosaur extinction. *Nat. Ecol. Evol.* **2017**, *1*, 1889–1895. [CrossRef] [PubMed]

145. Heesy, C.P.; Hall, M.I. The nocturnal bottleneck and the evolution of mammalian vision. *Brain Behav. Evol.* **2010**, *75*, 195–203. [CrossRef] [PubMed]

146. Malik, A.; Domankevich, V.; Lijuan, H.; Xiaodong, F.; Korol, A.; Avivi, A.; Shams, I. Genome maintenance and bioenergetics of the long-lived hypoxia-tolerant and cancer-resistant blind mole rat, Spalax: A cross-species analysis of brain transcriptome. *Sci. Rep.* **2016**, *6*, 38624. [CrossRef] [PubMed]

147. Klein, T.J.; Glazer, P.M. The tumor microenvironment and DNA repair. *Semin. Radiat. Oncol.* **2010**, *20*, 282–287. [CrossRef]

148. Gorbunova, V.; Hine, C.; Tian, X.; Ablaeva, J.; Gudkov, A.V.; Nevo, E.; Seluanov, A. Cancer resistance in the blind mole rat is mediated by concerted necrotic cell death mechanism. *Proc. Natl. Acad. Sci. USA* **2012**, *109*, 19392–19396. [CrossRef]

149. Ashur-Fabian, O.; Avivi, A.; Trakhtenbrot, L.; Adamsky, K.; Cohen, M.; Kajakaro, G.; Joel, A.; Amariglio, N.; Nevo, E.; Rechavi, G. Evolution of p53 in hypoxia-stressed Spalax mimics human tumor mutation. *Proc. Natl. Acad. Sci. USA* **2004**, *101*, 12236–12241. [CrossRef]

International Journal of
Molecular Sciences

Review

Mediators of Physical Activity Protection against ROS-Linked Skeletal Muscle Damage

Sergio Di Meo [1], Gaetana Napolitano [2] and Paola Venditti [1,*

[1]   Dipartimento di Biologia, Università di Napoli Federico II, Complesso Universitario Monte Sant'Angelo, Via Cinthia, I-80126 Napoli, Italy; serdimeo@unina.it
[2]   Dipartimento di Scienze e Tecnologie, Università degli Studi di Napoli Parthenope, via Acton n. 38-I-80133 Napoli, Italy; gaetana.napolitano@uniparthenope.it
*   Correspondence: venditti@unina.it; Tel.: +39-081-2535080; Fax: +39-081-679233

Received: 2 May 2019; Accepted: 17 June 2019; Published: 20 June 2019

Abstract: Unaccustomed and/or exhaustive exercise generates excessive free radicals and reactive oxygen and nitrogen species leading to muscle oxidative stress-related damage and impaired contractility. Conversely, a moderate level of free radicals induces the body's adaptive responses. Thus, a low oxidant level in resting muscle is essential for normal force production, and the production of oxidants during each session of physical training increases the body's antioxidant defenses. Mitochondria, NADPH oxidases and xanthine oxidases have been identified as sources of free radicals during muscle contraction, but the exact mechanisms underlying exercise-induced harmful or beneficial effects yet remain elusive. However, it is clear that redox signaling influences numerous transcriptional activators, which regulate the expression of genes involved in changes in muscle phenotype. The mitogen-activated protein kinase family is one of the main links between cellular oxidant levels and skeletal muscle adaptation. The family components phosphorylate and modulate the activities of hundreds of substrates, including transcription factors involved in cell response to oxidative stress elicited by exercise in skeletal muscle. To elucidate the complex role of ROS in exercise, here we reviewed the literature dealing on sources of ROS production and concerning the most important redox signaling pathways, including MAPKs that are involved in the responses to acute and chronic exercise in the muscle, particularly those involved in the induction of antioxidant enzymes.

Keywords: insulin resistance; cancer; cardiovascular disease; neurodegenerative disorders; exercise; mitochondria; oxidative stress; PGC-1; Nrf2; UCPs

---

1. Introduction

For several years the practice of physical activity has expanded in scope from competitive sports to disease prevention and health promotion. Therefore, physical activity has been widely recognized as a means for the primary prevention of chronic diseases as well as for patient treatment and rehabilitation [1]. Furthermore, regular physical activity (training) has beneficial effects on people's health and well-being.

The results of numerous studies have shown that regular physical activity reduces risk of several diseases including cardiovascular diseases, type 2 diabetes (T2DM), some types of cancer, osteoporosis, fall-related injuries, depression, and obesity [1–3].

Despite these clear benefits, little is known about the adaptive mechanisms involved in the protection offered by exercise even though to date accumulating evidence has allowed establishing that the production of free radicals represents a potential link between exercise and protection against diseases.

Currently, free radicals are recognized to play a crucial role in the regulation of critical physiological processes at both the cellular and system level, and be involved, as causal factors, in the development of pathological conditions. The regulatory role of free radicals is a relatively recent discovery, because for several decades they were thought to cause exclusively damaging effects and were gradually implicated in various pathologies, including cardiovascular disease, diabetes, rheumatoid arthritis, cancer, and neurodegenerative disorders [4].

The double role played by oxidants in living systems seems to be dependent on the extent of their production. Indeed, if produced in a massive extent, oxidants cause oxidative damage and tissue dysfunction whereas, when moderately produced, they serve as molecular signals activating adaptive responses that are useful for the organism.

A paradigmatic example is provided just by the exercise. Indeed, a single session of strenuous or prolonged exercise leads to the production of high amounts of radicals and other reactive oxygen species (ROS), which cause tissue damage and dysfunction. Conversely, the single sessions of a training program produce low amounts of ROS, which can induce adaptive responses beneficial for the organism [5].

Interestingly, the incidence of some ROS associated diseases, among which T2DM, rheumatic arthritis, heart disfunctions, Alzheimer and Parkinson diseases, is reduced by the execution of regular physical activity [1,6].

The balance of free radicals inside the skeletal muscle is very important, particularly in the context of exercise and sport as the main adaptations occur in the trained skeletal muscle, which can differ with the type of exercise but seem to be nevertheless dependent on ROS production. Thus, aerobic physical activity induces skeletal muscle adaptive responses [1] able to determine an increased resistance to conditions, among which prolonged or strenuous exercise, in which ROS production increase [7–9]. Conversely, heavy resistance exercise determines hypertrophy and increased strength production but does not change biochemical characteristics of muscle cells. On the other hand, metabolism of glucose and lipids in skeletal muscles during the resting state and insulin action in insulin-resistant individuals are improved by both aerobic [10] and resistance [11] exercises in skeletal muscle leading to decreased conversion rates to overt diabetes.

In recent times, much progress has been made in understanding the mechanisms underlying the adaptations evoked in skeletal muscle. However, other studies are needed to understand what factors lead ROS to become signal and/or stress agents and the molecular mechanisms through which ROS directly interact with critical signaling molecules to initiate signaling in skeletal muscle. This review, after examining the link between physical activity and ROS production, focuses on signaling pathways, such as MAPKS and transcription factors and cofactors, through which ROS produced in the regular physical activity elicit the adaptive responses implicated in increased antioxidant defenses effectiveness and mitochondrial content of skeletal muscle.

2. Reactive Oxygen and Nitrogen Species

Until about the mid-20th century free radicals, whose existence in chemical systems had been demonstrated by Gomberg's work [12], were still believed as too reactive species to exist in vivo [13]. Subsequently, when free radical existence in biological systems was recognized, they were thought to cause exclusively damaging effects and to be involved in the development of pathological conditions. In particular, ROS were thought to be involved in the general aging process and in many age-associated diseases [14].

This view was mainly supported by the finding that ROS, reacting with most biological macromolecules, cause their oxidative modification, which can result in the loss of their function [15]. In fact, ROS include both highly reactive species, such as the hydroxyl radical ($^\bullet$OH), which reacts soon after its formation, and less reactive species, among which are superoxide ($O_2^{\bullet-}$) and hydrogen peroxide (H_2O_2) [16]. Similarly, other reactive species containing nitrogen, named reactive nitrogen species (RNS), include both species not very reactive, such as nitric oxide (NO$^\bullet$), and species very reactive such as the peroxynitrite ($ONOO^-$), which originates from $NO^\bullet$ and is a very strong oxidant for biomolecules and can also undergo decomposition releasing small amounts of $^\bullet OH$ [17].

Many studies have reported that in normal conditions potentially toxic ROS and RNS are constantly produced at a low level in living systems. Aerobic organisms are equipped with an integrated system of antioxidant defenses to counteract the effects of ROS and RNS [18]. The antioxidant network consists of free radical scavengers of low molecular weight and a composite enzymatic system, which can scavenge free radicals, interrupt chain reactions, remove or repair the damaged components in the cells. Enzymatic antioxidants include a family of metalloenzymes called superoxide dismutases (SODs) [19], which convert $O_2^{\bullet-}$ to H_2O_2, catalase (CAT) an enzyme catalyzing the decomposition of H_2O_2 to H_2O and O_2 [20], glutathione peroxidases (GPXs) [21], a family of selenoproteins which decomposes H_2O_2 using as substrate the reduced glutathione (GSH) which is converted to oxidized glutathione (GSSG), glutathione reductase (GR) which reduces GSSG and restores GSH utilizing NADPH as a source of reducing equivalents [22] (Figure 1). In the cells the redox homeostasis is maintained other than the redox couple GSH/GSSG also by the thioredoxin proteins (Trxs), which are involved in the reduction of protein disulfide [23] and are regenerated by thioredoxin reductase (TrxR) and NADPH [24]. Trxs also collaborate with peroxiredoxins (Prxs) in the hydroperoxide removal as Prxs reduce both hydrogen peroxides and lipid hydroperoxides to water and alcohol with the help of the proteins containing thiol such as Trxs [25]. It is worth noting that Trxs, GPX, and SOD have also been recognized as potential systems able to remove RNS [26].

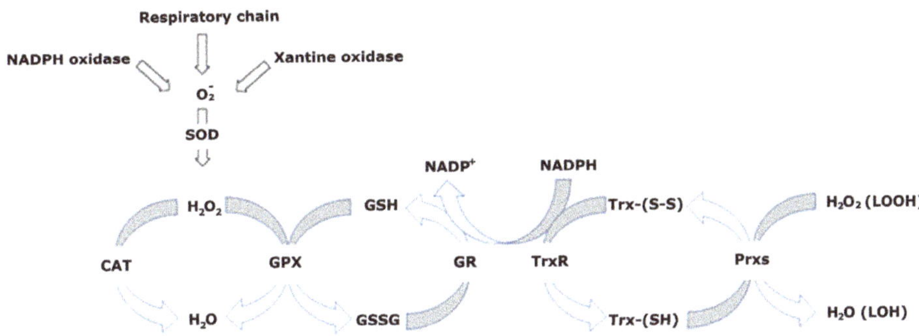

Figure 1. Reactions by which reactive oxygen species (ROS) are produced and removed by antioxidant defense system in skeletal muscle.

Normally, the antioxidant system rapidly removes ROS and RNS before they cause cellular dysfunction and eventual cell death. However, in the living systems generation slightly overcome the capacity of the antioxidant defense system to neutralize ROS, therefore a modest level of oxidative damage is always present. Probably, ROS are not all eliminated because they perform important roles, so that the challenge for the survival process was to evolve antioxidant defenses that allow such roles while minimizing damage.

However, when a greater imbalance occurs in favor of the ROS, oxidative stress ensues [27] characterized by widespread tissue damage, to which cells can adapt sometimes by upregulating the antioxidant system.

3. Exercise Induced Oxidative Damage

It is long known that physical activity promotes well-being and that in inactive subjects there is an increased incidence of several chronic diseases including obesity, diabetes, hypertension, osteoporosis and mood problems.

The observation that exercise, long-lasting in trained and of short term in non-trained subjects, induces damage seemed at odd with the idea of its beneficial effects. The damage was mainly observed in exercise in which the eccentric contractions were prevalent, and it included structural and functional alterations not only in skeletal muscles but also in other tissues [28–31].

ROS implication in tissue damage induced by acute exercise was reported as early as the late 1970s [32]. Some years later, Davies and collaborators [33], using electron spin resonance (ESR) spectroscopy method proved first that free radical signals were intensified in rat muscle after a bout of exhaustive running. These results were subsequently confirmed by Jackson et al. [34]. It was later observed that contracting muscles also produce $NO^\bullet$ and other RNS [35].

Since these early observations, many studies have confirmed that muscular exercise promotes the production of both ROS and RNS in skeletal muscle fibers. Jenkins et al. [36] found that the tert-butyl-hydroperoxide induced chemiluminescence, a marker of ROS production, was increased by an exhaustive run in the hindlimb muscles of rats. Furthermore, it was found that ROS production, measured using the intracellular probe 2',7'-dochlorofluorescein, increased in rat diaphragm muscle during contraction [37], in vastus lateralis after exhaustive exercise [38], and in single mature skeletal muscle fiber [39].

Important advances have also been made about to identification (principally) and quantification of reactive species. Indeed, a series of studies demonstrated that contracting skeletal muscle transiently overproduces parent reactive species, such as $O_2^{\bullet-}$ and $NO^\bullet$, and secondary reactive species, such as H_2O_2, $^\bullet OH$, $ONOO^-$, and lipid-derived oxygen (O_2)-centered alkoxyl radicals [39–43]. However, the direct mechanisms and sources of ROS and RNS production during exercise remain uncertain and they are likely to differ depending on the type of activity.

In the cell, $O_2^{\bullet-}$ is generated by the addition of a single electron to ground state oxygen in several sites including plasmalemma, cytosol, peroxisomes, mitochondria and endoplasmic reticulum [5].

It has been found that the mitochondria, the NADPH oxidase (NOX), and the enzyme xanthine oxidase (XO) are the main endogenous sources of ROS in skeletal muscle [44]. In mitochondria, superoxide production verifies mainly at complexes I and III of the electron transport chain [45], and it has often been assumed that it is the primary cell source of ROS in physiological and pathological conditions [46]. However, there is no convincing evidence that mitochondria are the main cellular source of ROS in contracting muscle fibers [47]. Conversely, using confocal microscopy with specific fluorescent probes it was observed that muscle contraction increases $O_2^{\bullet-}$ in cytosol and subsequently in mitochondria. This observation suggests that a ROS generator different from mitochondria could be the potential primary source of ROS production during muscle contraction [48]. Furthermore, the idea that mitochondria are the main source of ROS during the muscle contraction is also theoretically inconsistent. In fact, when muscle contraction begins mitochondrial respiration enters State 3 (active respiration). Since the reduction degree of the autoxidizable carriers, by which mitochondrial ROS production depends, decreases in State 3 [49], it is foreseeable that the rate of $O_2^{\bullet-}$ production also decreases during the muscle contraction.

Nevertheless, measurements of ROS release by muscle mitochondria isolated from exercised animals suggest that the rate of mitochondrial ROS release is increased by aerobic exercise. For example, ROS release by mitochondria isolated from rat exercised to swim until exhaustion increased during both State 4 and State 3 respiration [50,51]. This increase was accompanied by alterations in mitochondrial functionality as evidenced by enhanced State 4 and decreased State 3 respiration [50]. Thus, it is possible to hypothesize that during exercise a source other than mitochondria initially produces the ROS, which then damage the mitochondria altering their functionality and increasing their ROS release.

NOX, located within the sarcoplasmic reticulum, transverse tubules and sarcolemma, is considered a key ROS generator during muscle contractions. Indeed, both at rest and during contractile activity it appears to contribute more than mitochondria to cytosolic $O_2^{\bullet-}$ in skeletal muscle [52].

On the other hand, evidence also indicates that XO produces superoxide in the cytosol of contracting rat skeletal muscles [53] even if it has also been reported that the muscle cells do not contain large amounts of the enzyme [54]. However, this enzyme is present in associated endothelial cells and might contribute to exercise-induced muscle damage [55]. However, additional research is required to determine the role played by XO in exercise-induced ROS production in human skeletal muscle.

Long lasting, strenuous exercise also induces oxidative muscle damage via ROS production by phagocytic white blood cells, particularly neutrophils, which infiltrate the muscular tissue [56]. $NO^{\bullet}$ is produced from the conversion of L-arginine into L-citrulline by enzymes known as nitric oxide synthases (NOS) [57] of which three different isoforms have been defined: Type I neuronal (nNOS), type II inducible (iNOS) and type III endothelial (eNOS). Normally, skeletal muscle expresses nNOS and eNOS, whereas iNOS, is induced in response to infection, inflammation, or trauma [58]. Thus, it is well established that isolated skeletal muscle fibers produce low levels of $NO^{\bullet}$ during resting conditions while heavy muscle contraction results in increase of generation of $NO^{\bullet}$ which has many signaling functions but may also have some detrimental effect because of the danger linked to the formation of highly reactive $ONOO^{-}$ [42]. Passive stretching of the muscle has also been shown to increase $NO^{\bullet}$ release from rat skeletal muscle in vitro and to increase nNOS expression. Moreover, the use of inhibitors of putative generating pathways [42] and western blotting [43] indicates that contraction induced $NO^{\bullet}$ release is primarily from neuronal $NO^{\bullet}$ synthase enzyme.

4. Markers of Exercise-Induced Oxidative Damage

In addition to the ROS formation during exercise having been directly measured, other support exists to the idea that during exercise an increased ROS production verifies. Indeed, it is possible to determine the changes in tissue content of stable molecules arising from the reaction of free radicals with certain biomolecules. In fact, the increase in radical production often results in profound oxidative alterations of various biological substances, including lipids, proteins and nucleic acids, which are commensurate to the increase in free radical production. Thus, the measurement of the content of derivatives of these substances oxidatively damaged has been used to obtain information on ROS production in various physio-pathological conditions. Most commonly measured are the molecules derived from the oxidatively damaged lipids, proteins or DNA or the changes in the levels of antioxidant molecules such as GSH (Table 1).

Lipid peroxidation has been frequently considered as a marker of exercise-induced oxidative stress thanks to the extreme susceptibility of lipids to ROS and to the stability of lipid peroxidation byproducts. Therefore, different studies exist showing that lipid peroxidation increase in skeletal muscle after acute running [33,59–61] and swimming exercise [8,9,50].

In contrast to lipid peroxidation, few data are available on the effects of exercise on protein and DNA oxidation in skeletal muscle.

Protein oxidative damage generates several byproducts that originate from oxidative modifications of lateral chains of different amino acids. Protein carbonyls represent an irreversible form of protein modification and have been demonstrated to be relatively stable, so that they are considered an adequate marker of protein oxidation [62]. The first study on accumulation of protein-bound carbonyls was published by Reznick et al. [63], who reported that a single bout of exercise caused an increase in protein-bound carbonyl content in the rat skeletal muscle. Similar results were subsequently obtained in rat skeletal muscle subjected to exhaustive exercise [64]. Conversely, other researchers found that protein carbonyl formation in deep vastus lateralis [38] and in both fast and slow muscles [65] was unaffected by exhaustive exercise. More recently, oxidative damage to proteins in homogenates and mitochondria from skeletal muscle has been found after swimming exercise [50].

To establish DNA oxidative damage, the accumulation of 8-hydroxy-deoxyguanosine (8-OHdG) is normally determined. The levels of 8-OHdG in several tissues of dog, including skeletal muscle, showed no significant changes in tissues, except the colon, soon after exercise [66]. Similarly, no significant changes were found in the levels of 8-OHdG in the nuclear DNA of fast and slow muscles of rat because of acute exercise [65].

In contrast, increases in 8-OHdG were found in skeletal muscle from young and old subjects 24 h after a single bout of exercise [67]. It was suggested that the increase in DNA damage was due to a delayed effect of exercise, which results in activation of macrophages and neutrophils and involves massive ROS production.

Some explanations can be provided for the lack of 8-OHdG increase soon after the end of the exercise. First, it is likely that most of the ROS generated by various cellular sources during the exercise are intercepted by cytosolic antioxidants before they reach the nuclear DNA. Second, the DNA in the nucleus is protected by the histone proteins, which render the nuclear DNA less susceptible to ROS activity [68]. Finally, a specific enzyme, the 8-oxoguanine DNA glycosylase/lyase [69], is activated and rapidly repairs oxidatively damaged DNA [70].

GSH, a ubiquitous tripeptide thiol, is one of the most important scavengers of ROS, and its ratio with GSSG may be used as a marker of oxidative stress, because GSH is oxidized to GSSG, and the GSH/GSSG ratio decreases under oxidative conditions. Several studies have reported a decrease in muscle GSH/GSSG ratio in response to exercise. An early study by Lew and coauthors [71] showed that exhaustive exercise causes significant increases in GSSG and ratio between GSSG and total glutathione (GSH + GSSG) in rat skeletal muscle. Subsequently it was shown that GSSG was elevated to as high as 160% of the resting levels, whereas GSH/GSSG ratio fell significantly after an exhaustive bout of exercise [72]. More recent research has also shown that 6 h of swimming exercise reduces muscle GSH level and GSH/GSSG ratio [50].

Interestingly, like aerobic exercise, anaerobic exercise of enough intensity and duration increases oxidative modification of proteins, nucleic acids, and lipids [73]. However, aerobic exercise-induced ROS release and consequent oxidative damage depends on mitochondrial electron transport chain and on the enzyme NADPH oxidase, which is localized in the sarcoplasmic reticulum, the transverse tubules and on the muscle plasma membrane [74]. Conversely, the ROS production found during and after anaerobic physical activity can be due to other systems, among which xanthine oxidase [75].

In the whole, despite some disagreeing results, there seems to be little doubt that acute exercise results in both enhanced production of reactive species and oxidative damage to components of muscular cells.

An important consequence of the involvement of the free radicals in tissue damage caused by acute exercise is the possibility to reduce the radical effects by supplementation with antioxidants such as vitamins C and/or E, carotenoids, GSH or its precursor the N-acetylcysteine. It has been shown that antioxidant supplementation protects against the deleterious effects of intense exercise [76–78]. The involvement of free radical in the exercise-induced oxidative damage and the protective effect of antioxidants are further demonstrated by observation that low levels of vitamin E are associated with a high exercise induced lipid peroxidation [33].

5. Muscle Adaptations Induced by Training

Skeletal muscle is particularly responsive to training which induces adaptations such as potentiation of antioxidant system, increased mitochondrial content, increased sensitivity to insulin, ameliorating the muscle function and protecting against the onset of metabolic disorders [79].

An important concept developed over the past decade is that the responses to training are likely the result of the acute but cumulative effects of the responses to single exercise bouts [80]. Thus, each bout of exercise initiates acute and transient changes in gene transcription which are reinforced by repeated exercise stimuli, leading to altered, chronic expression of a variety of nuclear and mitochondrial DNA (mtDNA) gene products, that ultimately form the basis of skeletal muscle training adaptation and improvements in exercise capacity [81].

However, skeletal muscle responds to exercise in a training specific manner. Classically, training was distinguished in "endurance training" and "strength training" also referred as "resistance training". Endurance exercise (e.g., running, swimming, cycling) is generally characterized by high-frequency, long duration, and development of a relatively low force. Resistance exercise (e.g., weight lifting) is, in general, characterized by low frequency, short duration, and development of a relatively high force. These two training modalities represent the extremes of a continuum of exercise protocols of countless options that differ in terms of intensity, duration, frequency, and mode of contraction as well as any combination of these.

Endurance training enhances the muscle aerobic metabolism capacities but do not induce increases in muscle mass or capacity to develop strength. Indeed, training to endurance determines in skeletal muscles a transformation of fiber-type, increases the mitochondrial mass, the production of new blood vessels and other adaptations [82]. Muscle blood vessels increase is a necessary adaptation to the increased mitochondrial oxygen requests [83]. Mitochondrial density increases rapidly in muscle particularly when subjects are previously untrained. The increase in mitochondrial compartment is accompanied by enhancement in the content of the enzymes of both Krebs cycle and oxidative phosphorylation among which succinate dehydrogenase (SDH), citrate synthase (CS), and cytochrome *c* oxidase (COX) [84]. Due to these adaptations, and of the increased capillarization, in the endurance trained muscle oxidative capacities are greatly enhanced. Conversely, endurance exercise does not change the cross-section area of the fibers unless the muscle was preceded by immobilization or underuse [85].

The adaptations elicited by the endurance-type exercise increase the resistance to exercises of intensities that in the untrained state can be performed for shorter period.

Strength training induces muscle cells hypertrophy and increase strength production but does not affect biochemical composition. Classic strength training protocols predominantly impact on muscle and muscle fiber cross-sectional area. It is important to realize that, in terms of functional changes, significant strength gains can be obtained by changes in the nervous control of the muscle mainly at the onset of training session [86]. At the beginning, the functional adjustment can be obtained with low level of structural changes. Continuing the training of strength, the cross-sectional area increases, and this is more evident at the origin and insertion of the muscle [87].

It was initially hypothesized that the increase in the cross-section area was due to the expansion of the preexisting cells and not to the cell proliferation. Subsequently, it was shown that such a growth, was dependent on the enhanced content of myofibrils, and that the net increase in cross-section area was mainly due to the increase in the fast fibers of the type IIa and IIX in man [88]. However, evidence is now available that, in several animal species, eccentric strength training, during which muscle exerts force while lengthening, is capable of muscle hyperplasia with neoformation of muscle fibers even though muscle growth depends largely on fiber hypertrophy [89]. The expression of the heavy chain of myosin is changed by strength training in an extension and direction that apparently depends on the characteristics of the protocol of exercise.

In older adults particularly salutary is resistance training thanks to its capacity to reduce the sarcopenia that verifies with age [90]. Resistance training is advisable for all healthy adults for its beneficial effects in reducing blood pressure [91] and cardiovascular disease risk [92].

Early works suggested that strength training only marginally changes mitochondria and capillarization in muscle [93]. Indeed, mitochondrial volumes and capillary densities were found to be low in strength-trained human muscles; muscle metabolism remained dominantly carbohydrate-dependent such that the relative content of cytoplasm containing glycogen was increased [94].

However, more recent works indicate that strength training results in effects like those elicited by endurance training. Indeed, it can improve insulin action and glucose metabolism [11] and stimulate mitochondrial biogenesis [95]. Moreover, recent researches have challenged the view that endurance and strength training are distinct exercise modalities, which increase mitochondrial density [96] and myofibrillar units [88] of skeletal muscle, respectively. It was found that in lean sedentary adults both 10 weeks resistance training or aerobic training enhanced mitochondrial respiration in the skeletal muscle, and that the oxidative capacity increase was dependent on qualitative changes in mitochondria not being the mitochondrial density substantially modified [95]. This suggests that mitochondrial biogenesis is stimulated by both training modalities, although it is likely the two training modalities do not achieve the same outcome by identical mechanisms.

A subsequent study also showed that a long period (nine months) of resistance and endurance training induce muscle mitochondrial proliferation and that the combination of both training modalities induces a more marked reduction of oxidative damage to lipids and carbohydrates and a greater increase in mitochondria content and mitochondrial enzyme activities, suggesting that the two modes of training together are healthier by protecting against T2DM [97].

Interestingly, study performed on elderly muscle, which showed large energetic, but smaller structural, adaptations, demonstrated that only resistance training induced a rise in mitochondrial volume density and muscle size [98].

6. Mechanisms of Muscle Adaptive Responses to Training

Accumulating evidence has induced to think that the dual role of ROS in animal organisms can be responsible for the contrasting effects of acute and chronic exercise. Indeed, it is now well established that ROS can damage proteins, nucleic acids and membrane phospholipids leading to cellular dysfunction, but they also have essential physiological functions in the cells acting as signals for the regulation of transduction, proliferation and transcription [99]. Therefore, the different roles of ROS in training as signaling molecules for the induction of tissue adaptation, and in acute exercise as damaging molecules, can depend on differences in the extent and temporal pattern of ROS generation. The low levels of ROS produced intermittently for a short period of time during a training protocol program, activate intracellular signaling ways that promote cellular adaptations leading to increased capacities against subsequent stresses. Conversely, moderate levels of ROS generation for a long period of time, or high generation due to high intensity exercise, induces structural and functional damage.

In the past years, evidence has been obtained that during each session of a training program the low level of ROS regulates signaling cellular pathways that result in the induction of the training induced adaptations that are healthy for the organism [5].

In the subsequent parts of this review we will examine the literature concerning our current knowledge about some potential signaling pathways that link ROS to the remodeling that occurs in skeletal muscle during exercise training.

7. Muscle Performance

It was long accepted that the idea that training to physical exercise led the animals or human to successfully endure exercise loads of different intensities, types and durations. Training shows to be able to stem the homeostasis disturbance during an exercise bout, allowing the animals or humans to bear physical work for longer time before fatigue appears [100], but the factors that condition the physical performance are controversial.

8. ROS and Muscle Performance

In the past years, evidence was obtained that ROS affect the muscle capacity to generate force, being that the low levels of ROS in the resting (i.e., unfatigued) state is necessary for normal force production [101]. Therefore, in the muscle the excessive scavenging of ROS by the antioxidant is linked to reduced force generation [101–103], whereas a low ROS production increases force generation [104]. On the other hand, the ROS capacity to increase muscle force production is reversed when ROS levels are higher and force production is reduced with increased time of exposition and dose of ROS [104]. These results led to propose the existence of an ideal redox state in which the conditions for the force production by the muscle are optimal and that the removal from such an ideal state leads to reduced force production [104].

Endogenous production of $NO^\bullet$ can also modulate skeletal muscle force production. Indeed, studies using excised bundles of muscle fibers reveal that force production during submaximal tetanic contractions is depressed by $NO^\bullet$ donors and increased by NOS inhibitors and $NO^\bullet$ scavengers [74]. Conversely, a consensus of literature does not exist to support the notion that $NO^\bullet$ production promotes muscular fatigue [74]. The evidence that sports performance is impaired by redox imbalance through various mechanisms that compromise the structure and function of the muscle cells, suggested that ROS contribute to muscular fatigue during prolonged exercise. The process involved in exercise-induced muscle fatigue depends on several factors [105,106] and the specific causes of muscle fatigue vary depending on the type of exercise that produces it [107]. For example, the main factors contributing to fatigue during high-intensity contractions that take place during resistance exercises and that contributing to fatigue during low intensity exercise with continued contraction, are different. However, there is evidence indicating that free radical production in skeletal muscles contributes to fatigue during various types of exercise.

Indeed, by studying the contribution of oxidants to muscle fatigue using a variety of animal model evidence of a relationship between endurance and free radical generation was obtained. Several studies have demonstrated that antioxidant treatment can delay the fatigue, but other studies have shown that antioxidants are able to reduce the levels of oxidative stress markers but not the onset of fatigue, mainly in humans [108]. However, most human studies have been performed on athletes or trained subjects, in whom the antioxidant supplementation can have harmful effects that hinder the adaptive processes stimulated by ROS (see below). Indeed, endurance to isokinetic cycle exercise was increased in healthy untrained volunteers by diet supplementation with a whey-based cysteine donor [109]. Furthermore, the administration of vitamin E, which is known to reduce the exercise-induced oxidative damage [110], prolonged the endurance to physical exercise in mice [111]. Novelli et al. [112] also observed that directly administered GSH to mice resulted in an increase in swimming endurance. The most effective substance in inhibiting fatigue was N-acetylcysteine (NAC) [113], a nonspecific antioxidant and reduced thiol donor that is a precursor and upregulator of the synthesis of GSH [114]. The evidence that muscle performance is most consistently improved by antioxidants that oppose thiol oxidation leads to think that such antioxidants delay fatigue by helping to maintain thiol groups of myofibrillar proteins in a reduced state.

9. Training and Muscle Fatigue

In contrast to acute exercise, training does not increase muscle oxidative damage as demonstrated by the finding that levels of malondialdehyde (MDA), a product of lipid peroxidation, of white muscle are not modified whereas those of red muscle are decreased in rats trained to run [60]. Furthermore, levels of MDA and of lipid hydroperoxides, another product of lipid peroxidation, are not modified in gastrocnemius of young [8] and adult [9] rats trained to swim.

Training also exerts a protective effect against oxidative damage elicited by acute exercise since it prevents the appearance of some signs of exercise-induced free radical generation, such as the increases in muscle lipid oxidation normally elicited by acute run exercise [60]. In contrast, exhaustive swimming exercise gives rise to tissue damage irrespective of the trained state, as documented by similar levels of muscle lipid peroxidation and loss of SR and ER integrity found in exhausted trained and untrained rats [8]. However, exercise endurance capacity is greatly increased in trained rats indicating that lipid peroxidation and tissue damage are strongly slowed down.

10. Antioxidants Enzymes

An explanation for these training effects was provided by the finding that muscle activities of the key antioxidant enzymes were modified by chronic exercise protocols [115]. The results obtained in subsequent studies were somewhat variable, but in general they confirmed the idea that exercise training promotes an increase in key antioxidant enzymes in skeletal muscle (Table 2). Thus, glutathione peroxidase (GPX) activity was increased, whereas SOD was unmodified and CAT activity was decreased in skeletal muscle of rats trained to run [116]. GPX and GR activities were increased by swimming training in gastrocnemius muscles of young (four months) [8,117] and adult (12 months) rats [9]. Training increased total antioxidant capacity irrespective of age [8,9], but its effects on antioxidant enzymes were dependent on age. Indeed, training increased GPX and SOD activities in young rats, decreased GR and CAT activities in adult rats and CAT activity in old rats. Thus, it was suggested that exercise training, although increasing selective antioxidant enzymes in young rats, does not offer protection against oxidative stress in the senescent muscle [118]. This view was confirmed by the observation that run training increased the activities of CAT, GPX, and MnSOD, and did not modify CuZnSOD activity in young rats, whereas did not modify the activities of CAT, GPX and CuZnSOD in the soleus muscle and decreased the activity of MnSOD in aged rats, [119]. Moreover, swimming training enhanced the activity of GPX and CuZnSOD, but not that of MnSOD, in young mice, while it did not modify enzyme activities in old mice [120].

The magnitude of adaptation was in part dependent upon exercise intensity, so that higher training intensities induced greater changes in the antioxidant defense [121]. Moreover, the changes induced by training in muscle antioxidant enzymes were muscle fiber-specific because GPX and SOD activities increased in vastus lateralis, whereas GR activity declined and those of GPX and SOD remained unchanged in soleus [122].

Antioxidant adaptation is also dependent upon exercise duration. Indeed, GPX and GR activities, which were unchanged after the first six-week training period, increased after the 7th week of training, whereas SOD activities were unchanged after both training periods [54].

It is worth noting that chronic resistance training may provide a protective effect like aerobic exercise on redox homeostasis. Indeed, it was reported that resistance training reduces serum lipid peroxidation providing protection against oxidizing agents in vitro and against oxidative damage generated by aerobic exercise [123], perhaps mediated by improvements in the thiol portion of the antioxidant defense. Furthermore, resistance training reduces muscle DNA oxidative damage [124] and increases antioxidant defense in older adults [125]. Indeed, training results in a significant increase in CuZnSOD and CAT but not MnSOD activity in vastus lateralis muscle [125].

Table 1. Effect of exercise on markers of oxidative damage in skeletal muscle.

Species	Activity	Marker	Ref.
Rat (6 mo)	Exhaustive treadmill running (submaximal work intensity) (gastrocnemious, soleus, plantaris)	TBARS↑	[33]
Rat (2 mo	Exhaustive swimming (gastrocnemious)	HPs↑, MDA↑	[8]
Rat (12 mo)	Exhaustive swimming (gastrocnemious)	HPs↑, MDA↑	[9]
Rat (4 mo)	Acute swimming (6 h) (gastrocnemious)	HPs↑, MDA↑, C=O↑, GSH/GSSG↓, C=O (mit)↑	[50]
Rat	Moderate and high intensity running (red and white VL)	HPs↔, MDA ↑	[59]
Rat	Treadmill running (20 min)	MDA↑	[60]
Rat	Treadmill running (1 h) (20 m/min, 0% grade)	MDA (mit)↑	[61]
Rat	Exhaustive exercise (gastrocnemious)	C=O↑	[64]
Rat (8 mo, 24 mo)	Exhaustive treadmill running (25 m/min, 15 m/min, 5% grade)	MDA↑, C=O↔, GSH/GSSG↓	[38]
Rat (2 mo)	Exhaustive treadmill running (1.6 Km/h) (fast and slow muscle)	C=O↔, MDA↔, 8-oxodG↔	[65]
Dog	Treadmill running (7 h) (splenius, diaphragm, gastrocnemious)	8-oxodG↔	[66]
Men (~26, ~65 yr)	Exhaustive treadmill running (45 min, 75%VO$_{2max}$ and 45 min, 90% VO$_{2max}$)	8-oxodG↑	[67]
Rat	Exhaustive treadmill running	GSH/GSSG↓	[72]
Men (~68 yr)	Whole-body resistance exercise training (14 wk)	8-oxodG↓	[124]
Rat (2 mo)	Swim training (10 wk) (gastrocnemious)	MDA↔	[8]
Rat (12 mo)	Swim training (10 wk) (gastrocnemious)	MDA↔	[9]

↔ unchanged; ↓reduced; ↑increased; mo: months; yr: years; wk: weeks; mit: mitochondria; VL: vastus lateralis.

Table 2. Effect of training on antioxidant enzyme activity in skeletal muscle.

Species	Activity	Enzymes	Ref.
Rat (2 mo)	Swim training (1 h, 10 wk) (gastrocnemious)	GPX↑, GR↑	[8]
Rat (12 mo)	Swim training (1 h, 10 wk) (gastrocnemious)	GPX↑, GR↑	[9]
Rat (50 days)	Swim training (1 h, 10 wk) (gastrocnemious)	GPX↑, GR↑	[117]
Mouse (2 mo)	Swim Training (1 h, 6 wk)	GPX↑, GR↑, MnSOD↔, CuZnSOD↑	[120]
Mouse (26 mo)	Swim training (1 h, 6 wk)	MnSOD↔, CuZnSOD↔	[120]
Rat	Treadmill training (32 m/min, 8%, 2 h, 12 wk) (soleus, gastrocnemious)	CAT↓, GPX↑, SOD ↔	[116]
Rat (2 mo)	Treadmill training (1 h, 13 wk, 50–60% of maximal exercise capacity) (soleus)	CAT↑, GPX↑, MnSOD↑, CuZnSOD↔	[119]
Rat (21 mo)	Treadmill training (1 h, 13 wk, 50–60% of maximal exercise capacity) (soleus)	CAT↔, GPX↔, MnSOD↓, CuZnSOD↔	[119]
Rat	Treadmill training (25 m/min 10%, 2 h, 10 wk) (DVL)	SOD↑, GPX↑, GR↓	[122]
Rat	Treadmill training (25 m/min 10%, 2 h, 10 wk) (soleus)	SOD↔, GPX↔ GR↓	[122]
Rat (4 mo)	Treadmill training (25 m/min, 10%, 10 wk) (DVL)	GPX↑, MnSOD↔, CuZnSOD ↑	[126]
Rat (3 mo)	Treadmill training (27 m/min, 12% grade, 2 h, 10 wk) (SVL, soleus, plantaris)	GPX↔ CAT↔, MnSOD↔, CuZnSOD↔	[127]
Rat (3 mo)	Treadmill training (27 m/min, 12% grade, 2 h, 10 wk) (DVL)	GPX↑CAT↑, MnSOD↑, CuZnSOD↔	[127]
Men (~23 yr)	Maximal cycling sprint training (6 wk) (VL)	GPX↔ GR↔, SOD↔	[54]
Men (~23 yr)	Maximal cycling sprint training (7 wk) (VL)	GPX↑, GR↑, SOD↔	[54]
Men (~71 yr)	Unilateral resistance exercise training (12 wk) (VL)	CuZnSOD↑ MnSOD↔ CAT↑	[125]

↔ unchanged; ↓reduced; ↑increased; mo: months; yr: years; wk: weeks; DVL: deep vastus lateralis; VL: vastus lateralis.

Several experimental evidences suggest that the training-linked adaptations are induced by changes in gene expression with upregulation of both mRNA levels and protein expression. It must be underlined that the available data are few and debatable. For example, CuZnSOD activity and mRNA abundance were found to be higher in vastus lateralis (VL) muscle of trained compared to sedentary rats. However, the CuZnSOD protein content was not altered in the muscle, but increased in its superficial portion (type 2b). Moreover, the MnSOD protein content was higher in trained rats but its activity and mRNA abundance were not affected, whereas the GPX activity was increased without changing its mRNA abundance [126].

Training increased mitochondrial MnSOD activity in the deep portion of VL, and the MnSOD protein content in pooled superficial portion of VL and plantaris muscle [127]. The levels of the mRNA of MnSOD did not change in any muscle. The mRNA level, protein content and activity of CuZnSOD were not changed by training except for an increased protein content in pooled SVL. GPX and CAT activities were increased significantly by training only in the muscle DVL. Therefore, it was suggested that training induces adaptations of the antioxidant enzyme mainly in fibers of the type IIa, probably for the increased free radical production and for the low antioxidant capacity. The different training effects on mRNA, content of the enzyme protein and activity indicate that different cellular signals can affect the pre and post translational regulation of SOD.

It is worth noting that even acute bouts of exercise were found to be able to increase the activities of antioxidant enzymes, including SOD, CAT, and GPX [72,128,129] and GR [128], in skeletal muscle. The threshold and the greatness of the activation appeared different among enzymes and were fiber-specific because the enzyme activities increased in deep portion of the vastus lateralis but not in soleus [128]. The mechanisms by which antioxidant enzymes could be activated within a relatively short period of time during the exercise were largely unknown even though the fast activation suggested that enzymatic molecules underwent allosteric or covalent modifications. The rapid activation of antioxidant enzyme synthesis by oxidative stress through a transcriptional pathway had been shown in prokaryotes (*Salmonella and Escherichia*) [130], but there was no evidence that a similar mechanism existed in mammalian cells [131].

Subsequent studies showed muscle fiber specific upregulation of superoxide dismutase gene expression in skeletal muscle [132]. Indeed, increases in MnSOD mRNA levels were found in the DVL 0, 1, and 2 h after exercise, whereas MnSOD protein levels were not changed. MnSOD mRNA levels were not modified by exercise in SVL, whereas MnSOD protein levels were increased 10 and 24 h after exercise. CuZnSOD mRNA levels were not changed by exercise in DVL and SVL, whereas the CuZnSOD protein content was increased 48 h after exercise in both muscles. Activities of MnSOD, CuZnSOD and total SOD were not modified by exercise in either muscle [132]. The increases in the CuZnSOD protein content seen post-exercise, without increases in mRNA abundance in both DVL and SVL, suggested a translational mechanism in this SOD isoform [132].

Moreover, more recent studies showed that in monocytes, after individual exercise bouts, target genes of the nuclear transcription factor, peroxisome proliferator activated receptor-γ (PPARγ), were upregulated at the mRNA level up to 3 h after each exercise, and this effect persisted for less than 24 h [133,134]. In contrast, after an eight-week training program, increases in gene expression were observed at the protein level in samples taken 48 h after the previous bout of exercise [134].

Such a view was supported by subsequent studies which showed that increase in PPARγ target gene expression observed after single bouts of exercise were similar, but less pronounced, to that seen after eight-week training programs involving at least three exercise bouts per week [135]. Thus, a study using the mass spectrometry coupled with liquid chromatography (LC-MS/MS) to investigate the effect of an acute bout of endurance exercise on protein composition of human vastus lateralis (VL) muscle in endurance trained and untrained individuals, found that training altered the content of 92 structural and mitochondrial proteins. In contrast, a single bout of exercise (3 h) resulted in an alteration of the content of 44 proteins in untrained athletes [136].

These results suggest that the effects of each exercise bout can merge so that, after a training program, a more sustained effect is apparent. It can therefore be concluded that acute exercises, except for the most intense ones, which cause oxidative damage, result in small transient oxidative stresses which, in turn, induce redox-sensitive responses on local and systemic level, thus contributing to training adaptations and systemic health benefits.

11. ROS Production

Although it was apparent that training effects on muscle performance were associated to an increase in the effectiveness of antioxidant defense system, it was not known whether such an increase was the only change responsible for the slowdown of the peroxidative processes and the muscle fatigue. In fact, it was conceivable that a delayed onset of fatigue can also be dependent on adaptations involving a decrease in rate of production of reactive oxygen derivatives. Venditti et al. [137] first found that mitochondrial H_2O_2 release supported by succinate was lower in swimming trained than in untrained rats both in State 4 and State 3, whereas that supported by pyruvate/malate was lower only in State 4. The decrease in succinate-linked H_2O_2 release was such to compensate for the increase in mitochondrial protein content produced by training in the muscle [137]. Although the ROS produced by mitochondria are only a part of those produced in the cell, the above finding indicated that training can reduce their contribution to the intracellular steady state concentration of H_2O_2.

A subsequent study showed that eccentric training also leads to positive adaptations, decreasing mitochondrial H_2O_2 release when mitochondria are incubated with either pyruvate/malate or succinate [138]. It has also been shown that training decreases mitochondrial H_2O_2 release in healthy and diabetic subjects [139]. A recent research has shown that immobilization increases H_2O_2 emission, while subsequent aerobic training (supervised bicycle training) reverses these effects in young and older men [140]. A more recent study has shown that training decreases H_2O_2 production measured in freshly permeabilized soleus muscle in the presence of substrates of he ttricarboxylic acid cycle [141].

Surprisingly, another research has shown that training increases mitochondrial ROS production in old subjects with both normal and impaired glucose tolerance [142].

It remains unclear whether resistance training affects the mitochondrial ROS production in older adults, because a recent report [143] indicates that resistance training does not induce significant changes in mitochondrial ROS production in vastus lateralis muscle from older adults.

Although these in vitro models may not correspond to conditions in vivo, overall, the reported results seem to indicate a potential training-induced decrease for H_2O_2 release. However, the causes of the reduction of the mitochondrial ROS release induced by training in skeletal muscle are not well understood. One possibility is that the decrease in H_2O_2 release elicited by training is not due to a decrease in ROS production, but to a greater capacity of mitochondria to scavenge superoxide radical or hydrogen peroxide, thus limiting the generation of highly reactive hydroxyl radicals. However, the observation that regularly performed moderate exercise does not modify the total antioxidant capacity of mitochondria indicates that the reduction in mitochondrial H_2O_2 release is not due to a greater capacity of mitochondria to scavenge reduced oxygen intermediates [137]. This view is supported by the observation that the activities of mitochondrial antioxidant enzymes (MnSOD, CuZnSOD, GPX, GR, and CAT) are not affected by training [138].

Another possibility is that the chronic exercise leads to changes in factors, which can influence mitochondrial free radical production, such as mitochondrial membrane potential. In fact, evidence is available that training produces a drop in mitochondrial membrane potential, although the causes of such a drop remain still unknown [137]. In theory, the drop in mitochondrial membrane potential and, therefore, the decrease in H_2O_2 production could be due to increased uncoupling of the inner mitochondrial membrane. Such an uncoupling can be dependent on the levels of members of the mitochondrial transporters family, mitochondrial uncoupling proteins (UCPs), which contribute to energy dissipation as heat by uncoupling respiratory chain from ATP synthesis [144,145]. Of the five mammalian forms of UCPs, UCP4 and UCP5 are principally neuronally expressed [146]. The best characterized of these proteins, UCP1, is expressed exclusively in brown adipose tissue (BAT) [147], UCP2 is expressed in a large spectrum of tissues including skeletal and cardiac muscle, and UCP3 is mainly expressed in BAT and skeletal muscle [148] in which it is involved in decreasing ROS production [149].

Since peroxisome proliferator-activated receptor γ coactivator 1α (PGC-1α) can regulate the mRNA expression of Ucp2 and Ucp3 in the muscle cell culture [150], it was suggested that the increased PGC-1α expression induced by training could increase the uncoupling capacity of skeletal muscle mitochondria and thus decrease their ROS production by [151]. However, such an idea is not supported by experimental evidence since increases in UCP3 protein content in rat muscle have been found only after acute exercise [152] and short-term training (10 days) [153].

Rather, it has been hypothesized that the decreased ROS production observed with eccentric exercise is due to a mild uncoupling caused by an increase in the ratio between polyunsaturated and saturated fatty acids, and a decrease in the content in arachidonic acid and plasmalogens in mitochondrial membrane [138].

It is worth noting that the possibility that exercise training can lead to a decrease in skeletal muscle ROS production reducing the activity of other cellular sources has not been investigated. Only recently, evidence has been obtained that decrease in ROS production induced by aerobic exercise training in skeletal muscle is associated with reduced muscle NADPH oxidase activity [154].

12. Factors Regulating Protein Expression

Although the effects of aerobic exercise on antioxidant capabilities in skeletal muscle have been well described, the regulatory mechanisms underlying this adaptation are complex and incompletely understood.

In the 90s of the last century accumulating data indicated that cells exposed to ROS responded by inducing or repressing a wide variety of genes [155]. These effects seemed to be due to changes in the intracellular redox balance that influenced multiple signaling pathways leading to a modulation of the expression of some genes [156–158]. Several genes that could be differentially regulated by oxidative stress were characterized and included early response genes, genes for enzymes involved in antioxidant protection, and genes for specific stress and heat shock proteins (HSPs) [156–158].

At present, a complete answer to the question as to how changes in the redox status of muscle fibers regulate signaling pathways and gene expression is not yet available. However, it is known that redox signaling can affect numerous transcriptional activators leading to altered gene expression and modified muscle phenotype. An important mechanism by which redox signaling controls gene expression is the modulation of the phosphorylation state of transcriptional activating factors due to ROS ability to control the activities of many kinases and phosphatases [159,160].

Eukaryotic cells possess many families of kinases, but the family of mitogen-activated protein kinases (MAPKs) represents the main link between cell ROS levels and skeletal muscle adaptive responses.

12.1. MAPK

MAPK family is one of the main kinase families that are involved in the conversion of cell signals into cellular responses. MAPKs contribute to the regulation of life-and-death decisions taken in response to several stress signals including ROS [161]. The control exercised by MAPKs on a wide variety of pathways of cellular signaling is obtained through phosphorylation-mediated activation or deactivation of regulatory proteins [162].

All eukaryotic cells possess multiple MAPK pathways, and at least four major groups of MAPKs have been characterized in mammalian cells such as the c-Jun N-terminal kinases (JNK), the extracellular signal-related kinases (ERK1/2), the p38 kinase (p38), and the big MAP kinase 1 (BMK1/ERK5), which can be stimulated by cytokines, growth factors, and cellular stress even though their relative activation and the specific cellular response evoked depend on the different stimuli [163].

ERK, JNK, p38, and BMK1 are all proline-directed serine/threonine kinases, and the pathways in which they are activated also share similar homology. Indeed, all MAPKs are activated through a cascade of phosphorylation events, often referred to as the MAP kinase module, in which a MAP kinase kinase kinase (MAPKKK) phosphorylates and activates a MAP kinase kinase (MAPKK), which in turn phosphorylates and activates a MAPK [162,164].

A well-studied member of the MAPKKK family that is preferentially activated in response to various types of exogenous and endogenous cellular stress, including oxidative stress, is modulated by redox mechanisms, and mediates cell apoptosis, is apoptosis signal-regulating kinase-1 (ASK-1), a serine/threonine protein kinase that activates both p38 and JNK pathways [165]. In unstimulated conditions, ASK-1 binds to the repressor protein thioredoxin (Trx), a ubiquitously expressed redox regulatory protein, so that its kinase activity is inhibited. The binding of Trx to ASK-1 requires the presence of a reduced form of an intramolecular disulfide bridge between two cysteine residues in the catalytic site of Trx. This protein, after its oxidation by ROS molecules such as H_2O_2, dissociates from and liberates ASK-1, which is then activated by formation of an oligomeric complex and threonine autophosphorylation [165].

Several studies, which have revealed a direct link between ASK-1 and NOX, have also suggested that ASK-1 is an important effector of NOX in the redox signaling involved in cellular stress responses [166].

It has been shown that the three best-characterized MAPK subfamilies, JNK, p38 MAPK, and ERK, are activated by oxidative stress and could potentially be involved in pathways affecting the breakdown of muscle proteins or loss of nuclei via myonuclear apoptosis [160,167]. It has also been shown that H_2O_2 can elicit the activation of ERK, JNK, and p38 MAPK in skeletal myoblasts in a dose-and time-dependent manner [168]. On the other hand, it has been shown that exercise, as an intermittent form of cellular stress, is able to activate ERK1/2, p38, and JNK pathways in rat skeletal muscle [169].

Subsequent studies on the adaptations of muscle cells to exercise-linked oxidative stress have led to conclude that these distinct signaling pathways are partially dependent on the type, duration, and intensity of the contractile stimulus, and are critical cellular responses to maintain muscle homeostasis through upregulation of the expression of antioxidant enzymes and other cytoprotective proteins [170].

12.2. ERK

ERK is composed of two isoforms, ERK1 and ERK2, collectively referred to as ERK1/2 [162,164]. Several mitogens, including epidermal growth factor, platelet-derived growth factor, and ROS can activate ERK1/2 [162] The activation of ERK1/2 by oxidative stress is consistent with the idea that low, but adequate levels of ROS are mitogenic [164]. Once activated, the ERKs can phosphorylate different substrates, including other kinases and transcription factors, and are involved in mediating different responses, that depend on which ERK substrates the cell expresses [171].

Activation of ERK1/2 regulates the transcriptional activity of activator protein-1 (AP-1) [172], avian myelocytomatosis virus oncogene cellular homolog (c-Myc) and the cell survival protein B-cell lymphoma-2 (Bcl-2) [173]. Although it can be considered an oversimplification, in general, ERK1/2 activation seems to promote cellular adaptations that lead to survival [164].

ERK1/2 is phosphorylated rapidly and transiently in response to mechanical stress. Indeed, it was demonstrated that ERK1/2 is activated in skeletal muscle of rats running on a motorized treadmill for 10–60 min [169] and in a rat plantaris in situ preparation stimulated to contract for 5 min by electrical stimulation [174].

Early human studies showed an increase in ERK1/2 phosphorylation after endurance-type exercises, including acute submaximal cycling [175] and marathon running exercise [176]. Subsequent works also showed an increase in ERK1/2 phosphorylation in response to resistance exercise [177,178].

The magnitude of ERK1/2 phosphorylation during endurance exercise correlates with the intensity of the protocol [175]. Conversely, resistance exercise upregulates ERK1/2 signaling in a manner that does not seem to preferentially depend on exercise intensity [179].

Furthermore, one-legged cycling exercise leads to an increase in ERK1/2 phosphorylation only in the exercised legs, suggesting that phosphorylation is dependent on local rather than systemic factors [174,180]. This view is supported by observation that ERK1/2 phosphorylation increases in isolated rat [181,182] and mouse [183] skeletal muscles stimulated to contract in vitro.

The effect of chronic training has been studied on rats subjected to a program of either low- or moderate-to-high-intensity treadmill running. ERK1/2 phosphorylation was similar to sedentary values, whereas ERK1/2 expression was increased three- to fourfold irrespective of the prior training program in muscle sampled 48 h after the last exercise bout [184].

Contrary to previous studies, chronic endurance training does not greatly influence total MAPK protein expression and pERK/total-ERK in chronically trained runners [185]. Furthermore, total ERK1/2 content was lower in powerlifting and weightlifting trained subjects compared to their controls [186].

12.3. p38

p38 is activated in response to various physiological stresses, such as osmotic stress, endotoxins and ROS [164]. ASK-1 represents the link between oxidative stress and p38 activation, because it is activated in response to ROS, such as H_2O_2, and is required for phosphorylation-mediated activation of both p38 and JNK [161]. Five isoforms of p38 have been identified (p38α, p38β, p38β2, p38γ and p38δ), whose expression differs in the various tissues, with p38γ predominantly expressed in skeletal muscle [162].

Among the phosphorylation targets of p38 there are several important transcription factors, including tumor protein p53, a phosphoprotein crucial in prevention of cancer formation [187], nuclear factor κ-light-chain-enhancer of activated B cells (NF-κB), involved in the induction of antioxidant enzymes [188], and activating transcription factor 2 (ATF2), which regulates the transcription of various genes, including those involved in DNA damage response [189]. Of particular importance to apoptosis is the fact that activation of p53 results in the expression of the pro-apoptosis protein, bcl-2-like protein 4 (Bax), which can promote caspase-3 activation via a mitochondrion mediated pathway [190].

Although ERKs, JNKs and p38 are all activated by H_2O_2 treatment, following a time- and dose-dependent pattern in C2 skeletal myoblasts, the time-course of this activation differs among the MAPK subfamilies. Indeed, p38 activation is more rapid and displays a biphasic pattern, with a second peak obtained at 2 h of treatment [168]. This p38 re-activation could be attributed to a feedback mechanism, mediated by its either upstream activators or downstream targets, a phenomenon previously reported [163].

Phosphorylation of p38, like that of ERK1/2, increases during contraction of isolated skeletal muscles, implying a local activating factor [182,191]. Moreover, it is also increased by treadmill exercise in rodents [139] and cycling ergometry [174] and marathon running [176] in humans.

12.4. JNK

There are three isoforms of JNK (JNK1, JNK2, and JNK3) that are encoded by three different genes. JNK1 and JNK2 are ubiquitously expressed, while JNK3 is only expressed in brain, heart and testis [162]. Many of the stimuli that activate p38, including endotoxins, osmotic stress, and ROS, can also activate JNK. Moreover, like p38 activation, even activation of JNK induced by oxidative stress occurs via the ASK1 pathway [192]. The transcriptional factors AP-1, p53 and c-Myc and many other non-transcriptional factors, such as Bcl-2 family members, are among the specific molecular targets of JNK. About this, evidence has been obtained that JNK plays a major role in ROS-mediated apoptosis. Indeed, because ROS themselves are not able to activate caspases, JNK is required as another death-signaling pathway for oxidative stress-mediated apoptosis [161,192].

Signal transduction through the JNK pathway is also stimulated by intense exercise protocols and by those inducing muscular damage [193]. JNK phosphorylation increases linearly with increasing levels of muscular contraction force [174]. Therefore, JNK activity appears to be modulated by total muscle tension rather than duration of the contraction stimulus [194].

Summarizing, all three MAPK signaling pathways appear to be responsive to exercise even though their activation mechanisms (i.e., energetic/metabolic compared to mechanical) remain distinct. It is likely that the pattern of MAPK signaling have important implications in the different adaptive responses elicited by exercise. Indeed, MAPK may play an important role as a cellular intermediary able to couple perceived alteration in stress with adaptive changes, including the transcriptional regulation of redox state of the skeletal muscle.

12.5. MAPK and Modulation of Gene Expression

In order to execute their functions, the MAPKs phosphorylate hundreds of substrates, thus modulating their activities. MAPK substrates were identified in the cytoplasm, mitochondria, Golgi apparatus, endoplasmic reticulum, and particularly the nucleus where they modulate gene expression [195–197]. Indeed, stress responses, as well as other cellular processes, are mediated by MAPK cascade-dependent induction and regulation of *de-novo* gene expression [198]. For this to happen, it is necessary that the signals transmitted via the various cascades enter the nucleus where they modulate the activity of transcription factors, transcription suppressors, and chromatin remodeling proteins, in order to ensure the correct cellular responses [199].

In fact, inside the cell nucleus, the DNA is packed into the chromatin, a structure consisting of protein–DNA complexes [200]. This structure is very compact so that it is not accessible to other proteins, including transcription factors. Therefore, the transcription requires a "decompaction" and a change into active open euchromatin. Following various types of stimulations, several distinct processes become operative to induce chromatin remodeling and allow the access to the target genes. Such processes include histone acetylation, histone phosphorylation, poly ADP ribosylation, changes in DNA conformation, and binding of other proteins to the DNA. Some of them are regulated by the cascades of MAPKs, including in particular ERK1/2 and p38s, and are required for the correct transcription and the induction of processes dependent on MAPKs [201]. At present, among the numerous substrates of MAPKs, several transcription factors have been identified.

12.6. ROS Sensitive Transcription Factors

Over the past years, it was reported that MAPKs can regulate a wide range of transcription factors involved in response to oxidative stress elicited by exercise in skeletal muscle. On the other hand, based on the growing appreciation of the influence exerted by redox-sensitive signaling pathways on normal cellular processes, a reasonable hypothesis was that an important regulator of the adaptation in skeletal muscle in response to aerobic exercise may be ROS generated during the exercise.

ROS play a very important role to regulate several cell functions modulating the activity of preexisting proteins and inducing the expression of many genes via activation of specific redox-sensitive transcription factors [202]. Due to ROS involvement in almost all-important biological functions, it is difficult to define all the pathways and gene targets that redox signaling affects during exercise. Therefore, our examination will be limited to some of the most relevant factors that play critical roles in homeostatic regulation of muscle oxidant-antioxidant balance during exercise.

ROS are critical in the regulation of several transcription factors, including the activator protein-1 (AP-1) and the nuclear factor κ-light-chain-enhancer of activated B cells (NF-κB) [203–205] two transcription factors known to play crucial functions in proliferation, differentiation, and morphogenesis.

AP-1 and NF-kB response elements are located in the promoter regions of genes encoding CAT, GPX, Mn-SOD and CuZnSOD [204] and have been identified as the main factors that are both activated by exercise-produced ROS and directly implicated in the induction of the aforementioned antioxidant enzymes [206]. Moreover, combinations of AP-1 and NF-kB with other redox-sensitive transcription factors can determine which antioxidant enzyme is about to be induced and to what extent.

12.7. NF-κB

NF-κB, one of the most commonly investigated redox sensitive transcription factors, is a heterodimer composed of two related subunits, p65 and p50, which share a homologous region at the N-terminal end, necessary for DNA binding and dimerization.

The NF-κB/Rel transcription factors are normally sequestered in the cytoplasm in an inactive state, linked to the IκBα inhibitory protein. NF-κB is activated by several stimuli, including H_2O_2, proinflammatory cytokines, lipopolysaccharide (LPS), and phorbol esters, by the phosphorylation of IκBα at Ser-32 and -36 by IκB kinase (IKK). Phosphorylation of IκBα results in its dissociation from NF-κB and subsequent proteasomal degradation. NF-κB, once free, migrated into the nucleus where it binds to the corresponding DNA sequence of the target genes, including MnSOD and γ-glutamylcysteine synthetase (GCS) [207], the rate-limiting enzyme in the biosynthesis of glutathione.

In muscle cells, ROS such as H_2O_2 are able to induce degradation of the inhibitory IκB protein subunits bound to NF-κB subunits (p65, p50 and RelB), leading to the rapid migration of NF-kB to the nucleus and activation of the transcription of specific genes [203]. Zhou et al. [204] showed that the use of specific NF-κB inhibitors blocked the upregulation of the expression of Cat and Gpx induced by oxidative stress, thus confirming the hypothesis that ROS are able to modulate mRNA levels of antioxidant enzymes by activating redox-sensitive transcription factors, such as NF-κB.

In 1997, Sen et al. [208] were the first to bring NF-κB to the attention of exercise physiologists, demonstrating that NFκB activation in L6 muscle cells was responsive to H_2O_2 treatment and was controlled by intracellular GSH: GSSG status. Subsequently, Hollander et al. [134] reported that NF-κB (and AP-1) binding was significantly increased in rat skeletal muscle in a fiber-specific manner after an acute bout of prolonged exercise. Since NFκB binding was associated with an increase in MnSOD mRNA level and protein content, the authors hypothesized that NF-κB activation by ROS generated in contracting muscle may be the underlying mechanism for training adaptation and increase in antioxidant enzyme expression. Ji et al. [209], examining NFκB signaling cascades in response to exercise in rats, found that acute exercise increased NF-κB binding, IKK activity, IκBα phosphorylation and degradation, and P50 accumulation in the nucleus in rat DVL muscle. The exercise-induced activation of NF-κB was partially abolished by treatment with pyrrolidine dithiocarbamate, an inhibitor of the 26S proteosomes. Furthermore, the treatment with a high of t-butylhydroperoxide had scant effect on NF-κB, suggesting that the signaling was not induced by general oxidative stress but by specific chemical agents.

In a subsequent study, Ho et al. [210] found an increase in NF-κB activation accompanied by IKKα/β phosphorylation in the rat soleus (type 1) and red gastrocnemius (type 2a) muscles during 60 min of treadmill exercise. Peak IKKα/β activation was found early during exercise (15 min), whereas maximal NF-κB binding was found at 1–3 h. IKKα/β and IκB phosphorylation was also increased by the contraction of isolated extensor digitorum longus (EDL) muscles in vitro. Moreover, application of p38 and ERK inhibitors reduced IKKα/β activation, suggesting that MAPKs were upstream of NF-κB and could partially mediate stimulation of NF-κB activity by contraction.

Gomez-Cabrera et al. [53] found that an acute bout of treadmill running in rats activated ERK1/2 and p38 and the activation coincided with elevated gene expression of MnSOD and iNOS. Moreover, when an inhibitor a xanthine oxidase (XO), allopurinol, was used to partially block ROS generation, MnSOD and iNOS mRNA expression induced by exercise was severely hampered, and the activities of ERK, p38, and NFκB were decreased. Although it was not possible to conclude that attenuation of MAPK signaling was the reason for the decreased MnSOD and iNOS expression, these results suggested that MAPK proteins played a role in the signaling of antioxidant enzymes and that an integrated input from both the NFkB and MAPK signaling pathways was required to stimulate gene expression of these enzymes in the muscle fibers. These results also suggested that nonmitochondrial ROS were involved in the improvement of muscle antioxidant defense system.

12.8. AP-1

AP-1 is a heterodimer consisting of activating (c-Fos and c-Jun) and inhibitory (Fos-related antigen (Fra)-1 and 2) subunits, which can generate different heterodimers, thus modulating expression of target genes [211]. Depending on the cell type and cellular redox milieu, Fos and Jun can dimerize or interact with other transcription factors such as activating transcription factor (ATF), CCAAT enhancer binding protein (C/EBP), and proto-oncogene (Maf) leading to either activation or inhibition of gene transcription of antioxidant and immunoactive proteins [212,213].

AP-1 regulates the gene expression in response to signals generated by a wide variety of extracellular stimuli, among which growth factors, tumor promoters, neurotransmitters, UV light, and cytokines [214,215]. AP-1 can also be activated by ROS [216] and oxidative stress induces the binding of AP-1 complex proteins (c-Jun and c-Fos) to DNA [211]. According to this observation, an increase in AP-1 binding has also been found after a single bout of exercise [134].

Activation of various kinases which are involved in the MAPK signaling pathway can lead to the sequential phosphorylation of a variety of proteins, resulting in increased expression of c-Jun, a subunit of the transcription factor AP-1, which is an important DNA-binding site on many genes able to respond to oxidative stress [217].

It has been reported that c-Jun is regulated by JNKs to which it gave the name [218]. c-Jun is constantly expressed in both unstimulated and stimulated cells. Upon stimulation, c-Jun, to exert its activity, interacts with other transcription factors such as c-Fos, and ATF forming AP-1 [219]. The activation of c-Jun depends on phosphorylation of its transactivation domain by all JNK isoforms and to some extent by other MAPKs, which leads to induction of the full transcriptional activity within the AP-1 complex, independent of DNA binding [220].

12.9. Nrf2

Although the protection provided by NF-κB and AP-1 activation is important for cellular redox homeostasis, another pathway is the main regulator of cytoprotective responses to endogenous and exogenous stresses caused by electrophilic compounds and ROS [221]. The key signaling protein within the pathway is the transcription factor nuclear factor erythroid 2-related factor 2 (Nrf2) that can bind, together with small musculoaponeurotic fibrosarcoma (Maf) proteins, to a DNA sequence called antioxidant response element (ARE) in the regulatory regions of target genes. Nrf2 can also bind to Kelch ECH associating protein 1 (Keap1), a repressor protein very rich in cysteine residues most of which can be modified in vitro by different oxidants and electrophiles [222,223].

In unstressed conditions, the cellular concentration of Nrf2 protein is maintained at very low levels by its inhibitor Keap1, which sequesters Nrf2 in the cytosol and facilitates its ubiquitination through the Keap1/Cul3 ubiquitin ligase and rapid proteasomal degradation. Under conditions of stress or in the presence of Nrf2 activating compounds, this degradation is hampered because modification of reactive cysteine thiols of Keap1 and Nrf2 by inducers presumably alters the structure of the Nrf2/Keap1/Cul3 complex, leading to inhibition of Nrf2 ubiquitination Nrf2 release. Subsequently Nrf2, phosphorylated by protein kinases, moves into the nucleus where it forms heterodimers with Maf proteins. This, in turn, facilitates the binding of Nrf2 to the antioxidant response element (ARE), a cis-acting enhancer sequence (TCAG/CXXXGC) in the promoter region of Nrf2-regulated genes [224,225] (Figure 2).

Genome-wide search for Nrf2 target genes has led to identify an array of ARE-regulated genes, that lead to the production of phase II xenobiotic metabolizing enzymes, antioxidants, molecular chaperones, DNA repair enzymes, and anti-inflammatory response proteins [226]. They reduce reactive compounds such as electrophiles and free radicals to less toxic intermediates and increase cell capacity to repair any damage ensued.

An alternative model of the Nrf2-Keap1 pathway of gene regulation has also been proposed [227]. According to this model, Nrf2 is constitutively expressed in cells and moves directly into the nucleus to activate gene transcription. Nrf2 is then targeted for degradation by Keap1, a process that requires the transient Keap1 displacement in the nucleus. In cells under stress, the stabilization of Nrf2 in response to activating compounds is caused by mechanisms that prevent Nrf2 from binding to Keap1 and being degraded in the nucleus. The reduced degradation of Nrf2, together with its de novo synthesis, results in the accumulation and direct recruitment of Nrf2 to the ARE, so that transcription of its genes increases. This pathway of regulation of Nrf2 activity should allow it to exert its dual function of controlling gene expression constitutively and inducibly.

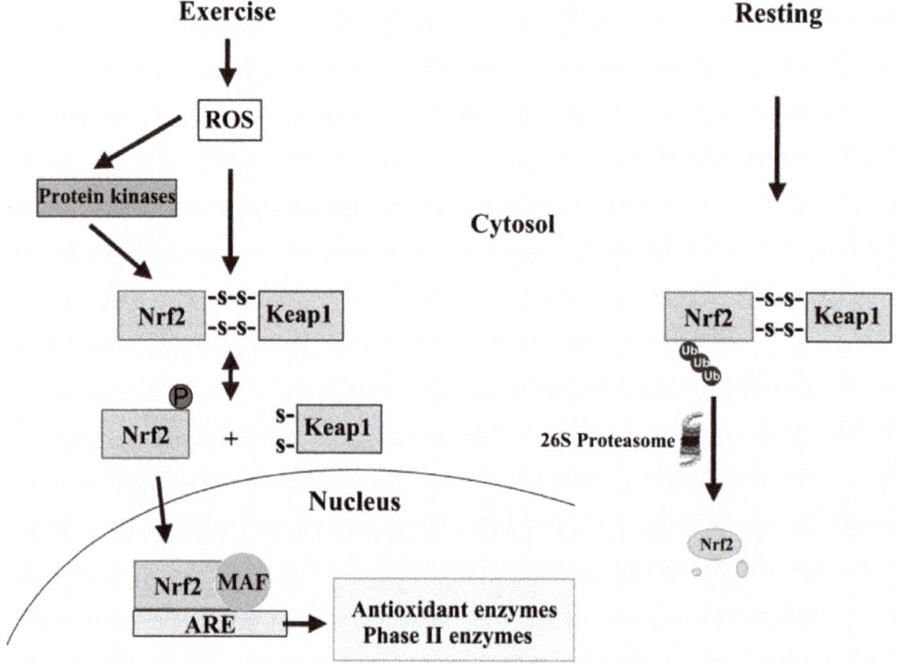

Figure 2. Schematic model of transcriptional activity of nuclear factor erythroid 2-related factor 2 (Nrf2) mediates by oxidants during exercise and Nrf2 degradation in resting condition. MAF, musculoaponeurotic fibrosarcoma protein; ARE, antioxidant response element; Keap1, Kelch ECH associating protein 1; Ub, ubiquitin.

Whatever the pathway of gene regulation may be, the transcription factor Nrf2 is certainly the master regulator of cellular antioxidant defense, because it regulates more than 200 cytoprotective genes in response to oxidative stress [228]. Nrf2 can regulate many antioxidative enzymes, including haem oxygenase-1, SOD, CAT, and NADPH quinone oxidoreductase [229]. Nrf2 also make sure that the antioxidant enzyme expression is coupled with the cofactor supply. Indeed, it controls the expression of GPX2 [230], which reduces peroxides producing GSSG, and GR1 [230], which reduces GSSG, thus allowing intracellular levels of GSH to remain constant. In addition to the GSH-based antioxidant system, Nrf2 also controls the expression of cytosolic Trx1 [231] TrxR1 [230,232] and sulfiredoxin (Srx1) (a cysteine sulfinic acid reductase) [233], all of which reduce oxidized protein thiols [234]. Antioxidant enzymes, such as GR1 and TrxR1, require NADPH as a cofactor so that it is notable that NADPH-generating enzymes such as glucose- 6-phosphate dehydrogenase, 6-phosphogluconate dehydrogenase, isocitrate dehydrogenase, and malic enzyme are all regulated by Nrf2 [235].

Interestingly, Nrf2 also contributes to the maintenance of metabolic homeostasis since Nrf2 induction in pancreatic β cells markedly suppresses oxidative-stress-mediated dysfunction [236].

Evidence is available that the Nrf2 pathway also plays a key role in how oxidative stress mediates the exercise beneficial effects. Increases in ROS production induced by bouts of acute exercise stimulate Nrf2 activation and when they are applied repeatedly, as with regular physical activity, this may lead to upregulation of endogenous antioxidant defenses and overall greater capacity to counteract the oxidative damage of biological molecules.

Cell culture study using C2C12 skeletal muscle cells provided evidence that Nrf2 is activated by ROS and this activation is suppressed when antioxidants, such as N-acetylcysteine, are added to the culture medium [237]. Subsequent study showed an increase in Nrf2 protein expression after myotube treatment of myotubes with H_2O_2 [238].

A single bout of acute exercise in wild-type mice has been shown to increase Nrf2 gene expression [238,239], Nrf2 protein abundance in skeletal muscle [239], and Nrf2-dependent phase II enzymes [238,240]. Conversely, no change in Nrf2 activity was observed in Nrf2$^{-/-}$ mice after acute bout of exercise [239].

The exercise increased oxidative stress and activated Ref1/Nrf2 signaling in a time-dependent manner, with a linear correlation between the mitochondrial H_2O_2 content and Ref1/Nrf2 expressions.

The effect of regular exercise training on the Nrf2 response has been studied extensively and it has been found that, regardless of duration or training regimen, regular aerobic exercise in rodent models activates Nrf2 signaling across multiple tissues including skeletal [241] and cardiac muscle [242,243].

Taken together, the studies demonstrate that regular exercise upregulates Nrf2 protein abundance and phase II and antioxidant enzyme amounts. Furthermore, emerging evidence suggests that an active lifestyle can conserve skeletal muscle cellular redox status via activation of Nrf2 –Keap1 signaling in elderly. Indeed, a cross-sectional study comparing Nrf2 and Keap1 protein content from a single muscle biopsy in sedentary and active older adults has shown the age-associated decline in antioxidant response is due, at least in part, to dysfunction in Nrf2–Keap1 redox signaling, which is preserved in the skeletal muscle of older adults thus maintaining cellular redox homeostasis [244]. However, it is not known whether an exercise program can restore redox balance in individuals who already display a Nrf2 signaling impairment, even though moderate exercise training has been shown to be able to restore Nrf2 signaling in cardiac muscle in older age [243].

12.10. PGC-1α

Although the molecular mechanisms of the adaptive response to exercise remain to be fully elucidated, PGC-1α, a transcriptional coactivator, is currently considered a major regulator of phenotypic adaptation induced by exercise.

PGC-1α was first identified as a transcriptional coactivator of the peroxisome proliferator-activated receptor (PPAR)-γ in brown fat cells [245]. Subsequently, it was found in other mitochondria-rich tissues, including skeletal and cardiac muscle, as well as in kidney, liver, and brain [246] in which it influences numerous aspects of metabolism [247]. PGC-1α and its homolog PGC-1β are also co-activators for PPARα and PPARδ (involved in adipocyte differentiation and thermogenesis), and for a variety of other transcription factors [248,249]. Furthermore, PGC-1α promotes upregulation of itself by an interaction with myocyte enhancer factor 2 (MEF2) on its own promoter [248].

PGC-1α can interact with nuclear receptors and transcription factors activating transcription of their target genes, and its activity is responsive to a wide variety of stimuli including calcium ion, ROS, insulin, thyroid and estrogen hormones, hypoxia, ATP demand, and cytokines [249] (Figure 3).

For a long time PGC-1α has been considered to be exclusively a master regulator of mitochondrial biogenesis by coactivating numerous transcription factors that, in turn, bind to the promoters of distinct sets of nuclear-encoded mitochondrial genes [250]. However, more recent studies have shown that PGC-1α is also able to stimulate the expression of endogenous antioxidant proteins. Reduced mRNA levels of CuZnSod, MnSod, and/or Gpx1 [251], as well as MnSOD protein content [252,253], were found in skeletal muscle from PGC-1α knockout mice compared to wild type, while PGC-1α overexpressing mice showed an increase in MnSOD protein content [254]. PGC-1α KO fibroblasts exhibited a decrease in MnSod, Cat, and Gpx1 mRNA content relative to wild-type fibroblasts and PGC-1α KO mice were more vulnerable to oxidative stress [255]. Furthermore, PGC-1α is able to regulate RNA expression of UCP2 and UCP3 in cell culture [150], suggesting that PGC-1α may increase the uncoupling capacity of mitochondria, thus reducing their ROS production. PGC-1α also promotes mSIRT3 gene expression, which is mediated by an ER-α binding element mapped to the SIRT3 promoter region [256]. SIRT3, in turn, binds to mitochondrial enzymes, including MnSOD, and activates them by deacylation [257,258]. Taken together, PGC-1 α appears to play a role in reducing cell oxidative damage by upregulating antioxidant gene expression and activity.

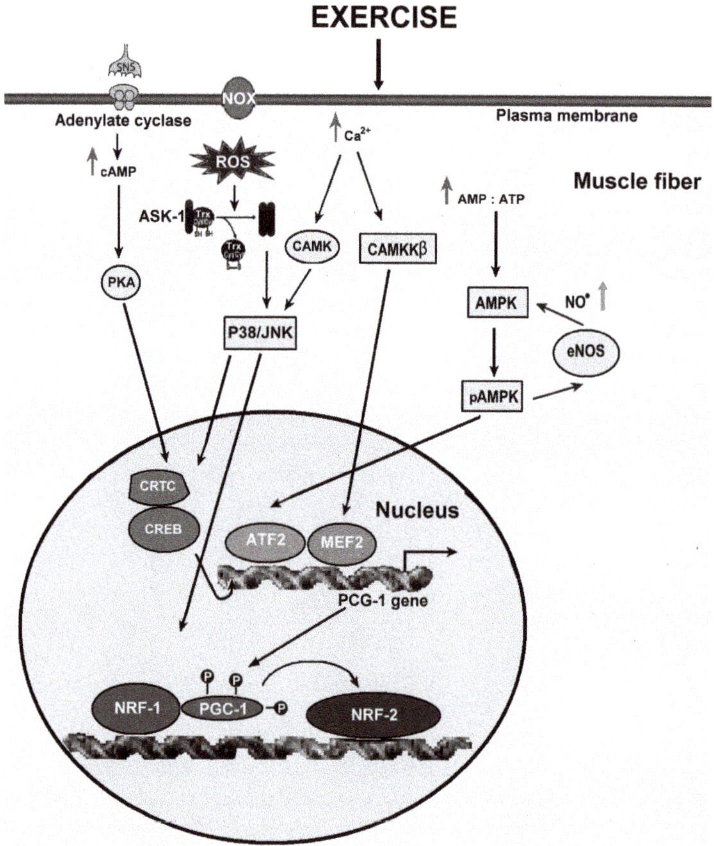

Figure 3. Schematic representation of the signalling pathways that mediate the exercise- induced PGC-1 expression and mitochondrial biogenesis in skeletal muscle. PGC-1, peroxisome proliferator–activated receptor coactivator 1; NRF-1, nuclear respiratory factor 1; NRF-2, nuclear respiratory factor 2; ATF2, activating transcription factor 2; MEF2, myocyte enhancer factor-2; cAMP, cyclic adenosine monophosphate; CREB, cAMP response element-binding protein; CRTC, cAMP-regulated transcriptional co-activators; AMPK, AMP-activated protein kinase; PKA, protein kinase A; NO•, nitric oxide; eNOS, endothelial nitric oxide synthase; CAMK, Ca2+/calmodulin-dependent protein kinase; p38, p38 mitogen-activated protein kinases; JNK, c-Jun N-terminal kinase; ASK-1, apoptosis signal-regulating kinase-1.

Interestingly, recent report has shown that PGC-1 is also necessary for the activation of the signaling network called unfolded protein responses (UPR) during pharmacologically induced endoplasmic reticulum stress and exercise training [259].

PGC-1 protein expression increases rapidly in muscle fibers stimulated to contract [260]. Moreover, Pgc-1 gene expression increases in rat skeletal muscle after a single bout of exercise [261] and in human skeletal muscle after endurance training [262]. Increased levels of PGC-1α protein expression were also found in rat skeletal muscle after 10 weeks of training to swimming [263].

It is worth noting that several initiating stimuli, activated during exercise, can contribute to induction of the PGC-1 gene response. First, acute exercise leads to rapid activation of p38 [264], which in turn activates PGC-1α by phosphorylation [265] and produces the increase in its expression [266].

Other stimuli activated by exercise, that are able to induce Pgc-1 gene response include: (i) Increased concentration of cytosolic calcium, which activates several signaling pathways regulated by the calcineurin phosphatase and the calmodulin-modulated kinase, (ii) decreased levels of high-energy phosphates, which lead to AMPK activation of, (iii) stimulation of the adrenergic system, which leads to cyclic AMP synthesis, and activation of various kinases, including protein kinase A [151] (Figure 3).

However, it is worth noting that the regulation of PGC-1α is not limited to variations in its expression but is also dependent on covalent modifications including phosphorylation, acetylation, methylation and ubiquitination [267]. Indeed, in vitro experiments have shown that PGC-1α phosphorylation by p38 MAPK and AMPK produces a more active protein [151].

Most studies point toward H_2O_2 as an important molecule for PGC-1α upregulation in skeletal muscle. ROS involvement in contraction-induced increases in Pgc-1α expression is supported by the observation that the increase in Pgc-1α mRNA, induced by electrical stimulation in cell culture of rat skeletal muscle, is prevented by antioxidant incubation [268]. Thus, the idea that the upregulation of ROS-removing enzymes in response to increases in ROS can be in part mediated by PGC-1α is supported by the observation that the increase in the Sod, Cat and Gpx mRNA content induced by H_2O_2 in Pgc-1α KO fibroblasts is lower than that in wild-type fibroblasts [255].

Furthermore, the observation that treatment of cultured muscle myotubes with exogenous H_2O_2 activates AMPK and increases Pgc-1α expression [268] suggests that H_2O_2 can promote Pgc-1α expression through AMPK. Moreover, the sensitivity of PGC-1α to the redox status is confirmed by the observation that the antioxidant *N*-acetylcysteine inhibits Pgc-1α upregulation [269].

Pharmacological inhibition of xanthine oxidase with allopurinol also suppresses the upregulation of PGC-1α induced by a single bout of anaerobic exercise in parallel to blunted activation (i.e., phosphorylation) of p38 MAPK in rat vastus lateralis muscle [75]. This finding suggests that ROS generated in response to in vivo contraction are involved in p38 MAPK activation and subsequent regulation of PGC-1α expression [75]. Moreover, evidence that allopurinol treatment also reduces the exercise-induced increases in levels of transcription factors, such as nuclear respiratory factor 1 (NRF-1) and factor of transcription A (Tfam), which are involved in mitochondrial biogenesis, indicates that ROS arising from nonmitochondrial sources play a major role in stimulating mitochondrial biogenesis [75].

Although in literature there are conflicting results [270], ROS have also been shown to be functionally important for PGC-1α expression and adaptive responses induced by endurance exercise in skeletal muscle. Indeed, several experimental studies have reported that antioxidant supplementation attenuates the increase in Pgc-1 gene expression [271,272] and PGC-1 protein content [263,273] elicited by endurance training.

Antioxidant supplementation also prevent health-promoting effects of physical exercise, including mitochondrial biogenesis [263,271,273], endurance performance (running to exhaustion) [273], and insulin sensitivity [271]. These results strongly support the view that the ROS generated during each session of exercise can cause beneficial effects functioning as signals regulating molecular events critical for muscle adaptive responses to training.

It is worth noting that PGC-1α expression in muscle can be regulated by a variety of stimuli associated with muscular exercise, which, however, seem to be dependent on ROS production. Thus, the finding that the human PGC-1α promoter contains a binding site for NF-κB suggests that the expression of PGC-1α may also be regulated by NF-κB [274]. Analysis of the human PGC-1α promoter has revealed a variety of consensus transcription binding sites to the following transcription factors: Specificity protein 1 (SP1), cAMP response element binding protein (CREB), CREB related family member, activating transcription factor 2 (ATF2), forkhead transcription factor (FKHR), p53, EBox binding proteins, GATA and muscle enhancer factor 2 (MEF2) [274]. Many of these transcription factors have been shown to be ROS-sensitive, which indicates numerous potential possibilities for redox control of PGC-1α expression.

Additionally, RNS, particularly NO$^\bullet$, may also be involved in the regulation of PGC-1α. The idea that NO$^\bullet$ mediates the upregulation of PGC-1a, thus modulating mitochondrial function and biogenesis, is supported by the evidence that low levels of NO$^\bullet$ induce mitochondrial biogenesis, PGC-1a and GLUT4 expression in cultured muscle cells [275] and that a genetic deletion of NOS or their pharmacological inhibition prevents PGC-1a induction that is triggered by endurance exercise [276].

It has also been observed that administration to humans of inorganic nitrate (which can be converted into NO$^\bullet$ in the body) significantly improves energy metabolism during exercise [277]. Reports showing that NOS activity is involved in mitochondrial biogenesis induced by AMPK and CaMK and PGC-1α expression in L6 myotubes [278] and that AMPK phosphorylates and activates both eNOS and nNOS [279], led to propose that there is a positive feedback loop between NO$^\bullet$ production and AMPK activity in skeletal muscle [280]. The evidence that NO$^\bullet$ production promotes PGC-1α expression via NO-mediated activation of AMPK (i.e., AMPKα1 isoform) demonstrated that the proposed model of synergistic interaction between AMPK and NOS is crucial to maintain metabolic function in skeletal muscle cells [280]. Moreover, it suggested that both ROS and RNS can contribute to PGC-1α expression via a common signaling pathway (i.e., AMPK activation).

13. Regulation of Cellular Phosphatases by ROS

It has been previously pointed out that changes in the phosphorylation status of signaling molecules play an important role in the control of cellular adaptation. In this regard, it is necessary to note that the phosphorylation status of regulatory proteins and/or transcriptional activators is regulated not only by kinase activity but also by changes in phosphatase activity.

In general, phosphatases are divided into two major classes (i.e., serine/threonine phosphatases and phosphotyrosine phosphatases), both of which are known to be redox sensitive in many different cell types, including skeletal muscle. Serine/threonine phosphatases contain metal ions that are susceptible to oxidation, which leads to their inactivation. Similarly, the phosphotyrosine phosphatases (PTPs) are susceptible to oxidation-induced inactivation. The PTPs contain a cysteine residue in their active site, and oxidation of this cysteine inactivates the enzyme [281]. A subclass of PTPs, called dual-specificity phosphatases (DUSPs), can remove phosphates from both tyrosine and serine/threonine residues. The DUSPs contain two cysteines in their active sites, leading to inactivation of the enzyme during oxidizing conditions. DUSP family include 10 members, which differ in their substrate specificity, subcellular localization, tissue expression, and inducibility by extracellular stimuli [282].

MAPKs including ERK1/2 are dephosphorylated on both the threonine/serine and tyrosine residues by MAPK phosphatases (MKPs) belonging to the DUSP family. It has been shown that, in human skeletal muscle, ERK1/2 phosphorylation is increased in an intensity-dependent manner by acute contractions, but after exercise this phosphorylation is rapidly reduced, and resting levels are restored within 60 min [175,283]. Recent study has demonstrated that two ERK1/2-specific MKPs, dual specificity phosphatase 5 (DUSP5) and DUSP6, are the most regulated MKPs in skeletal muscle after acute exercise [284]. DUSP5 expression is nine-fold higher immediately after exercise and returns to pre-exercise level within 2 h, whereas DUSP6 expression is reduced by 43% just after exercise and remains below pre-exercise level after 2 h recovery. It has also been proposed a hypothetical interplay between ERK1/2 signaling, DUSP5, and DUSP6 in skeletal muscle before and after exercise. Before exercise, basal phosphorylation of MEK (the kinase phosphorylating ERK1/2) and ERK1/2 is low and inactive ERK1/2 is bound to inactive MEK and DUSP6 in cytoplasm. During exercise MEK is activated leading to increased phosphorylation and translocation into the nucleus of ERK1/2, which enhances expression of its target genes, including DUSP5, which, in turn, increases dephosphorylation and trapping of ERK1/2 in the nucleus and reduces ERK1/2 recycling to cytoplasm. A higher proportion of cytoplasmic ERK1/2 is available for phosphorylation by MEK due to the reduced level of DUSP6. During recovery, MEK and ERK1/2 activities are reduced to the basal level, normalizing the DUSP5 level. ERK1/2 is translocated back to the cytoplasm and most of it is bound to MEK, whereas DUSP6 level is still low [284].

14. Conclusions

The idea is now widely shared that the utilization of oxygen by aerobic organisms exposes them to the attack of reactive oxygen and nitrogen species, which can initiate chain reactions leading to oxidative damage of important biological molecules. Aerobic organisms are provided with an efficient antioxidant defense system that allows them to neutralize the oxidative effects of reactive metabolites of oxygen and nitrogen. However, when reactive species production exceeds the cellular antioxidant capacity, oxidative stress develops, potentially leading to cell structural and functional alterations and to the development of many pathological conditions. A single session of strenuous or prolonged exercise leads to the production of high number of radicals and other reactive oxygen species (ROS), which cause tissue damage and dysfunction. On the other hand, regular exercise appears to decrease the incidence of a wide range of ROS-associated diseases because the single sessions of a training program produce low amounts of ROS, which can induce adaptive responses beneficial for the organism. Cells may adapt to the stress by upregulation of systems of defense against and repair of oxidative damage so that they are then resistant to higher levels of oxidative stress imposed subsequently.

Although the effects of exercise on antioxidant capabilities in skeletal muscle have been well described, the regulatory mechanisms underlying this adaptation are complex and incompletely understood. To date it is known that cells exposed to ROS are brought to respond by inducing or repressing a remarkable variety of target genes [155]. These effects appear to be due to modification of the intracellular redox balance resulting in the activation of several signaling pathways ultimately leading to a modulation of gene expression. A comprehensive answer to the question as to how changes in the redox status of muscle fibers regulate signaling pathways and gene expression is not yet available. However, it is clear that a major mechanism by which redox signaling is able to alter gene expression and modify muscle phenotype involves changes in phosphorylation status of transcriptional activators due to ROS capacity to control the activities of many kinases and phosphatases. Evidence that ROS play a pivotal role in adaptive response elicited by exercise training is provided by researches indicating that antioxidant supplementation blunts benefits of regular physical activity on skeletal muscle.

In conclusion, ROS generated during muscle activity, from both mitochondrial and nonmitochondrial sources, may play a pivotal role in muscle adaptive responses to exercise-induced oxidative stress by activating redox-sensitive signal transduction. Important signaling pathways that can be activated in response to ROS stimulation include NFκB, Nrf2, and MAPK. Moreover, the existence of multiple redox-sensitive binding sites on antioxidant genes suggests that the fidelity of gene expression requires the synergistic activation and interaction of several transcription factors. These regulatory mechanisms may control not only the effectiveness of antioxidant defense system through upregulation of antioxidant enzyme expression but also other biological activities in skeletal muscle, including mitochondrial biogenesis.

Funding: This research received no external funding.

Conflicts of Interest: The authors declare no conflict of interest.

References

1. Warburton, D.E.R.; Nicol, C.W.; Bredin, S.S.D. Health benefits of physical activity: The evidence. *Can. Med. Assoc. J.* **2006**, *174*, 801–809. [CrossRef] [PubMed]
2. Simoni, C.; Zauli, G.; Martelli, A.M.; Vitale, M.; Sacchetti, G.; Gonelli, A.; Neri, L.M. Oxidative stress: Role of physical exercise and antioxidant nutraceuticals adulthood and aging. *Oncotarget* **2018**, *9*, 17181–17192. [CrossRef] [PubMed]

3. Kruk, J. Physical activity in the prevention of the most frequent chronic diseases: An analysis of the recent evidence. *Asian Pac. J. Cancer Prev.* **2007**, *8*, 325–338. [PubMed]

4. Valko, M.; Leibfritz, D.; Moncol, J.; Cronin, M.T.; Mazur, M.; Telser, J. Free radicals and antioxidants in normal physiological functions and human disease. *Int. J. Biochem. Cell Biol.* **2007**, *39*, 44–84. [CrossRef] [PubMed]

5. Di Meo, S.; Reed, T.T.; Venditti, P.; Victor, M.V. Role of ROS and RNS sources in physiological and pathological conditions. *Oxid. Med. Cell. Longev.* **2016**, 1245049. [CrossRef]

6. Pedersen, B.K.; Saltin, B. Exercise as medicine—evidence for prescribing exercise as therapy in 26 different chronic diseases. *Scand. J. Med. Sci. Sports* **2015**, *2*, 1–72. [CrossRef] [PubMed]

7. Ebbeling, C.B.; Clarkson, P.M. Exercise-induced muscle damage and adaptation. *Sports Med.* **1989**, *7*, 207–234. [CrossRef]

8. Venditti, P.; Di Meo, S. Antioxidants, tissue damage, and endurance in trained and untrained young male rats. *Arch. Biochem. Biophys.* **1996**, *331*, 63–68. [CrossRef]

9. Venditti, P.; Di Meo, S. Effect of training on antioxidant capacity, tissue damage, and endurance of adult male rats. *Int. J. Sports Med.* **1997**, *1*, 497–502. [CrossRef]

10. Colberg, S.R. Physical activity, insulin action, and diabetes prevention and control. *Curr. Diab. Rev.* **2007**, *3*, 176–184. [CrossRef]

11. Wojtaszewski, J.F.; Richter, E.A. Effects of acute exercise and training on insulin action and sensitivity: Focus on molecular mechanisms in muscle. *Essays Biochem.* **2006**, *42*, 31–46. [CrossRef] [PubMed]

12. Gomberg, M. An instance of trivalent carbon: Triphenylmethyl. *J. Am. Chem. Soc.* **1900**, *22*, 757–771. [CrossRef]

13. Pryor, W.A. Organic free radicals. *Chem. Eng. News* **1968**, *46*, 70–89. [CrossRef]

14. Harman, D. Aging: A theory based on free radical and radiation chemistry. *J. Gerontol.* **1956**, *11*, 298–300. [CrossRef]

15. Bartosz, G, Reactive oxygen species: Destroyers or messengers? *Biochem. Pharmacol.* **2009**, *77*, 1303–1315. [CrossRef] [PubMed]

16. Halliwell, B. Oxidants and human disease: Some new concepts. *FASEB J.* **1987**, *1*, 358–364. [CrossRef]

17. Radi, R. Peroxynitrite, a stealthy biological oxidant. *J. Biol. Chem.* **2013**, *288*, 26464–26472. [CrossRef]

18. Yu, B.P. Cellular defenses against damage from reactive oxygen species. *Physiol. Rev.* **1994**, *74*, 139–162. [CrossRef]

19. Fridovich, I. Superoxide radical and superoxide dismutases. *Annu. Rev. Biochem.* **1995**, *64*, 97–112. [CrossRef]

20. Chelikani, P.; Fita, I.; Loewen, P.C. Diversity of structures and properties among catalase. *Cell. Mol. Life Sci.* **2004**, *61*, 192–208. [CrossRef]

21. Brigelius-Flohé, R.; Maiorino, M. Glutathione peroxidases. *Biochim. Biophys. Acta* **2013**, *1830*, 3289–3303. [CrossRef] [PubMed]

22. Mannervik, B. The enzymes of glutathione metabolism: An overview. *Biochem. Soc. Trans.* **1987**, *15*, 717–718. [CrossRef] [PubMed]

23. Nordberg, J.; Arnér, E.S.J. Reactive oxygen species, antioxidants, and the mammalian thioredoxin system. *Free Radic. Biol. Med.* **2001**, *31*, 1287–1312. [CrossRef]

24. Nakamura, H.; Nakamura, K.; Yodoi, J. Redox regulation of cellular activation. *Annu. Rev. Immunol.* **1997**, *15*, 351–369. [CrossRef] [PubMed]

25. Rhee, S.G.; Chae, H.Z.; Kim, K. Peroxiredoxins: A historical overview and speculative preview of novel mechanisms and emerging concepts in cell signaling. *Free Radic. Biol. Med.* **2005**, *38*, 1543–1552. [CrossRef] [PubMed]

26. Martínez, M.C.; Andriantsitohaina, R. Reactive Nitrogen Species: Molecular Mechanisms and Potential Significance in Health and Disease. *Antioxid. Redox Signal.* **2009**, *11*, 669–702. [CrossRef] [PubMed]

27. Sies, H. *Oxidative Stress, Oxidants and Antioxidants*; Academic Press: London, UK, 1991.

28. Gollnick, P.D.; King, D.W. Effect of exercise and training on mitochondria of rat skeletal muscle. *Am. J. Physiol.* **1969**, *216*, 1502–1509. [CrossRef]

29. King, D.W.; Gollnick, P.D. Ultrastructure of rat heart and liver after exhaustive exercise. *Am. J. Physiol.* **1970**, *218*, 1150–1155. [CrossRef]

30. McCutcheon, L.J.; Byrd, S.K.; Hodgson, D.R. Ultrastructural changes in skeletal muscle after fatiguing exercise. *J. Appl. Physiol.* **1992**, *72*, 1111–1117. [CrossRef]

31. Clarkson, P.M. Eccentric exercise and muscle damage. *Int. J. Sports Med.* **1997**, *18*, S314–S317. [CrossRef]

32. Dillard, C.J.; Litov, R.E.; Savin, W.M.; Dumclin, E.E.; Tapple, A.L. Effect of exercise, vitamin E, and ozone on pulmonary function and lipid peroxidation. *J. Appl. Physiol.* **1978**, *45*, 927–932. [CrossRef] [PubMed]

33. Davies, K.J.; Quintanilha, A.T.; Brooks, G.A.; Packer, L. Free radicals and tissue damage produced by exercise. *Biochem. Biophys. Res. Commun.* **1982**, *107*, 1198–1205. [CrossRef]

34. Jackson, M.J.; Edwards, R.H.; Symons, M.C. Electron spin resonance studies of intact mammalian skeletal muscle. *Biochim. Biophys. Acta* **1985**, *847*, 185–190. [CrossRef]

35. Balon, T.W.; Nadler, J.L. Nitric oxide release is present from incubated skeletal muscle preparations. *J. Appl. Physiol.* **1994**, *77*, 2519–2521. [CrossRef]

36. Jenkins, R.R.; Krause, K.; Schofield, L.S. Influence of exercise on clearance of oxidant stress products and loosely bound iron. *Med. Sci. Sports Exerc.* **1993**, *25*, 213–217. [CrossRef]

37. Reid, M.B.; Haack, K.E.; Franchek, K.M.; Valberg, P.A.; Kobzik, L.; West, M.S. Reactive oxygen in skeletal muscle: Intracellular oxidants kinetics and fatigue in vitro. *J. Appl. Physiol.* **1992**, *73*, 1797–1804. [CrossRef] [PubMed]

38. Bejma, J.; Ji, L.L. Aging and acute exercise enhance free radical generation in rat skeletal muscle. *J. Appl. Physiol.* **1999**, *87*, 465–470. [CrossRef]

39. Palomero, J.; Pye, D.; Kabayo, T.; Spiller, D.G.; Jackson, M.J. In situ detection and measurement of intracellular reactive oxygen species in single isolated mature skeletal muscle fibers by real time fluorescence microscopy. *Antioxid. Redox Signal.* **2008**, *10*, 1463–1474. [CrossRef]

40. Gomez-Cabrera, M.C.; Close, G.L.; Kayani, A.; McArdle, A.; Viña, J.; Jackson, M.J. Effect of xanthine oxidase-generated extracellular superoxide on skeletal muscle force generation. *Am. J. Physiol. Regul. Integr. Comp. Physiol.* **2010**, *298*, R2–R8. [CrossRef]

41. Cheng, A.J.; Bruton, J.D.; Lanner, J.T.; Westerblad, H. Antioxidant treatments do not improve force recovery after fatiguing stimulation of mouse skeletal muscle fibres. *J. Physiol.* **2015**, *593*, 457–472. [CrossRef]

42. Pattwell, D.M.; McArdle, A.; Morgan, J.E.; Patridge, T.A.; Jackson, M.J. Release of reactive oxygen and nitrogen species from contracting skeletal muscle cells. *Free Radic. Biol. Med.* **2004**, *37*, 1064–1072. [CrossRef] [PubMed]

43. Tidball, J.G.; Lavergne, E.; Lau, K.S.; Spencer, M.J.; Stull, J.T.; Wehling, M. Mechanical loading regulates NOS expression and activity in developing and adult skeletal muscle. *Am. J. Physiol. Cell Physiol.* **1998**, *275*, C260–C266. [CrossRef] [PubMed]

44. Steinbacher, P.; Eckl, P. Impact of oxidative stress on exercising skeletal muscle. *Biomolecules* **2015**, *5*, 356–377. [CrossRef] [PubMed]

45. Barja, G. Mitochondrial oxygen radical generation and leak: Sites of production in states 4 and 3, organ specificity, and relation to aging and longevity. *J. Bioenerg. Biomembr.* **1999**, *31*, 347–366. [CrossRef] [PubMed]

46. Lambert, A.J.; Brand, M.D. Reactive oxygen species production by mitochondria. *Methods Mol. Biol.* **2009**, *554*, 165–181.

47. Powers, S.K.; Talbert, E.; Adhihetty, P.J. Reactive oxygen and nitrogen species as intracellular signals in skeletal muscle. *J. Physiol.* **2011**, *589*, 2129–2138. [CrossRef] [PubMed]

48. Pearson, T.; Kabayo, T.; Ng, R.; Chamberlain, J.; McArdle, A.; Jackson, M.J. Skeletal muscle contractions induce acute changes in cytosolic superoxide, but slower responses in mitochondrial superoxide and cellular hydrogen peroxide. *PLoS ONE* **2014**. [CrossRef]

49. Di Meo, S.; Venditti, P. Mitochondria in exercise- induced oxidative stress. *Biol. Signals Recept.* **2001**, *10*, 125–140. [CrossRef]

50. Carfagna, G.; Napolitano, G.; Barone, G.; Pinto, A.; Pollio, A.; Venditti, P. Dietary supplementation with the microalga Galdieria sulphuraria (Rhodophyta) reduces prolonged exercise induced oxidative stress in rat tissues. *Oxid. Med. Cell. Longev.* **2015**. [CrossRef]

51. Venditti, P.; Bari, A.; Di Stefano, L.; Di Meo, S. Role of mitochondria in exercise-induced oxidative stress in skeletal muscle from hyperthyroid rats. *Arch. Biochem. Biophys.* **2007**, *463*, 12–18. [CrossRef]

52. Sakellariou, G.K.; Vasilaki, A.; Palomero, J.; Kayani, A.; Zibrik, L.; McArdle, A.; Jackson, M.J. Studies of mitochondrial and nonmitochondrial sources implicate nicotinamide adenine dinucleotide phosphate oxidase(s) in the increased skeletal muscle superoxide generation that occurs during contractile activity. *Antioxid. Redox Signal.* **2013**, *18*, 603–621. [CrossRef]

53. Gomez-Cabrera, M.-C.; Borrás, C.; Pallardo, F.V.; Sastre, J.; Ji, L.L.; Viña, J. Decreasing xanthine oxidase-mediated oxidative stress prevents useful cellular adaptations to exercise in rats. *J. Physiol.* **2005**, *567*, 113–120. [CrossRef]

54. Hellsten, Y.; Apple, F.S.; Sjödin, B. Effect of sprint cycle training on activities of antioxidant enzymes in human skeletal muscle. *J. Appl. Physiol.* **1996**, *81*, 1484–1487. [CrossRef] [PubMed]

55. Duarte, J.A.R.; Appell, H.-J.; Carvalho, F.; Bastos, M.L.; Soares, J.M.C. Endothelium-derived oxidative stress may contribute to exercise-induced muscle damage. *Int. J. Sports Med.* **1993**, *14*, 440–443. [CrossRef] [PubMed]

56. Aoi, W.; Naito, Y.; Takanami, Y.; Kawai, Y.; Sakuma, K.; Ichikawa, H.; Yoshida, N.; Yoshikawa, T. Oxidative stress and delayed-onset muscle damage after exercise. *Free Radic. Biol. Med.* **2004**, *37*, 480–487. [CrossRef] [PubMed]

57. Mayer, B.; Hemmens, B. Biosynthesis and action of nitric oxide in mammalian cells. *Trends Biochem. Sci.* **1997**, *22*, 477–481. [CrossRef]

58. Moylan, J.S.; Reid, M.B. Oxidative stress, chronic disease, and muscle wasting. *Muscle Nerve* **2007**, *35*, 411–429. [CrossRef]

59. Alessio, H.M.; Goldfarb, A.H.; Cutler, R.G. MDA content increases in fast- and slow-twitch skeletal muscle with intensity of exercise in a rat. *Am. J. Physiol.* **1988**, *255*, C874–C877. [CrossRef]

60. Alessio, H.M.; Goldfarb, A.H. Lipid peroxidation and scavenger enzymes during exercise: Adaptive response to training. *J. Appl. Physiol.* **1988**, *64*, 1333–1336. [CrossRef]

61. Ji, L.L.; Stratman, F.W.; Lardy, H.A. Antioxidant enzyme system in rat liver and skeletal muscle: Influences of selenium deficiency, acute exercise and chronic training. *Arch. Biochem. Biophys.* **1988**, *263*, 150–160. [CrossRef]

62. Weber, D.; Davies, M.J.; Grune, T. Determination of protein carbonyls in plasma, cell extracts, tissue homogenates, isolated proteins: Focus on sample preparation and derivatization conditions. *Redox Biol.* **2015**, *5*, 367–380. [CrossRef] [PubMed]

63. Reznick, A.Z.; Witt, E.; Matsumoto, M.; Packer, L. Vitamin E inhibits protein oxidation in skeletal muscle of resting and exercised rat. *Biochem. Biophys. Res. Commun.* **1992**, *189*, 801–806. [CrossRef]

64. Atalay, M.; Laaksonen, D.E.; Khanna, S.; Kaliste-Korhonen, E.; Hänninen, O.; Sen, C.K. Vitamin E regulates changes in tissue antioxidants induced by fish oil and acute exercise. *Med. Sci. Sports Exerc.* **2000**, *32*, 601–607. [CrossRef] [PubMed]

65. Liu, J.; Yeo, H.C.; Overvik-Douki, E.; Hagen, T.; Doniger, S.J.; Chyu, D.W.; Brooks, G.A.; Ames, B.N. Chronically and acutely exercised rats: Biomarkers of oxidative stress and endogenous antioxidants. *J. Appl. Physiol.* **2000**, *89*, 21–28. [CrossRef] [PubMed]

66. Okamura, K.; Doi, T.; Sakurai, M.; Hamada, K.; Yoshioka, Y.; Sumida, S.; Sugawa-Katayama, Y. Effect of endurance exercise on the tissue 8-hydroxy-deoxyguanosine content in dogs. *Free Radic. Res.* **1997**, *26*, 523–528. [CrossRef] [PubMed]

67. Radak, Z.; Bori, Z.; Koltai, E.; Fatouros, I.G.; Jamurtas, A.Z.; Douroudos, I.I.; Terzis, G.; Nikolaidis, M.G.; Chatzinikolaou, A.; Sovatzidis, A.; et al. Age-dependent changes in 8-oxoguanine-DNA-glycosylase activity is modulated by adaptive responses to physical exercise in human skeletal muscle. *Free Radic. Biol. Med.* **2011**, *51*, 417–423. [CrossRef] [PubMed]

68. Richter, C.; Park, J.W.; Ames, B.N. Normal oxidative damage to mitochondrial and nuclear DNA is extensive. *Proc. Natl. Acad. Sci. USA* **1988**, *85*, 6465–6467. [CrossRef]

69. Lu, R.; Nash, H.M.; Verdine, G.L.A. A mammalian DNA repair enzyme that excises oxidatively damaged guanines maps to a locus frequently lost in lung cancer. *Curr. Biol.* **1997**, *7*, 397–407. [CrossRef]

70. Yamaguchi, R.; Hirano, T.; Asami, S.; Sugita, A.; Kasai, H. Increase in 8-hydroxyguanine repair activity in the rat kidney after the administration of a renal carcinigen, ferric nitroloacetate. *Environ. Health Perspec.* **1996**, *104*, 651–653.

71. Lew, H.; Pyke, S.; Quintanilha, A. Changes in the glutathione status of plasma, liver and muscle following exhaustive exercise in rats. *FEBS Lett.* **1985**, *185*, 262–266. [CrossRef]

72. Ji, L.L.; Fu, R. Responses of glutathione system and antioxidant enzymes to exhaustive exercise and hydroperoxide. *J. Appl. Physiol.* **1992**, *72*, 549–554. [CrossRef] [PubMed]

73. Bloomer, R.J.; Goldfarb, A.H. Anaerobic exercise and oxidative stress: A review. *Can. J. Appl. Physiol.* **2004**, *29*, 245–263. [CrossRef] [PubMed]

74. Powers, S.K.; Jackson, M.J. Exercise-induced oxidative stress: Cellular mechanisms and impact on muscle force production. *Physiol. Rev.* **2008**, *88*, 1243–1276. [CrossRef]

75. Kang, C.; O'Moore, K.M.; Dickman, J.R.; Ji, L.L. Exercise activation of muscle peroxisome proliferator-activated receptor-gamma coactivator-1alpha signaling is redox sensitive. *Free Radic. Biol. Med.* **2009**, *47*, 1394–1400. [CrossRef] [PubMed]

76. Witt, E.H.; Reznick, A.Z.; Viguie, C.A.; Starke-Reed, P.; Packer, L. Exercise, oxidative damage and effects of antioxidant manipulation. *J. Nutr.* **1992**, *122* (Suppl. 3), 766–773. [CrossRef] [PubMed]

77. Kanter, M. Free radicals, exercise and antioxidant supplementation. *Proc. Nutr. Soc.* **1998**, *57*, 9–13. [CrossRef] [PubMed]

78. Clarkson, P.M.; Thompson, H.S. Antioxidants: What role do they play in physical activity and health? *Am. J. Clin. Nutr.* **2000**, *72*, 637S–646S. [CrossRef] [PubMed]

79. Egan, B.; Zierath, J.R. Exercise metabolism and the molecular regulation of skeletal muscle adaptation. *Cell Metab.* **2013**, *17*, 162–184. [CrossRef]

80. Perry, C.G.; Lally, J.; Holloway, G.P.; Heigenhauser, G.J.; Bonen, A.; Spriet, L.L. Repeated transient mRNA bursts precede increases in transcriptional and mitochondrial proteins during training in human skeletal muscle. *J. Physiol.* **2010**, *588*, 4795–4810. [CrossRef]

81. Hawley, J.A. Adaptations of skeletal muscle to prolonged, intense endurance training. *Clin. Exp. Pharmacol. Physiol.* **2002**, *29*, 218–222. [CrossRef]

82. Yan, Z.; Okutsu, M.; Akhtar, Y.N.; Lira, V.A. Regulation of exercise-induced fiber type transformation, mitochondrial biogenesis, and angiogenesis in skeletal muscle. *J. Appl. Physiol.* **2011**, *110*, 264–274. [CrossRef] [PubMed]

83. Prior, B.M.; Yang, H.T.; Terjung, R.L. What makes vessels grow with exercise training? *J. Appl. Physiol.* **2004**, *97*, 1119–1128. [CrossRef] [PubMed]

84. Reichmann, H.; Hoppeler, H.; Mathieu-Costello, O.; von Bergen, F.; Pette, D. Biochemical and ultrastructural changes of skeletal muscle mitochondria after chronic electrical stimulation in rabbits. *Pflugers Arch.* **1985**, *404*, 1–9. [CrossRef] [PubMed]

85. Roesler, K.M.; Conley, K.E.; Claassen, H.; Howald, H.; Hoppeler, H.; Gehr, P. Transfer effects in endurance exercise: Adaptations in trained and untrained muscles. *Eur. J. Appl. Physiol.* **1985**, *54*, 355–362. [CrossRef]

86. Sale, D.G. Neural adaptation to resistance training. *Med. Sci. Sports Exerc.* **1988**, *20*, S135–S145. [CrossRef] [PubMed]

87. Narici, M.V.; Hoppeler, H.; Kayser, B.; Landoni, L.; Claassen, H.; Gavardi, C.; Conti, M.; Cerretelli, P. Human quadriceps cross-sectional area, torque and neural activation during 6 months strength training. *Acta Physiol. Scand.* **1996**, *157*, 175–186. [CrossRef] [PubMed]

88. Goneya, W.J.; Sale, D. Physiology of weight-lifting exercise. *Arch. Phys. Med. Rehabil.* **1982**, *63*, 235–237.

89. Kelley, G. Mechanical overload and skeletal muscle fiber hyperplasia: A meta-analysis. *J. Appl. Physiol.* **1996**, *81*, 1584–1588. [CrossRef]

90. Doherty, T.J. Invited review: Aging and sarcopenia. *J. Appl. Physiol.* **2003**, *95*, 1717–1727. [CrossRef]

91. Cornelissen, V.A.; Fagard, R.H. Effect of resistance training on resting blood pressure: A meta-analysis of randomized controlled trials. *J. Hypertens.* **2005**, *23*, 251–259. [CrossRef]

92. Tanasescu, M.; Leitzmann, M.F.; Rimm, E.B.; Willett, W.C.; Stampfer, M.J.; Hu, F.B. Exercise type and intensity in relation to coronary heart disease in men. *Jama* **2002**, *288*, 1994–2000. [CrossRef] [PubMed]

93. Luethi, J.M.; Howald, H.; Claassen, H.; Roesler, K.; Vock, P.; Hoppeler, H. Structural changes in skeletal muscle tissue with heavy-resistance exercise. *Int. J. Sports Med.* **1986**, *7*, 123–127. [CrossRef] [PubMed]

94. MacDougall, J.D.; Sale, D.G.; Elder, G.C.; Sutton, J.R. Muscle ultrastructural characteristics of elite powerlifters and bodybuilders. *Europ. J. Appl. Physiol.* **1982**, *48*, 117–126. [CrossRef]

95. Pesta, D.; Hoppel, F.; Macek, C.; Messner, H.; Faulhaber, M.; Kobel, C.; Parson, W.; Burtscher, M.; Schocke, M.; Gnaiger, E. Similar qualitative and quantitative changes of mitochondrial respiration following strength and endurance training in normoxia and hypoxia in sedentary humans. *Am. J. Physiol.* **2011**, *301*, R1078–R1087. [CrossRef] [PubMed]

96. Davies, K.J.A.; Packer, L.; Brooks, G.A. Biochemical adaptation of mitochondria, muscle, and whole-animal respiration to endurance training. *Arch. Biochem. Biophys.* **1981**, *209*, 539–554. [CrossRef]

97. Sparks, L.M.; Johannsen, N.M.; Church, T.S.; Earnest, C.P.; Moonen-Kornips, E.; Moro, C.; Hesselink, M.K.; Smith, S.R.; Schrauwen, P. Nine months of combined training improves ex vivo skeletal muscle metabolism in individuals with type 2 diabetes. *J. Clin. Endocrinol. Metab.* **2013**, *98*, 1694–1702. [CrossRef] [PubMed]

98. Jubrias, S.A.; Esselman, P.C.; Price, L.B.; Cress, M.E.; Conley, K.E. Large energetic adaptations of elderly muscle to resistance and endurance training. *J. Appl. Physiol.* **2001**, *90*, 1663–1670. [CrossRef] [PubMed]

99. Dröge, W. Free radicals in the physiological control of cell function. *Physiol. Rev.* **2002**, *82*, 47–95. [CrossRef]

100. Booth, F.W.; Thomason, D.B. Molecular and cellular adaptation of muscle in response to exercise: Perspectives of various models. *Physiol. Rev.* **1991**, *71*, 541–585. [CrossRef]

101. Reid, M.B.; Khawli, F.A.; Moody, M.R. Reactive oxygen in skeletal muscle. III. Contractility of unfatigued muscle. *J. Appl. Physiol.* **1993**, *75*, 1081–1087. [CrossRef]

102. Reid, M.B.; Moody, M.R. Dimethyl sulfoxide depresses skeletal muscle contractility. *J. Appl. Physiol.* **1994**, *76*, 2186–2190. [CrossRef] [PubMed]

103. Coombes, J.S.; Powers, S.K.; Rowell, B.; Hamilton, K.L.; Dodd, S.L.; Shanely, R.A.; Sen, C.K.; Packer, L. Effects of vitamin E and alpha-lipoic acid on skeletal muscle contractile properties. *J. Appl. Physiol.* **2001**, *90*, 1424–1430. [CrossRef] [PubMed]

104. Reid, M.B. Invited Review: Redox modulation of skeletal muscle contraction: What we know and what we don't. *J. Appl. Physiol.* **2001**, *90*, 724–731. [CrossRef] [PubMed]

105. Fitts, R.H. Cellular mechanisms of muscle fatigue. *Physiol. Rev.* **1994**, *74*, 49–94. [CrossRef] [PubMed]

106. Fitts, R.H. Muscle fatigue: The cellular aspects. *Am. J. Sports Med.* **1996**, *24* (Suppl. 6), S9–S13. [CrossRef]

107. Ament, W.; Verkerke, G.J. Exercise and fatigue. *Sports Med.* **2009**, *39*, 389–422. [CrossRef] [PubMed]

108. Ferreira, L.F.; Reed, M.B. Muscle–derived ROS and thiol regulation in muscle fatigue. *J. Appl. Physiol.* **2008**, *104*, 853–860. [CrossRef]

109. Lands, L.C.; Grey, V.L.; Smountas, A.A. Effect of supplementation with a cysteine donor on muscular performance. *J. Appl. Physiol.* **1999**, *87*, 1381–1385. [CrossRef]

110. Brady, P.S.; Brady, L.J.; Ulrey, D.E.J. Selenium, vitamin E and the response to swimming stress in the rat. *Nutrition* **1979**, *109*, 1103–1109. [CrossRef]

111. Novelli, G.P.; Bracciotti, G.; Falsini, S. Spin-trappers and vitamin E prolong endurance to muscle fatigue in mice. *Free Radic. Biol. Med.* **1990**, *8*, 9–13. [CrossRef]

112. Novelli, G.P.; Falsini, S.; Bracciotti, G. Exogenous glutathione increases endurance to muscle effort in mice. *Pharmacol. Res.* **1991**, *23*, 149–155. [CrossRef]

113. Shindoh, C.; Di Marco, A.; Thomas, A.; Manubay, P.; Supinski, G. Effect of N-acetylcysteine on diaphragm fatigue. *J. Appl. Physiol.* **1990**, *68*, 2107–2113. [CrossRef] [PubMed]

114. Moldeus, P.; Derw, R.; Berggren, M. Lung protection by a thiol-containing antioxidant: N-acetyl-cysteine. *Respiration* **1986**, *50*, 31–42. [CrossRef] [PubMed]

115. Quintanilha, A.T.; Packer, L. Vitamin E, physical exercise and tissue oxidative damage. *Ciba Found. Symp.* **1983**, *101*, 56–69. [PubMed]

116. Laughlin, M.H.; Simpson, T.; Sexton, W.L.; Brown, O.R.; Smith, J.K.; Korthuis, R.J. Skeletal muscle oxidative capacity, antioxidant enzymes, and exercise training. *J. Appl. Physiol.* **1990**, *68*, 2337–2343. [CrossRef] [PubMed]

117. Venditti, P.; Bari, A.; Di Stefano, L.; Di Meo, S. Effect of T3 on metabolic response and oxidative stress in skeletal muscle from sedentary and trained rats. *Free Radic. Biol. Med.* **2009**, *46*, 360–366. [CrossRef] [PubMed]

118. Leeuwenburgh, C.; Fiebig, R.; Chandwaney, R.; Ji, L.L. Aging and exercise training in skeletal muscle: Responses of glutathione and antioxidant enzyme systems. *Am. J. Physiol.* **1994**, *267*, R439–R445. [CrossRef]

119. Lambertucci, R.H.; Levada-Pires, A.C.; Rossoni, L.V.; Curi, R.; Pithon-Curi, T.C. Effects of aerobic exercise training on antioxidant enzyme activities and mRNA levels in soleus muscle from young and aged rats. *Mech. Ageing Dev.* **2007**, *128*, 267–275. [CrossRef]

120. Oh-ishi, S.; Toshinai, K.; Kizaki, T.; Haga, S.; Fukuda, K.; Nagata, N.; Ohno, H. Effects of aging and/or training on antioxidant enzyme system in diaphragm of mice. *Respir. Physiol.* **1996**, *105*, 195–202. [CrossRef]

121. Powers, S.K.; Criswell, D.; Lawler, J.; Ji, L.L.; Martin, D.; Herb, R.A.; Dudley, G. Influence of exercise and fiber type on antioxidant enzyme activity in rat skeletal muscle. *Am. J. Physiol.* **1994**, *266*, R375–R380. [CrossRef] [PubMed]

122. Leeuwenburgh, C.; Hollander, J.; Leichtweis, S.; Griffiths, M.; Gore, M.; Ji, L.L. Adaptations of glutathione antioxidant system to endurance training are tissue and muscle fiber specific. *Am. J. Physiol.* **1997**, *272*, R363–R369. [CrossRef] [PubMed]

123. Vincent, K.R.; Vincent, H.K.; Braith, R.W.; Lennon, S.L.; Lowenthal, D.T. Resistance exercise training attenuates exercise-induced lipid peroxidation in the elderly. *Eur. J. Appl. Physiol.* **2002**, *87*, 416–423. [CrossRef] [PubMed]

124. Parise, G.; Brose, A.N.; Tarnopolsky, M.A. Resistance exercise training decreases oxidative damage to DNA and increases cytochrome oxidase activity in older adults. *Exp. Gerontol.* **2005**, *40*, 173–180. [CrossRef] [PubMed]

125. Parise, G.; Phillips, S.M.; Kaczor, J.J.; Tarnopolsky, M.A. Antioxidant enzyme activity is up-regulated after unilateral resistance exercise training in older adults. *Free Radic. Biol. Med.* **2005**, *39*, 289–295. [CrossRef] [PubMed]

126. Gore, M.; Fiebig, R.; Hollander, J.; Leeuwenburgh, C.; Ohno, H.; Ji, L.L. Endurance training alters antioxidant enzyme gene expression in rat skeletal muscle. *Can. J. Physiol. Pharmacol.* **1998**, *76*, 1139–1145. [CrossRef] [PubMed]

127. Hollander, J.; Fiebig, R.; Gore, M.; Bejma, J.; Ookawara, T.; Ohno, H.; Ji, L.L. Superoxide dismutase gene expression in skeletal muscle: Fiber-specific adaptation to endurance training. *Am. J. Physiol.* **1999**, *277*, R856–R862. [CrossRef]

128. Ji, L.L.; Fu, R.G.; Mitchell, E. Glutathione and antioxidant enzyme in skeletal muscle: Effect of fiber type and exercise intensity. *J. Appl. Physiol.* **1992**, *73*, 1854–1859. [CrossRef] [PubMed]

129. Ji, L.L. Antioxidant enzyme response to exercise and aging. *Med. Sci. Sports Exer.* **1993**, *25*, 225–231. [CrossRef]

130. Storz, G.; Tartaglia, L.A.; Ames, B.N. Transcriptional regulator of oxidative stress-inducible genes: Direct activation by oxidation. *Science* **1990**, *48*, 189–194. [CrossRef]

131. Harris, A.D. Regulation of antioxidant enzymes. *FASEB J.* **1992**, *6*, 2675–2683. [CrossRef]

132. Davies, N.A.; Watkeys, L.; Butcher, L.; Potter, S.; Hughes, M.G.; Moir, H.; Morris, K.; Thomas, A.W.; Webb, R. The contributions of oxidative stress, oxidised lipoproteins and AMPK towards exercise-associated PPAR signalling within human monocytic cells. *Free Radic. Res.* **2015**, *49*, 45–56. [CrossRef] [PubMed]

133. Thomas, A.W.; Davies, N.A.; Moir, H.; Watkeys, L.; Ruffino, J.-S.; Isa, S.A.; Butcher, L.R.; Hughes, M.G.; Morris, K.; Webb, R. Exercise-associated generation of PPAR ligands activates PPAR signalling events and upregulates genes related to lipid metabolism. *J. Appl. Physiol.* **2012**, *112*, 806–815. [CrossRef] [PubMed]

134. Hollander, J.; Fiebig, R.; Gore, M.; Ookawara, T.; Ohno, H.; Ji, L.L. Superoxide dismutase gene expression is activated by a single bout of exercise in rat skeletal muscle. *Pflugers Arch.* **2001**, *442*, 426–434. [CrossRef] [PubMed]

135. Webb, R.; Hughes, M.G.; Thomas, A.W.; Morris, K. The Ability of Exercise-Associated Oxidative Stress to Trigger Redox-Sensitive Signalling Responses. *Antioxidants* **2017**, *6*, 63. [CrossRef] [PubMed]

136. Schild, M.; Ruhs, A.; Beiter, T.; Zügel, M.; Hudemann, J.; Reimer, A.; Krumholz-Wagner, I.; Wagner, C.; Keller, J.; Eder, K.; et al. Basal and exercise induced label-free quantitative protein profiling of m. vastus lateralis in trained and untrained individuals. *J. Proteom.* **2015**, *122*, 119–132. [CrossRef] [PubMed]

137. Venditti, P.; Masullo, P.; Di Meo, S. Effect of training on H_2O_2 release by mitochondria from rat skeletal muscle. *Arch. Biochem. Biophys.* **1999**, *372*, 315–320. [CrossRef] [PubMed]

138. Molnar, A.M.; Servais, S.; Guichardant, M.; Lagarde, M.; Macedo, D.V.; Pereira-Da-Silva, L.; Sibille, B.; Favier, R. Mitochondrial H_2O_2 production is reduced with acute and chronic eccentric exercise in rat skeletal muscle. *Antioxid. Redox Signal.* **2006**, *8*, 548–558. [CrossRef]

139. Hey-Mogensen, M.; Højlund, K.; Vind, B.F.; Wang, L.; Dela, F.; Beck-Nielsen, H.; Fernström, M.; Sahlin, K. Effect of physical training on mitochondrial respiration and reactive oxygen species release in skeletal muscle in patients with obesity and type 2 diabetes. *Diabetologia* **2010**, *53*, 1976–1985. [CrossRef]

140. Gram, M.; Vigelsø, A.; Yokota, T.; Helge, J.W.; Dela, F.; Hey-Mogensen, M. Skeletal muscle mitochondrial H_2O_2 emission increases with immobilization and decreases after aerobic training in young and older men. *J. Physiol.* **2015**, *593*, 4011–4027. [CrossRef]

141. Binsch, C.; Jelenik, T.; Pfitzer, A.; Dille, M.; Müller-Lühlhoff, S.; Hartwig, S.; Karpinski, S.; Lehr, S.; Kabra, D.G.; Chadt, A.; et al. Absence of the kinase S6k1 mimics the effect of chronic endurance exercise on glucose tolerance and muscle oxidative stress. *Mol. Metab.* **2017**, *6*, 1443–1453. [CrossRef]

142. Ghosh, S.; Lertwattanarak, R.; Lefort, N.; Molina-Carrion, M.; Joya-Galeana, J.; Bowen, B.P.; Garduno-Garcia Jde, J.; Abdul-Ghani, M.; Richardson, A.; DeFronzo, R.A.; et al. Reduction in reactive oxygen species production by mitochondria from elderly subjects with normal and impaired glucose tolerance. *Diabetes* **2011**, *60*, 2051–2060. [CrossRef]

143. Flack, K.D.; Davy, B.M.; DeBerardinis, M.; Boutagy, N.E.; McMillan, R.P.; Hulver, M.W.; Frisard, M.I.; Anderson, A.S.; Savla, J.; Davy, K.P. Resistance exercise training and in vitro skeletal muscle oxidative capacity in older adults. *Physiol. Rep.* **2016**, *4*, e12849. [CrossRef]

144. Klingeberg, M. Mechanism and evolution of the uncoupling protein of brown adipose tissue. *Trends Biol. Sci.* **1990**, *15*, 108–112. [CrossRef]

145. Silva, J.E.; Rabelo, R. Regulation of the uncoupling protein gene expression. *Eur. J. Endocrinol.* **1997**, *136*, 251–264. [CrossRef]

146. Ramsden, D.B.; Ho, P.W.; Ho, J.W.; Liu, H.F.; So, D.H.; Tse, H.M.; Chan, K.H.; Ho, S.L. Human neuronal uncoupling proteins 4 and 5 (UCP4 and UCP5): Structural properties, regulation, and physiological role in protection against oxidative stress and mitochondrial dysfunction. *Brain Behav.* **2012**, *2*, 468–478. [CrossRef]

147. Ricquier, D. Uncoupling protein 1 of brown adipocytes, the only uncoupler: A historical perspective. *Front Endocrinol.* **2011**, *2*, 85. [CrossRef]

148. Nedergaard, J.; Cannon, B. The 'novel' 'uncoupling' proteins UCP2 and UCP3: What do they really do? Pros and cons for suggested functions. *Exp. Physiol.* **2003**, *88*, 65–84. [CrossRef]

149. Toime, L.J.; Brand, M.D. Uncoupling protein-3 lowers reactive oxygen species production in isolated mitochondria. *Free Radic. Biol. Med.* **2010**, *49*, 606–611. [CrossRef]

150. St-Pierre, J.; Lin, J.; Krauss, S.; Tarr, P.T.; Yang, R.; Newgard, C.B.; Spiegelman, B.M. Bioenergetic analysis of peroxisome proliferator-activated receptor γ coactivators 1α and 1β (PGC-1α and PGC-1β) in muscle cells. *J. Biol. Chem.* **2003**, *278*, 26597–26603. [CrossRef]

151. Kang, C.; Ji, L.L. Role of PGC-1α signaling in skeletal muscle health and disease. *Ann. N. Y. Acad. Sci.* **2012**, *1271*, 110–117. [CrossRef]

152. Zhou, M.; Lin, B.Z.; Coughlin, S.; Vallega, G.; Pilch, P.F. UCP-3 expression in skeletal muscle: Effects of exercise, hypoxia, and AMP-activated protein kinase. *Am. J. Physiol. Cell Physiol.* **2000**, *279*, E622–E629. [CrossRef] [PubMed]

153. Jones, T.E.; Baar, K.; Ojuka, E.; Chen, M.; Holloszy, J.O. Exercise induces an increase in muscle UCP3 as a component of the increase in mitochondrial biogenesis. *Am. J. Physiol. Cell Physiol.* **2003**, *284*, E96–E101. [CrossRef] [PubMed]

154. Cunha, T.F.; Bechara, L.R.; Bacurau, A.V.; Jannig, P.R.; Voltarelli, V.A.; Dourado, P.M.; Vasconcelos, A.R.; Scavone, C.; Ferreira, J.C.; Brum, P.C. Exercise training decreases NADPH oxidase activity and restores skeletal muscle mass in heart failure rats. *J. Appl. Physiol.* **2017**, *122*, 817–827. [CrossRef] [PubMed]

155. McArdle, A.; Pattwell, D.; Vasilaki, A.; Griffiths, R.D.; Jackson, M.J. Contractile activity-induced oxidative stress: Cellular origin and adaptive responses. *Am. J. Physiol. Cell Physiol.* **2001**, *280*, C621–C627. [CrossRef] [PubMed]

156. Ammendola, R.; Fiore, F.; Esposito, F.; Caserta, G.; Mesuraca, M.; Russo, T.; Cimino, F. Differentially expressed mRNAs as a consequence of oxidative stress in intact cells. *FEBS Lett.* **1995**, *371*, 209–213.

157. Storz, G.; Polla, B.S. Transcriptional regulators of oxidative stress inducible genes in prokaryotes and eukaryotes. *EXS* **1996**, *77*, 239–254. [PubMed]

158. Jackson, M.J.; Papa, S.; Bolaños, J.; Bruckdorfer, R.; Carlsen, H.; Elliott, R.M.; Flier, J.; Griffiths, H.R.; Heales, S.; Holst, B.; et al. Antioxidants, reactive oxygen and nitrogen species, gene induction and mitochondrial function. *Mol. Asp. Med.* **2002**, *23*, 209–285. [CrossRef]

159. Chiarugi, P.; Cirri, P. Redox regulation of protein tyrosine phosphatases during receptor tyrosine kinase signal transduction. *Trends Biochem. Sci.* **2003**, *28*, 509–514. [CrossRef]

160. Torres, M.; Forman, H.J. Redox signaling and the MAP kinase pathways. *Biofactors* **2003**, *17*, 287–296. [CrossRef]

161. Matsuzawa, A.; Ichijo, H. Stress-responsive protein kinases in redox-regulated apoptosis signaling. *Antioxid. Redox Signal.* **2005**, *7*, 472–481. [CrossRef]

162. Cuschieri, J.; Maier, R.V. Mitogen-activated protein kinase (MAPK). *Crit. Care Med.* **2005**, *33*, S417–S419. [CrossRef]

163. Kyriakis, J.M.; Avruch, J. Mammalian mitogen-activated protein kinase signal transduction pathways activated by stress and inflammation. *Physiol. Rev.* **2001**, *81*, 807–869. [CrossRef]

164. Chen, Z.; Gibson, T.B.; Robinson, F.; Silvestro, L.; Pearson, G.; Xu, B.; Wright, A.; Vanderbilt, C.; Cobb, M.H. MAP kinases. *Chem. Rev.* **2001**, *101*, 2449–2476. [CrossRef]

165. Matsukawa, J.; Matsuzawa, A.; Takeda, K.; Ichijo, H. The ASK1-MAP kinase cascades in mammalian stress response. *J. Biochem.* **2004**, *136*, 261–265. [CrossRef]

166. Jiang, F.; Zhang, Y.; Dusting, G.J. NADPH oxidase-mediated redox signaling: Roles in cellular stress response, stress tolerance, and tissue repair. *Pharmacol. Rev.* **2011**, *63*, 218–242. [CrossRef]

167. Li, Y.P.; Chen, Y.; John, J.; Moylan, J.; Jin, B.; Mann, D.L.; Reid, M.B. TNF-α acts via p38 MAPK to stimulate expression of the ubiquitin ligase atrogin1/MAFbx in skeletal muscle. *FASEB J.* **2005**, *19*, 362–370. [CrossRef]

168. Kefaloyianni, E.; Gaitanaki, C.; Beis, I. ERK1/2 and p38-MAPK signalling pathways, through MSK1, are involved in NF-kappaB transactivation during oxidative stress in skeletal myoblasts. *Cell Signal.* **2006**, *18*, 2238–2251. [CrossRef]

169. Goodyear, L.J.; Chang, P.Y.; Sherwood, D.J.; Dufresne, S.D.; Moller, D.E. Effects of exercise and insulin on mitogen-activated protein kinase signaling pathways in rat skeletal muscle. *Am. J. Physiol. Endocrinol. Metab.* **1996**, *271*, E403–E408. [CrossRef]

170. Kramer, H.F.; Goodyear, L.J. Exercise, MAPK, and NF-B signaling in skeletal muscle. *J. Appl. Physiol.* **2007**, *103*, 388–395. [CrossRef]

171. McCubrey, J.A.; LaHair, M.M.; Franklin, R.A. Reactive Oxygen Species-Induced Activation of the MAP Kinase Signaling Pathways. *Antioxid. Redox Signal.* **2006**, *8*, 1775–1789. [CrossRef]

172. Whitmarsh, A.J.; Davis, R.J. Transcription factor AP-1 regulation by mitogen-activated protein kinase signal transduction pathways. *J. Mol. Med.* **1996**, *74*, 589–607. [CrossRef] [PubMed]

173. Qi, M.; Elion, E.A. MAP kinase pathways. *J. Cell Sci.* **2005**, *118*, 3569–3572. [CrossRef] [PubMed]

174. Martineau, L.C.; Gardiner, P.F. Insight into skeletal muscle mechanotransduction: MAPK activation is quantitatively related to tension. *J. Appl. Physiol.* **2001**, *91*, 693–702. [CrossRef] [PubMed]

175. Widegren, U.; Wretman, C.; Lionikas, A.; Hedin, G.; Henriksson, J. Influence of exercise intensity on ERK/MAP kinase signalling in human skeletal muscle. *Pflug. Arch.* **2000**, *441*, 317–322. [CrossRef]

176. Yu, M.; Blomstrand, E.; Chibalin, A.V.; Krook, A.; Zierath, J.R. Marathon running increases ERK1/2 and p38 MAP kinase signalling to downstream targets in human skeletal muscle. *J. Physiol.* **2001**, *536*, 273–282. [CrossRef] [PubMed]

177. Karlsson, H.K.; Nilsson, P.A.; Nilsson, J.; Chibalin, A.V.; Zierath, J.R.; Blomstrand, E. Branched-chain amino acids increase p70S6k phosphorylation in human skeletal muscle after resistance exercise. *Am. J. Physiol.* **2004**, *287*, E1–E7.

178. Creer, A.; Gallagher, P.; Slivka, D.; Jemiolo, B.; Fink, W.; Trappe, S. Influence of muscle glycogen availability on ERK1/2 and Akt signaling after resistance exercise in human skeletal muscle. *J. Appl. Physiol.* **2005**, *99*, 950–956. [CrossRef] [PubMed]

179. Taylor, L.W.; Wilborn, C.D.; Kreider, R.B.; Willoughby, D.S. Effects of resistance exercise intensity on extracellular signal-regulated kinase 1/2 mitogen-activated protein kinase activation in men. *J. Strength Cond. Res.* **2012**, *26*, 599–607. [CrossRef] [PubMed]

180. Aronson, D.; Violan, M.A.; Dufresne, S.D.; Zangen, D.; Fielding, R.A.; Goodyear, L.J. Exercise stimulates the mitogen-activated protein kinase pathway in human skeletal muscle. *J. Clin. Investig.* **1997**, *99*, 1251–1257. [CrossRef]

181. Ryder, J.W.; Fahlman, R.; Walleberb-Henriksson, H.; Alessi, D.R.; Krook, A.; Zierath, J.R. Effect of contraction on mitogen-activated protein kinase signal transduction in skeletal muscle. Involvement of the mitogen- and stress-activated protein kinase 1. *J. Biol. Chem.* **2000**, *275*, 1457–1462. [CrossRef]

182. Wretman, C.; Lionikas, A.; Widegren, U.; Lännergren, J.; Westerblad, H.; Henriksson, J. Effects of concentric and eccentric contractions on phosphorylation of MAPK(erk1/2) and MAPK(p38) in isolated rat skeletal muscle. *J Physiol.* **2001**, *535*, 155–164. [CrossRef]

183. Leng, Y.; Steiler, T.L.; Zierath, J.R. Effects of insulin, contraction, and phorbol esters on mitogen-activated protein kinase signaling in skeletal muscle from lean and ob/ob mice. *Diabetes* **2004**, *53*, 1436–1444. [CrossRef]

184. Lee, J.S.; Bruce, C.R.; Spurrell, B.E.; Hawley, J.A. Effect of training on activation of extracellular signal-regulated kinase 1/2 and p38 mitogen-activated protein kinase pathways in rat soleus muscle. *Clin. Exp. Pharmacol. Physiol.* **2002**, *29*, 655–660. [CrossRef]

185. Nicoll, J.X.; Fry, A.C.; Galpin, A.J.; Thomason, D.B.; Moore, C.A. Resting MAPK expression in chronically trained endurance runners. *Eur. J. Sport Sci.* **2017**, *17*, 1194–1202. [CrossRef]

186. Galpin, A.J.; Fry, A.C.; Nicoll, J.X.; Moore, C.A.; Schilling, B.K.; Thomason, D.B. Resting extracellular signal-regulated protein kinase 1/2 expression following a continuum of chronic resistance exercise training paradigms. *Res. Sports Med.* **2016**, *24*, 298–303. [CrossRef]

187. Surget, S.; Khoury, M.P.; Bourdon, J.C. Uncovering the role of p53 splice variants in human malignancy: A clinical perspective. *Onco Targets Ther.* **2013**, *7*, 57–68. [CrossRef]

188. Morgan, M.J.; Liu, Z.G. Crosstalk of reactive oxygen species and NF-κB signaling. *Cell Res.* **2011**, *21*, 103–115. [CrossRef]

189. Bhoumik, A.; Lopez-Bergami, P.; Ronai, Z. ATF2 on the double - activating transcription factor and DNA damage response protein. *Pigment Cell Res.* **2007**, *20*, 498–506. [CrossRef]

190. Primeau, A.J.; Adhihetty, P.J.; Hood, D.A. Apoptosis in heart and skeletal muscle. *Can. J. Appl. Physiol.* **2002**, *27*, 349–395. [CrossRef]

191. Boppart, M.D.; Hirshman, M.F.; Sakamoto, K.; Fielding, R.A.; Goodyear, L.J. Static stretch increases c-Jun NH2-terminal kinase activity and p38 phosphorylation in rat skeletal muscle. *Am. J. Physiol. Cell Physiol.* **2001**, *280*, C352–C358. [CrossRef]

192. Shen, H.M.; Liu, Z.G. JNK signaling pathway is a key modulator in cell death mediated by reactive oxygen and nitrogen species. *Free Radic. Biol. Med.* **2006**, *40*, 928–939. [CrossRef] [PubMed]

193. Aronson, D.; Boppart, M.D.; Dufresne, S.D.; Fielding, R.A.; Goodyear, L.J. Exercise stimulates c-Jun NH2 kinase activity and c-Jun transcriptional activity in human skeletal muscle. *Biochem. Biophys. Res. Commun.* **1998**, *251*, 106–110. [CrossRef] [PubMed]

194. Russ, D.W.; Lovering, R.M. Influence of activation frequency on cellular signalling pathways during fatiguing contractions in rat skeletal muscle. *Exp. Physiol.* **2006**, *91*, 957–966. [CrossRef] [PubMed]

195. Yoon, S.; Seger, R. The extracellular signal-regulated kinase: Multiple substrates regulate diverse cellular functions. *Growth Factors* **2006**, *24*, 21–44. [CrossRef] [PubMed]

196. Turjanski, A.G.; Vaqué, J.P.; Gutkind, J.S. MAP kinases and the control of nuclear events. *Oncogene* **2007**, *26*, 3240–3253. [CrossRef] [PubMed]

197. Yao, Z.; Seger, R. The ERK signaling cascade–views from different subcellular compartments. *Biofactors* **2009**, *35*, 407–416. [CrossRef] [PubMed]

198. Whitmarsh, A.J. Regulation of gene transcription by mitogen-activated protein kinase signaling pathways. *Biochim. Biophys. Acta.* **2007**, *1773*, 1285–1298. [CrossRef] [PubMed]

199. Zehorai, E.; Yao, Z.; Plotnikov, A.; Seger, R. The subcellular localization of MEK and ERK–a novel nuclear translocation signal (NTS) paves a way to the nucleus. *Mol. Cell. Endocrinol.* **2010**, *314*, 213–220. [CrossRef]

200. Fischle, W.; Wang, Y.; Allis, C.D. Histone and chromatin cross-talk. *Curr. Opin. Cell Biol.* **2003**, *15*, 172–183. [CrossRef]

201. Brami-Cherrier, K.; Roze, E.; Girault, J.A.; Betuing, S.; Caboche, J. Role of the ERK/MSK1 signalling pathway in chromatin remodelling and brain responses to drugs of abuse. *J. Neurochem.* **2009**, *108*, 1323–1335. [CrossRef]

202. Dalton, T.P.; Shertzer, H.G.; Puga, A. Regulation of gene expression by reactive oxygen. *Annu. Rev. Pharmacol. Toxicol* **1999**, *39*, 67–101. [CrossRef]

203. Müller, J.M.; Rupec, R.A.; Baeuerle, P.A. Study of gene regulation by NF-kappa B and AP-1 in response to reactive oxygen intermediates. *Methods* **1997**, *11*, 301–312. [CrossRef]

204. Zhou, L.Z.; Johnson, A.P.; Rando, T.A. NF kappa B and AP-1 mediate transcriptional responses to oxidative stress in skeletal muscle cells. *Free Radic. Biol. Med.* **2001**, *31*, 1405–1416. [CrossRef]

205. Catani, M.V.; Savini, I.; Duranti, G.; Caporossi, D.; Ceci, R.; Sabatini, S.; Avigliano, L. Nuclear factor kappaB and activating protein 1 are involved in differentiation-related resistance to oxidative stress in skeletal muscle cells. *Free Radic. Biol. Med.* **2004**, *37*, 1024–1036. [CrossRef]

206. Gomez-Cabrera, M.C.; Domenech, E.; Viña, J. Moderate exercise is an antioxidant: Upregulation of antioxidant genes by training. *Free Radic. Biol. Med.* **2008**, *44*, 126–131. [CrossRef]

207. Baeuerle, P.A.; Henkel, T. Function and activation of NF-κB in the immune system. *Annu. Rev. Immunol.* **1994**, *12*, 141–179. [CrossRef]

208. Sen, C.K.; Khanna, S.; Reznick, A.Z.; Roy, S.; Packer, L. Glutathione regulation of tumor necrosis factor-alpha-induced NF-kappa B activation in skeletal muscle-derived L6 cells. *Biochem. Biophys. Res. Commun.* **1997**, *237*, 645–649. [CrossRef]

209. Ji, L.L.; Gomez-Cabrera, M.-C.; Steinhafel, N.; Vina, J. Acute exercise activates nuclear factor (NF) κB signaling pathway in rat skeletal muscle. *FASEB J.* **2004**, *18*, 1499–1506. [CrossRef]

210. Ho, R.C.; Hirshman, M.F.; Li, Y.; Cai, D.; Farmer, J.R.; Aschenbach, W.G.; Witczak, C.A.; Shoelson, S.E.; Goodyear, L.J. Regulation of IkappaB kinase and NF-kappaB in contracting adult rat skeletal muscle. *Am. J. Physiol. Cell Physiol.* **2005**, *289*, C794–C801. [CrossRef]

211. Gomez del Arco, P.; Martinez-Martinez, S.; Calvo, V.; Armesilla, A.L.; Redondo, J.M. Antioxidants and AP-1 activation: A brief overview. *Immunobiology* **1997**, *198*, 273–278. [CrossRef]

212. Bergelson, S.; Pinkus, R.; Daniel, V. Induction of AP-1 (Fos/Jun) by chemical agents mediates activation of glutathione S-transferase and quinone reductase gene expression. *Oncogene* **1994**, *9*, 565–571. [PubMed]

213. Sekhar, K.R.; Meredith, M.J.; Kerr, L.D.; Soltaninassab, S.R.; Spitz, D.R.; Xu, Z.Q.; Freeman, M.L. Expression of glutathione and gammaglutamylcysteine synthetase mRNA is Jun dependent. Biochem. *Biophys. Res. Commun.* **1997**, *234*, 588–593. [CrossRef] [PubMed]

214. Karin, M. The regulation of AP-1 activity by mitogen-activated protein kinases. *Bio. Chem.* **1995**, *270*, 16483–16486. [CrossRef] [PubMed]

215. Angel, P.; Karin, M. The role of Jun, Fos and the AP-1 complex in cell proliferation and transformation. *Biochem. Biophys. Acta* **1991**, *1072*, 129–157. [CrossRef]

216. Hollander, M.C.; Fornace, A.J.J. Induction of fos RNA by DNA-damaging agents. *Cancer Res.* **1989**, *49*, 1687–1692. [PubMed]

217. Pulverer, B.J.; Kyriakis, J.M.; Avruch, J.; Nikolakaki, E.; Woodgett, J.R. Phosphorylation of c-jun mediated by MAP kinases. *Nature* **1991**, *353*, 670–674. [CrossRef] [PubMed]

218. Hibi, M.; Lin, A.; Smeal, T.; Minden, A.; Karin, M. Identification of an oncoprotein- and UV-responsive protein kinase that binds and potentiates the c-Jun activation domain. *Genes Dev.* **1993**, *11*, 2135–2148. [CrossRef]

219. Shaulian, E. AP-1–The Jun proteins: Oncogenes or tumor suppressors in disguise? *Cell Signal.* **2010**, *6*, 894–899. [CrossRef]

220. Shaulian, E.; Karin, M. AP-1 as a regulator of cell life and death. *Nat. Cell Biol.* **2002**, *5*, E131–E136. [CrossRef]

221. Kansanen, E.; Jyrkkänen, H.K.; Levonen, A.L. Activation of stress signaling pathways by electrophilic oxidized and nitrated lipids. *Free Radic. Biol. Med.* **2012**, *52*, 973–982. [CrossRef]

222. Kansanen, E.; Kivela, A.M.; Levonen, A.L. Regulation of Nrf2-dependent gene expression by 15-deoxy-Delta12,14-prostaglandin J2. *Free Radic. Biol. Med.* **2009**, *47*, 1310–1317. [CrossRef] [PubMed]

223. Bryan, H.K.; Olayanju, A.; Goldring, C.E.; Park, B.K. The Nrf2 cell defence pathway: Keap1-dependent and-independent mechanisms of regulation. *Biochem. Pharmacol.* **2013**, *85*, 705–717. [CrossRef] [PubMed]

224. Nguyen, T.; Sherratt, P.J.; Pickett, C.B. Regulatory mechanisms controlling gene expression mediated by the antioxidant response element. *Annu. Rev. Pharmacol. Toxicol.* **2003**, *43*, 233–260. [CrossRef] [PubMed]

225. Itoh, K.; Chiba, T.; Takahashi, S.; Ishii, T.; Igarashi, K.; Katoh, Y. An Nrf2/small Maf heterodimer mediates the induction of phase II detoxifying enzyme genes through antioxidant response elements. *Biochem. Biophys. Res. Commun.* **1997**, *236*, 313–322. [CrossRef] [PubMed]

226. Hayes, J.D.; McLellan, L.I. Glutathione and glutathione-dependent enzymes represent a co-ordinately regulated defence against oxidative stress. *Free Radic. Res.* **1999**, *31*, 273–300. [CrossRef] [PubMed]

227. Nguyen, T.; Sherratt, P.J.; Nioi, P.; Yang, C.S.; Pickett, C.B. Nrf2 controls constitutive and inducible expression of ARE-driven genes through a dynamic pathway involving nucleocytoplasmic shuttling by Keap1. *J. Biol. Chem.* **2005**, *280*, 32485–32492. [CrossRef] [PubMed]

228. Tebay, L.E.; Robertson, H.; Durant, S.T.; Vitale, S.R.; Penning, T.M.; Dinkova-Kostova, A.T.; Hayes, J.D. Mechanisms of activation of the transcription factor Nrf2 by redox stressors, nutrient cues, and energy status and the pathways through which it attenuates degenerative disease. *Free Radic. Biol. Med.* **2015**, *88*, 108–146. [CrossRef] [PubMed]

229. Liu, Y.; Zhang, L.; Liang, J. Activation of the Nrf2 defense pathway contributes to neuroprotective effects of phloretin on oxidative stress injury after cerebral ischemia/reperfusion in rats. *J. Neurol. Sci.* **2015**, *351*, 88–92. [CrossRef]

230. Agyeman, A.S.; Chaerkady, R.; Shaw, P.G.; Davidson, N.E.; Visvanathan, K.; Pandey, A.; Kensler, T.W. Transcriptomic and proteomic profiling of KEAP1 disrupted and sulforaphane-treated human breast epithelial cells reveals common expression profiles. *Breast Cancer. Res. Treat.* **2012**, *132*, 175–187. [CrossRef]

231. Hawkes, H.J.K.; Karlenius, T.C.; Tonissen, K.F. Regulation of the human thioredoxin gene promoter and its key substrates: A study of functional and putative regulatory elements. *Biochim. Biophys. Acta* **2013**, *1840*, 303–314. [CrossRef]

232. Malhotra, D.; Portales-Casamar, E.; Singh, A.; Srivastava, S.; Arenillas, D.; Happel, C.; Shyr, C.; Wakabayashi, N.; Kensler, T.W.; Wasserman, W.W.; et al. Global mapping of binding sites for Nrf2 identifies novel targets in cell survival response through ChIP-Seq profiling and network analysis. *Nucleic Acids Res.* **2010**, *38*, 5718–5734. [CrossRef] [PubMed]

233. Abbas, K.; Breton, J.; Planson, A.G.; Bouton, C.; Bignon, J.; Seguin, C.; Riquier, S.; Toledano, M.B.; Drapier, J.C. Nitric oxide activates an Nrf2/sulfiredoxin antioxidant pathway in macrophages. *Free Radic. Biol. Med.* **2011**, *51*, 107–114. [CrossRef] [PubMed]

234. Lu, J.; Holmgren, A. The thioredoxin antioxidant system. *Free Radic. Biol. Med.* **2014**, *66*, 75–87. [CrossRef] [PubMed]

235. Wu, K.C. Beneficial role of Nrf2 in regulating NADPH generation and consumption. *Toxicol. Sci.* **2011**, *123*, 590–600. [CrossRef] [PubMed]

236. Yagishita, Y.; Fukutomi, T.; Sugawara, A.; Kawamura, H.; Takahashi, T.; Pi, J.; Uruno, A.; Yamamoto, M. Nrf2 protects pancreatic β-cells from oxidative and nitrosative stress in diabetic model mice. *Diabetes* **2014**, *63*, 605–618. [CrossRef]

237. Horie, M.; Warabi, E.; Komine, S.; Oh, S.; Shoda, J. Cytoprotective role of Nrf2 in electrical pulse stimulated C2C12 myotube. *PLoS ONE* **2015**, *10*, e0144835. [CrossRef]

238. Merry, T.L.; Ristow, M. Nuclear factor erythroid-derived 2-like 2 (NFE2L2, Nrf2) mediates exercise-induced mitochondrial biogenesis and antioxidant response in mice. *J. Physiol.* **2016**, *594*, 5195–5207. [CrossRef]

239. Wang, P.; Li, C.G.; Qi, Z.; Cui, D.; Ding, S. Acute exercise stress promotes Ref/Nrf signaling and increases mitochondrial antioxidant activity in skeletal muscle. *Exp. Physiol.* **2015**, *101*, 410–420. [CrossRef]

240. Narasimhan, M.; Hong, J.; Atieno, N.; Muthusamy, V.R.; Davidson, C.J.; Abu-Rmaileh, N.; Richardson, R.S.; Gomes, A.V.; Hoidal, J.R.; Rajasekaran, N.S. Nrf2 deficiency promotes apoptosis and impairs PAX7/MyoD expression in aging skeletal muscle cells. *Free Radic. Biol. Med.* **2014**, *71*, 402–414. [CrossRef]

241. Kumar, R.; Negi, P.S.; Singh, B.; Ilavazhagan, G.; Bhargava, K.; Sethy, N.K. Cordycepssinensis promotes exercise endurance capacity of rats by activating skeletal muscle metabolic regulators. *J. Ethnopharmacol.* **2011**, *136*, 260–266. [CrossRef]

242. Muthusamy, V.R.; Kannan, S.; Sadhaasivam, K.; Gounder, S.S.; Davidson, C.J.; Boeheme, C.; Hoidal, J.R.; Wang, L.; Rajasekaran, N.S. Acute exercise stress activatesNrf2/ARE signaling and promotes antioxidant mechanisms in the myocardium. *Free Radic. Biol. Med.* **2012**, *52*, 366–376. [CrossRef] [PubMed]

243. Gounder, S.S.; Kannan, S.; Devadoss, D.; Miller, C.J.; Whitehead, K.J.; Odelberg, S.J.; Firpo, M.A.; Paine, R., 3rd; Hoidal, J.R.; Abel, E.D.; et al. Impaired transcriptional activity of Nrf2 in age-related myocardial oxidative stress is reversible by moderate exercise training. *PLoS ONE* **2012**, *7*, e45697. [CrossRef]

244. Safdar, A.; deBeer, J.; Tarnopolsky, M.A. Dysfunctional Nrf2-Keap1 redox signaling in skeletal muscle of the sedentary old. *Free Radic. Biol. Med.* **2010**, *49*, 1487–1493. [CrossRef] [PubMed]

245. Puigserver, P.; Wu, Z.C.; Park, W.; Graves, R.; Wright, M.; Spiegelman, B.M. A cold-inducible coactivator of nuclear receptors linked to adaptive thermogenesis. *Cell* **1998**, *92*, 829–839. [CrossRef]

246. Finck, B.N.; Kelly, D.P. PGC-1 coactivators: Inducible regulators of energy metabolism in health and disease. *J. Clin. Investig.* **2006**, *116*, 615–622. [CrossRef] [PubMed]

247. Lin, J.; Wu, H.; Tarr, P.T. Transcriptional co-activator PGC-1α drives the formation of slow-twitch muscle fibres. *Nature* **2002**, *418*, 797–801. [CrossRef] [PubMed]

248. Handschin, C.; Rhee, J.; Lin, J.; Tarr, P.T.; Spiegelman, B.M. An autoregulatory loop controls peroxisome proliferator activated receptor γ coactivator 1α expression in muscle. *Proc. Natl. Acad. Sci. USA* **2003**, *100*, 7111–7116. [CrossRef] [PubMed]

249. Puigserver, P.; Spiegelman, B.M. Peroxisome proliferators-activated receptor gamma coactivator 1α (PGC-1α): Transcriptional coactivator and metabolic regulator. *Endocr. Rev.* **2003**, *24*, 78–90. [CrossRef]

250. Scarpulla, R.C. Transcriptional paradigms in mammalian mitochondrial biogenesis and function. *Physiol. Rev.* **2008**, *88*, 611–638. [CrossRef]

251. Leick, L.; Wojtaszewsk, I.J.F.; Johansen, S.T. PGC-1_ is not mandatory for exercise- and training-induced adaptive gene responses inmouse skeletal muscle. *Am. J. Physiol.* **2008**, *294*, E463–E474.

252. Geng, T.; Li, P.; Okutsu, M. PGC-1_ plays a functional role in exercise-induced mitochondrial biogenesis and angiogenesis but not fiber-type transformation inmouse skeletal muscle. *Am. J. Physiol. Cell Physiol.* **2010**, *298*, 572–579. [CrossRef] [PubMed]

253. Leick, L.; Lyngby, S.S.; Wojtasewski, J.F.; Pilegaard, H. PGC-1_ is required for training-induced prevention of age associated decline in mitochondrial enzymes in mouse skeletal muscle. *Exp. Gerontol.* **2010**, *45*, 336–342. [CrossRef] [PubMed]

254. Wenz, T.; Rossi, S.; Rotundo, R.L. Increased muscle PGC-1_ expression protects from sarcopenia and metabolic disease during aging. *Proc. Natl. Acad. Sci. USA* **2009**, *106*, 20405–20410. [CrossRef] [PubMed]

255. St-Pierre, J.; Drori, S.; Uldry, M. Suppression of reactive oxygen species and neurodegeneration by the PGC-1 transcriptional coactivators. *Cell* **2006**, *127*, 397. [CrossRef] [PubMed]

256. Kong, X.; Wang, R.; Xue, Y. Sirtuin 3, a new target of PGC-1_, plays an important role in the suppression of ROS and mitochondrial biogenesis. *PLoS ONE* **2010**, *5*, e11707. [CrossRef] [PubMed]

257. Shi, T.; Wang, F.; Stieren, E.; Tong, Q. SIRT3, a mitochondrial sirtuin deacetylase, regulates mitochondrial function and thermogenesis in brown adipocytes. *J. Biol. Chem.* **2005**, *280*, 13560–13567. [CrossRef] [PubMed]

258. Bellizzi, D. A novel VNTR enhancer within the SIRT3 gene, a human homologue of SIR2, is associated with survival at oldest ages. *Genomics* **2005**, *85*, 258–263. [CrossRef] [PubMed]

259. Wu, J.; Ruas, J.L.; Estall, J.L.; Rasbach, K.A.; Choi, J.H.; Je, L.; Bostrom, P.; Tyra, H.M.; Crawford, R.W.; Campbell, K.P.; et al. The unfolded protein response mediates adaptation to exercise in skeletal muscle trough a PGC1α/ATF6α complex. *Cell Metab.* **2011**, *13*, 160–169. [CrossRef] [PubMed]

260. Irrcher, I.; Adhihetty, P.J.; Sheehan, T.; Joseph, A.M.; Hood, D.A. PPARγ coactivator-1α expression during thyroid hormone-and contractile activity-induced mitochondrial adaptations. *Am. J. Physiol.* **2003**, *284*, C1669–C1677. [CrossRef]

261. Baar, K.; Wende, A.R.; Jones, T.E.; Marison, M.; Nolte, L.A.; Chen, M.; Kelly, D.P.; Holloszy, J.O. Adaptations of skeletal muscle to exercise: Rapid increase in the transcriptional coactivator PGC-1. *FASEB J.* **2002**, *16*, 1879–1886. [CrossRef]

262. Wang, L.; Mascher, H.; Psilander, N.; Blomstrand, E.; Sahlin, K. Resistance exercise enhances the molecular signaling of mitochondrial biogenesis induced by endurance exercise in human skeletal muscle. *J. Appl. Physiol.* **2011**, *111*, 1335–1344. [CrossRef]

263. Venditti, P.; Napolitano, G.; Barone, D.; Di Meo, S. Vitamin E supplementation modifies adaptive responses to training, in rat skeletal muscle. *Free Radic. Res.* **2014**, *48*, 1179–1189. [CrossRef] [PubMed]

264. Boppart, M.D.; Asp, S.; Wojtaszewski, J.F.P.; Fielding, R.A.; Mohr, T.; Goodyear, L.J. Marathon running transiently increases c-Jun NH2-terminal kinase and p38 activities in human skeletal muscle. *J. Physiol.* **2000**, *526*, 663–669. [CrossRef] [PubMed]

265. Puigserver, P.; Rhee, J.; Lin, J.; Wu, Z.; Yoon, J.C.; Zhang, C.Y.; Krauss, S.; Mootha, V.K.; Lowell, B.B.; Spiegelman, B.M. Cytokine stimulation of energy expenditure through p38 MAP kinase activation of PPARgamma coactivator-1. *Mol. Cell.* **2001**, *8*, 971–982. [CrossRef]

266. Knutti, D.; Kressler, D.; Kralli, A. Regulation of the transcriptional coactivator PGC-1 via MAPK sensitive interaction with a repressor. *Proc. Natl. Acad. Sci. USA* **2001**, *98*, 9713–9718. [CrossRef]

267. Fernandez-Marcos, P.J.; Auwerx, J. Regulation of PGC-1a, a nodal regulator of mitochondrial biogenesis. *Am. J. Clin. Nutr.* **2011**, *93*, 884S–890S. [CrossRef] [PubMed]

268. Silveira, L.R.; Pilegaard, H.; Kusuhara, K.; Curi, R.; Hellsten, Y. The effect of reactive oxygen species and antioxidants on basal and contraction-induced gene expression of PGC-1α, UCP3 and HKII in primary rat skeletal muscle cells. *BBA* **2006**, *1763*, 969–976. [CrossRef]

269. Irrcher, I.; Ljubicic, V.; Hood, D.A. Interactions between ROS and AMP kinase activity in the regulation of PGC-1α transcription in skeletal muscle cells. *Am. J. Physiol. Cell Physiol.* **2009**, *296*, C116–C123. [CrossRef]

270. Higashida, K.; Kim, S.H.; Higuchi, M.; Holloszy, J.O.; Han, D.H. Normal adaptations to exercise despite protection against oxidative stress. *Am. J. Physiol. Cell Physiol.* **2011**, *301*, E779–E784. [CrossRef]

271. Ristow, M.; Zarsea, K.; Oberbach, A.; Klöting, N.; Birringer, M.; Kiehntopf, M.; Stumvoll, M.; Kahn, C.R.; Blüher, M. Antioxidants prevent health-promoting effects of physical exercise in humans. *Proc. Natl. Acad. Sci. USA* **2009**, *106*, 8665–8670. [CrossRef]

272. Paulsen, G.; Cumming, K.T.; Holden, G.; Hallén, J.; Rønnestad, B.R.; Sveen, O.; Skaug, A.; Paur, I.; Bastani, N.E.; Østgaard, H.N.; et al. Vitamin C and E supplementation hampers cellular adaptation to endurance training in humans: A double-blind, randomised, controlled trial. *J. Physiol.* **2014**, *592*, 1887–1901. [CrossRef] [PubMed]

273. Gomez-Cabrera, M.C.; Domenech, E.; Romagnoli, M.; Arduini, A.; Borras, C.; Pallardo, F.V.; Sastre, J.; Viña, J. Oral administration of vitamin C decreases muscle mitochondrial biogenesis and hampers training-induced adaptations in endurance performance. *Am. J. Clin. Nutr.* **2008**, *87*, 142–149. [CrossRef] [PubMed]

274. Irrcher, I.; Ljubicic, V.; Kirwan, A.F.; Hood, D.A. AMP-Activated Protein Kinase-Regulated Activation of the PGC-1α Promoter in Skeletal Muscle Cells. *PLoS ONE* **2008**, *3*, e3614. [CrossRef] [PubMed]

275. Nisoli, E.; Falcone, S.; Tonello, C.; Cozzi, V.; Palomba, L.; Fiorani, M.; Pisconti, A.; Brunelli, S.; Cardile, A.; Francolini, M.; et al. Mitochondrial biogenesis by NO yields functionally active mitochondria in mammals. *Proc. Natl. Acad. Sci. USA* **2004**, *101*, 16507–16512. [CrossRef] [PubMed]

276. Wadley, G.D.; Choate, J.; McConell, G.K. NOS isoform specific regulation of basal but not exercise-induced mitochondrial biogenesis in mouse skeletal muscle. *J. Physiol.* **2007**, *585*, 253–262. [CrossRef] [PubMed]

277. Carlstrom, M.; Larsen, F.J.; Nystrom, T.; Hezel, M.; Borniquel, S.; Weitzberg, E.; Lundberg, J.O. Dietary inorganic nitrate reverses features of metabolic syndrome in endothelial nitric oxide synthase-deficient mice. *Proc. Natl. Acad. Sci. USA* **2010**, *107*, 17716–17720. [CrossRef]

278. McConell, G.K.; Ng, G.P.; Phillips, M.; Ruan, Z.; Macaulay, S.L.; Wadley, G.D. Central role of nitric oxide synthase in AICAR and caffeine-induced mitochondrial biogenesis in L6 myocytes. *J. Appl. Physiol.* **2010**, *108*, 589–595. [CrossRef]

279. Chen, Z.P.; McConell, G.K.; Michell, B.J.; Snow, R.J.; Canny, B.J.; Kemp, B.E. AMPK signaling in contracting human skeletal muscle: Acetyl-CoA carboxylase and NO synthase phosphorylation. *Am. J. Physiol. Endocrinol. Metab.* **2000**, *279*, E1202–E1206. [CrossRef]

280. Lira, V.A.; Brown, D.L.; Lira, A.K.; Kavazis, A.N.; Soltow, Q.A.; Zeanah, E.H.; Criswell, D.S. Nitric oxide and AMPK cooperatively regulate PGC-1 in skeletal muscle cells. *J. Physiol.* **2010**, *588*, 3551–3566. [CrossRef]

281. Tonks, N.K. Redox redux: Revisiting PTPs and the control of cell signaling. *Cell* **2005**, *121*, 667–670. [CrossRef]

282. Camps, M.; Nichols, A.; Arkinstall, S. Dual specificity phosphatases: A gene family for control of MAP kinase function. *FASEB J.* **2000**, *14*, 6–16. [CrossRef] [PubMed]

283. Krook, A.; Widegren, U.; Jiang, X.J.; Henriksson, J.; Wallberg-Henriksson, H.; Alessi, D.; Zierath, J.R. Effects of exercise on mitogen- and stress-activated kinase signal transduction in human skeletal muscle. *Am. J. Physiol.* **2000**, *279*, R1716–R1721. [CrossRef] [PubMed]

284. Pourteymour, S.; Hjorth, M.; Lee, S.; Holen, T.; Langleite, T.M.; Jensen, J.; Birkeland, K.I.; Drevon, C.A.; Eckardt, K. Dual specificity phosphatase 5 and 6 are oppositely regulated in human skeletal muscle by acute exercise. *Physiol. Rep.* **2017**, *5*, e13459. [CrossRef] [PubMed]

Review

Interplay Between Mitochondrial Peroxiredoxins and ROS in Cancer Development and Progression

Tayaba Ismail [†], Youni Kim [†], Hongchan Lee, Dong-Seok Lee and Hyun-Shik Lee *

KNU-Center for Nonlinear Dynamics, CMRI, School of Life Sciences, BK21 Plus KNU Creative BioResearch Group, College of Natural Sciences, Kyungpook National University, Daegu 41566, Korea
* Correspondence: leeh@knu.ac.kr; Tel.: +82-53-950-7367; Fax: +82-53-943-2762
† These authors contributed equally to this work.

Received: 1 July 2019; Accepted: 5 September 2019; Published: 7 September 2019

Abstract: Mitochondria are multifunctional cellular organelles that are major producers of reactive oxygen species (ROS) in eukaryotes; to maintain the redox balance, they are supplemented with different ROS scavengers, including mitochondrial peroxiredoxins (Prdxs). Mitochondrial Prdxs have physiological and pathological significance and are associated with the initiation and progression of various cancer types. In this review, we have focused on signaling involving ROS and mitochondrial Prdxs that is associated with cancer development and progression. An upregulated expression of Prdx3 and Prdx5 has been reported in different cancer types, such as breast, ovarian, endometrial, and lung cancers, as well as in Hodgkin's lymphoma and hepatocellular carcinoma. The expression of Prdx3 and Prdx5 in different types of malignancies involves their association with different factors, such as transcription factors, micro RNAs, tumor suppressors, response elements, and oncogenic genes. The microenvironment of mitochondrial Prdxs plays an important role in cancer development, as cancerous cells are equipped with a high level of antioxidants to overcome excessive ROS production. However, an increased production of Prdx3 and Prdx5 is associated with the development of chemoresistance in certain types of cancers and it leads to further complications in cancer treatment. Understanding the interplay between mitochondrial Prdxs and ROS in carcinogenesis can be useful in the development of anticancer drugs with better proficiency and decreased resistance. However, more targeted studies are required for exploring the tumor microenvironment in association with mitochondrial Prdxs to improve the existing cancer therapies and drug development.

Keywords: mitochondria; peroxiredoxins; tumorigenesis; reactive oxygen species; ROS scavengers

1. Introduction

Mitochondria, with their multifaceted roles, are recognized as indispensable organelles in eukaryotic cells and are considered as the energy currency of the cell because of adenosine triphosphate (ATP) production [1]. In addition to being energy sources, they are involved in heme synthesis, metabolism of amino acids, and the regulation of the redox state of cells [2,3]. These multiple functions of mitochondria make them prerequisites for cellular life of eukaryotes [4]. Apart from these functions that are related to cell survival, a large body of evidence has exhibited that this subcellular organelle also has crucial functions in the cell death program [5,6]. Mitochondria play significant roles in apoptosis by regulating the release of pro-apoptotic factors [7]. In addition, they are also critical participants in necrosis and autophagy [8,9].

Mitochondria continuously communicate with other cells by signal transduction to carry out this diverse array of functions [10,11]. The generation of reactive oxygen species (ROS) is one of the way of mitochondrial signal transduction and these generated (ROS) function as secondary messengers [12,13] and play a significant role in cellular signal transduction [14]. ROS are basically short-lived species having unpaired electrons [15,16], and they are endogenously generated as a byproduct during mitochondrial energy production as well as produced as a consequence of fatty acid β-oxidation and exposure to radiation, light, metals, and redox drugs [17]. Mitochondria being the largest contributors of ROS in mammalian cells convert approximately 1% of their consumed oxygen to superoxide anion (O^{2-}) [18], and they have up to ten sites with ROS generation ability [19]. ROS serve as secondary messengers by interacting with a variety of molecules; they are important for many biologically significant processes, such as adaptive immunity, cell differentiation, and oxygen cell sensing [13,15,16,20,21]. Mitochondrial ROS, in addition to their cellular signaling properties, also have cell damaging roles [22].

Therefore, ROS homeostasis is essential for steady state functions of cells [23,24]. The accumulation of ROS resulting from any imbalance in ROS production and metabolism induces oxidative stress [25]. In other words, oxidative stress is the result of imbalance between two opposite and opposing forces, i.e., ROS production and antioxidation, and it can lead to pathological defects in living organisms, such as cancer, atherosclerosis, neurological diseases, aging, and diabetes, as well as can harm the cellular components, such as DNA, RNA, lipids, and proteins [26,27]. To protect cells from the oxidative stress, there are many enzymatic and non-enzymatic defense systems in mitochondria [28,29]. The non-enzymatic defense system includes flavonoids, vitamins (A, C, and E), and glutathione [28]. Superoxide dismutase (SOD), superoxide reductase, catalase, glutathione peroxidase, glutathione reductase, peroxiredoxins (Prdxs), and thioredoxins (Trx) are important enzymatic antioxidants that are involved in the regulation of mitochondrial ROS [30,31]. ROS production in mitochondria and their subsequent fate involve interactions between various molecules in normal and pathological conditions [32] (Figure 1).

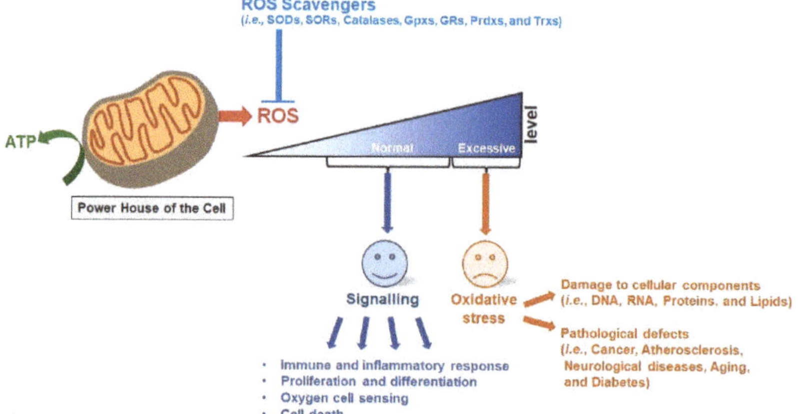

Figure 1. Mitochondrial reactive oxygen species (ROS) scavengers and cancer development. Eukaryotic cells depend upon the mitochondria for energy production, and reactive oxygen species (ROS) are produced as byproducts during adenosine triphosphate (ATP) generation by the mitochondria. The amount of mitochondrial ROS is balanced by ROS scavengers in the mitochondria of normal eukaryotic cells and these ROS regulate different cellular processes such as cell proliferation and differentiation by acting as signaling molecules. The loss of this specific balance between ROS and ROS scavengers leads to an outburst growth of cells, resulting in cancer initiation and development.

The aim of the present review is to focus on the signaling for ROS regulation involving mitochondrial Prdxs in normal as well as disease conditions. Mitochondrial Prdxs are recently discovered antioxidants presenting a variety of roles in eukaryotic organisms. Recently, mitochondrial Prdxs have gathered much interest due to their diversified functions and interactions in living organisms.

In the present review, we describe the factors that are associated with mitochondrial Prdxs in ROS scavenging, as well as its signaling pattern in normal and pathological cellular conditions.

2. Mitochondrial Prdxs

Prdxs denote the family of proteins having the ability to efficiently scavenge peroxides in the form of hydrogen peroxide, alkyl hydro peroxide, and peroxynitrite [33]. Prdxs are classified into six subfamilies, according to PeroxiRedoxin classification index (PREX) and, in mammals, there are six Prdxs, among which Prdx 1–4 belongs to Prdx1 subfamily, Prdx5 is a member of Prdx5 subfamily, and Prdx6 represent Prdx6 subfamily [34,35]. Prdxs are also classified as 1-Cys containing only Prdx6, while typical 2-Cys comprised of Prdx1–4 and Prdx5 belongs to atypical 2-Cys on the basis of their cysteine residues and their catalytic mechanism [36]. The catalytic activity of Prdxs is highly dependent upon the conserved peroxidatic cysteine (Cp) in the amino terminal region of the protein. In addition to this conserved Cp, there is an additional conserved cysteine residue that is present in the carboxyl terminal portion of the five out of six mammalian Prdxs, termed as resolving cysteine (Cr). [37]. Prdxs belonging to typical 2-Cys class are typical homodimers containing two identical active sites that brings peroxidatic and resolving cysteines (Cp and Cr) into juxtaposition. Peroxidatic cysteine of typical 2-Cys is oxidized by peroxide, resulting in the formation of sulfenic acid (C-SOH), which in turn condenses with Cr of opposite subunit, and results in the formation of intermolecular disulfide bond. The oxidized Prdxs are reduced by appropriate electron donor to complete the catalytic cycle. The oxidized Cp of atypical 2-Cys condenses with Cr within same polypeptide to form intramolecular disulfide bond and in turn reduced by the Trx2. In contrast to typical and atypical 2-Cys Prdxs, 1-Cys Prdxs has only one cysteine residue and depends on glutathione (GSH) to complete the catalytic peroxidatic cycle [38]. Members of typical 2-Cys class have the ability to form decamers and do-decamers in their reduced or hyperoxidized state. This feature enables them to function as chaperons, enzyme activators, and redox sensors, in addition to their antioxidative abilities [39]. While, atypical 2-Cys Prdxs can undergo protein-protein interaction in combination with their antioxidative properties [40]. In contrast, 1-Cys Prdxs cannot form decamers, and mainly functions as antioxidant rather than molecular chaperons, although their antioxidation catalytic mechanism is similar to typical 2-Cys [41].

Mammalian Prdxs also differ in their cellular location and are distributed in cytosol, mitochondria, endoplasmic reticulum, peroxisomes, and nuclei [42]. Based on subcellular distribution, Prdx3 and 5 are categorized as mitochondrial Prdxs [43]. Studies on mitochondrial Prdxs were initiated in 1989 with the cloning of the *mer5* gene [44]. This gene is expressed in murine erythroleukemia [45]. Prdx3 was thought to play a significant role in erythrocyte differentiation [46]. This gene was further investigated to have considerable sequence identity with bacterial alkyl hydroperoxide reductase (AhpC) [47]. Meanwhile, substrates of mitochondrial ATP-dependent protease, later identified as Lon protease, were investigated, leading to the discovery of a 22-kDa substrate named as SP-22 [48]. This SP-22 was found to be a highly abundant protein in the mitochondrial matrix, sharing a sequence homology of 90% with Mer5 and having considerable identity with AhpC. It is now recognized as "Prdx3". Prdx3 was also termed as an antioxidant protein 1 (AoP1) [48]. Simultaneously, when the *mer5* gene was cloned, a novel gene, *PMP20*, was identified in *Candida boidinii*, encoding a peroxisomal membrane-associated protein [49]. The homologue of *PMP20* was discovered 10 years later in humans with thiol-specific antioxidant properties [50]. In the same year, a novel gene, named "AEB166", having a sequence identity of 65% with bacterial Prdx, was identified [51]. This gene was later termed as "Prdx5" [52].

3. Characteristics of mitochondrial Prdxs

Prdx3 is located on chromosome 10 (q25–q26) and it is transcribed from seven exons in humans. It is a 256-amino acid containing protein with a 61-amino acid long mitochondrial targeting sequence at the N-terminal. The cleavage of this mitochondrial sequence results in a 21.5-kDa protein that resides in the mitochondrial matrix [37]. Peroxidatic cysteine (Cp) is located at the residue 47 and it is a highly conserved region of the protein. Prdx3 belongs to "typical 2-Cys" class of Prdxs, i.e., it is a homodimer organized in a head to tail manner and during catalysis, utilizes Cp and resolving cysteine (Cr) residues on opposite subunits [43] (Figure 2).

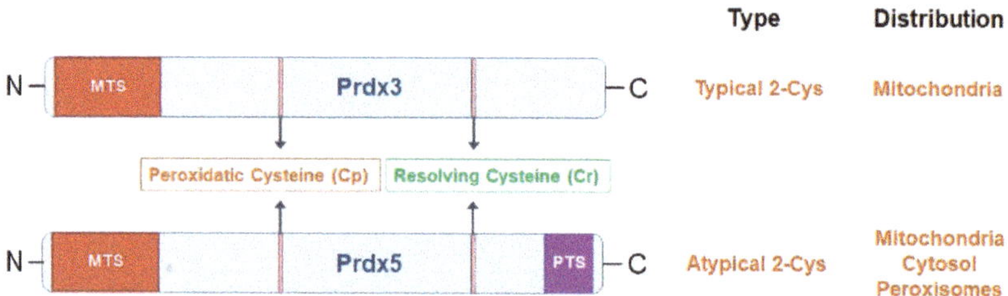

Figure 2. Structural comparison of mitochondrial peroxiredoxins (Prdxs). Peroxiredoxin 3 (Prdx3) and Peroxiredoxin 5 (Prdx5) contain a mitochondrial targeting sequence (MTS) at their amino terminal (N-terminal) and this enables them to reside in the mitochondria. Prdx5 in addition to MTS also contains a peroxisomal targeting sequence (PTS) at its carboxyl terminal (C-terminal) that is responsible for the entry of Prdx5 in the peroxisomes. Both mitochondrial Prdxs comprise peroxidatic cysteine (Cp) and resolving cysteine (Cr) which play an essential role in their catalytic mechanisms. In Prdx3, Cp is located at residue 47, whereas, in Prdx5, it is located at residue 48 [35].

In contrast to Prdx3, human Prdx5 is located on chromosome 11 (q13) and it is transcribed from six exons. It contains a 52-amino acid-long mitochondrial targeting leader sequence at its N-terminal and it also has a C-terminal SQL sequence that is specific for peroxisomes [52] (Figure 2). This peroxisomal targeting sequence enables this protein to localize in both peroxisomes and mitochondria [53]. The cleavage of mitochondrial targeting sequence yields a 17-kDa protein having Cp at residue 48 of the mature protein. The active site of Prdx5 has several sequences in common with other Prdx family members [54]. The human Prdx5 has a divergent sequence sharing a sequence homology of 28–30% with Prdx1, Prdx2, and Prdx3. Prdx5 is the only mammalian Prdx that belongs to "atypical 2-Cys" subfamily of Prdxs [55](Figure 2).

4. ROS and Mitochondrial Prdxs

ROS are extraordinarily active oxygen species involving superoxide anion (O_2^-), hydroxyl radical (OH·), and hydrogen peroxide (H_2O_2) [56]. They are generated by the partial reduction of oxygen, and comprise radical and non-radical oxygen species [56]. Endogenous sources of ROS include mitochondrial electron transport linked phosphorylation, P450 metabolism, peroxisome metabolism, and activation of inflammatory cells [13]. Mitochondrial ROS are produced in mitochondrial electron transport chain during ATP generation [13,15].

Previously, ROS were considered to be damaging to cells because of their association with oxidative stress, which leads to the induction of pathological responses [14]. However, in the past few decades, ROS have been recognized as signaling molecules that regulate different processes of biological and physiological importance [57]. H_2O_2 is the ROS that directly serves as a secondary messenger and is majorly produced by the mitochondria [58,59]. Redox signal transduction involves the oxidation of Cys residues in the proteins that is mediated by H_2O_2 [58]. At physiological pH, Cys residues exist in the form of thiolate anions (Cys-S$^-$) and are more receptive to oxidation as compared to protonated Cys thiols (Cys-SH) [60]. During the process of redox signaling, Cys-S$^-$ is oxidized to its sulfenic form (Cys-SOH) by H_2O_2, which leads to allosteric alterations in the protein, and ultimately changes its function [61,62]. Thioredoxin and glutaredoxins can reduce back this sulfenic form of Cys to thiolate anion and bring back the protein to its original form [63]. Thus, the oxidation of Cys residues in proteins involves a reversible signal transduction mechanism that mostly occurs at nanomolar concentrations of H_2O_2, whereas at higher H_2O_2 concentrations, thiolate anions are further oxidized to sulfinic (SO_2H) and sulfonic (SO_3H) species [62,63]. Sulfiredoxin (Srx) can slowly reduce back the SO_2H form of Prdxs to its native enzyme [64]. In contrast to Cys-SOH and SO_2H species, the oxidation to sulfonic species can be irreversible, which results in damage to the protein (oxidative stress) [61,65]. Mitochondria and cells are equipped with professional enzymes to prevent the accumulation of H_2O_2, particularly Prdxs and glutaredoxins [66].

Mitochondrial Prdxs play a significant role in H_2O_2 scavanging and to maintain a tight balance of H_2O_2 [67]. Mitochondrial Prdxs that belong to the typical and atypical 2-Cys subfamilies contain a redox sensitive cysteine at the active site that is oxidized by H_2O_2 and then reduced by thioredoxins to complete the catalytic cycle [68]. The oxidized thioredoxins are reduced by thioredoxin reductase at expense of NADPH [69]. The discriminative characteristic of Prdxs is their ability to undergo reversible hyperoxidation by a second molecule of H_2O_2, which results in the formation of sulfinic acid that consequently leads to the transient inactivation of the protein [70]. The reduction of the hyperoxidized form of Prdxs to SO^- is catalyzed by sulfredoxins through an ATP-dependent mechanism [70].

5. ROS and Mitochondrial Antioxidants in Oncogenic Signaling

Many vital biological processes, such as cellular signaling, metabolism, and epigenetics, are significantly regulated by ROS and, consequently, these biologically active oxygen species are involved in disease initiation and progression [15,30]. Mitochondria being one of the major source of ROS are associated with the roles of ROS in the living organisms and are significantly involved in the regulation of ROS to maintain a steady state cellular environment and prevent cells from oxidative damage [71]. ROS, particularly H_2O_2, is a byproduct of mitochondrial energy generation and is regulated by mitochondrial antioxidants [72]. Altered cellular metabolism is associated with the production of cancerous cells [73]. Cancerous cells are characterized by a high rate of proliferation and active metabolism [73]. These cancerous cells require higher energy levels for aberrant cellular proliferation and to maintain the high growth rate, these cells hijack the machinery of normal cells [74–76]. Growth-related pathways are constitutively activated in cancer cells and, consequently, these cells take up excess amount of nutrients, endure stress, and multiply rapidly [77,78]. This results in hyper metabolism, which leads to an excessive generation of ROS by the mitochondria, endoplasmic reticulum, and the action of NADPH oxidases [79]. Mitochondrial ROS are required for the proliferation of cancerous cells driven by K-ras oncogenes [80,81]. Mutations in the mitochondria that result in dysfunctions of TCA cycle/electron transport chain produce excess amount of ROS to trigger tumor associated signaling pathways, such as PI3K and MAP kinase signaling pathways [82–84]. In addition to these signaling pathways, ROS also target the transcription factor NF-kB (earliest discovered transcription factor responsive to ROS) that is associated with the survival of tumor cells [85,86] (Figure 3).

Studies have shown that the association of ROS with cancer progression and suppression is dependent on their intracellular levels [87]. At low levels, ROS augment cancer proliferation by working as signal transducers or by prompting the genomic DNA alteration or damage to DNA [88]. For example, ROS can stimulate cyclin D1 expression [89,90], lead to the promotion of extracellular-signal-related kinase (ERK) Jun N–terminal kinase (JNK) phosphorylation [91,92], and trigger mitogen-activated protein kinase (MAPK) activation [93], all of which are associated with carcinogenesis and the survival of cancerous cells [94]. In addition to the functions of ROS in the pathways associated with cell proliferation, ROS have been recognized to inversely debilitate tumor suppressors, such as protein tyrosine phosphatases (PTPs) and phosphatase and tensin homolog (PTEN), because of the presence of redox-sensitive cysteine in their catalytic sites [95]. PTPs also function to regulate signaling pathways by inducing antioxidant enzyme expression and decreasing ROS levels [96,97]. Moreover, ROS act as regulators of normal stem cell renewal and differentiation. Cancerous stem cells (CSCs) exhibit similar characteristics; however, there is limited knowledge regarding their association with maintenance of cellular redox balance [98]. Recent studies have exhibited low ROS levels in liver and breast cancer stem cells, consequently resulting in the upregulation of ROS-scavenging signaling proteins expression [99]. If growth of CSCs is indispensable for tumor induction, maintenance of low ROS levels might be a prerequisite for pre-neoplastic foci tolerance [98]. Chemo and radiotherapies stimulate ROS production and they are useful for demolishing most cancerous cells [98,99]. Certain patients are unable to cure by the chemo and radio therapies because of increased endurance of CSCs to conditions of high ROS levels by augmented antioxidant levels [100]. Some adverse effects of anticancer drugs are debatably mediated by ROS; thus, CSCs may be appropriately released and vigorously chosen by the factors, depending on increased ROS levels [101]. Moreover, further oxidative stress that is mediated by these factors may result in additional mutations and DNA damage, ultimately leading to enlargement of drug-resistant cancer cells [98,101]. Previously, it was assumed that increased ROS generation in cancerous cells leads to genomic instability, and consequently promotes tumor formation [102]. However, genomic instability is associated with a loss of p53 and other processes that are related to aneuploidy [103,104]. ROS-dependent oncogenesis driven by Myc oncogenes in cancer cells exhibits no chromosomal instability [105,106].

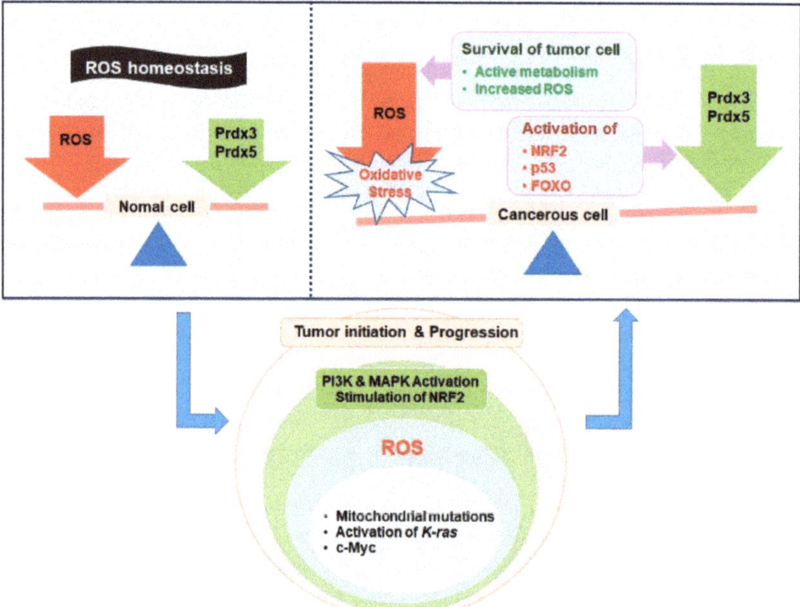

Figure 3. Role of antioxidants in cancerous cells. Mutations in mitochondria and mitochondrial dysfunctions lead to the production of increased amount of ROS by the activation of c-Myc and K-ras factors. The increased production of ROS leads to an imbalance between ROS and its scavengers. This imbalance activates a number of signaling pathways such as PI3K and mitogen-activated protein kinase (MAPK), resulting in tumor initiation and progression. The increased amount of ROS in cancerous cells requires an upregulated expression of antioxidants to prevent cell death. Thus, the activation of certain transcription factors (NRF2 and FOXO) and tumor suppressors (p53) induces the upregulation of mitochondrial Prdxs to compensate the increased production of ROS. Thus, mitochondrial Prdxs along with other scavengers act to protect the cancerous cells from ROS-mediated cell death.

ROS at high levels mediate cell death and cause adverse damage to the cells [107]. Therefore, cancerous cells must control increased ROS levels, specifically during the initial stages of cellular proliferation [108]. In cancerous cells with elevated ROS levels, redox balance is maintained by equally high amounts of antioxidants [109]. These antioxidants control the ROS levels and prevent the cells from cell death mediated by oxidative stress [110]. Recently, it has been discovered that conditions that facilitate oxidative stress also induce particular pressure on pre-neoplastic cells to trigger effective antioxidant mechanisms [111]. This characteristic is relevant, particularly during metastasis, which is characterized by the spread of cancer cells to distant organs [112]. Thus, cancer cells have extraordinary antioxidant capabilities to circumvent elevated ROS levels and they are accustomed to the diverse biological functions of cells [113]. The antioxidant capabilities of cancer cells are quite high when compared with those of normal cells [21,73,114]. The elevated antioxidant level in cancerous cells is achieved by the activation of transcription factors, such as nuclear factor erythroid 2 related factor 2 (NRF2) [115,116], that interacts with kelch-like ECH-associated proteins (KEAP1) and targets proteosomal degradation [117]. ROS sensitive Cys residues of KEAP1 are oxidized by elevated ROS levels, leading to KEAP1 dissociation from NRF2 [118]. The translocation of dissociated NRF2 to nucleus and its heterodimerization occurs with a small MAF protein [119]. Subsequently, it binds with antioxidant-responsive elements (AREs) of various antioxidative genes [119,120]. Although NRF2 protects the cells from cancer causing agents, it also stimulates tumor formation and cancer development by safeguarding cancer cells from ROS and the resulting DNA damage [108,115,121]. In some tumor cells, KEAP1 mutations lead to the constitutive NRF2 activation [122]. The loss of NRF2 induces oxidative stress in cancerous cells, and ultimately inhibits tumor formation [123]. The loss of NRF2 inhibits multiple antioxidant signaling pathways, resulting in damage to the cancer cells [124] (Figure 3).

Elevated ROS are associated with tumor formation and progression. The tumor suppressive genes can serve as antioxidants for scavenging ROS and to maintain them at the levels that do not promote tumor formation [125]. Tumor suppressor p53 regulates the expression of many antioxidant genes [126,127]. Moreover, mice that are deficient in p53 have demonstrated a reduction in tumor formation by dietary supplement NAC, which suggests that, in certain cancers, the primary suppressive function of p53 is to reduce ROS levels [128]. However, it is also suggested that a major tumor suppressive function of p53 is to regulate the antioxidative and metabolic genes [129]. One mechanism by which p53 regulates the metabolism to control antioxidant function is the induction of p53 target TIGAR expression [130]. TIGAR, also known as 2, 6-fructose bisphosphatase, reduces the levels of fructose 2, 6-bisphosphate (a positive regulator of phosphofructokinase 1) [131]. Consequently, it leads to a reduction in glycolytic flux by shunting the glycolytic carbons to the pentose phosphate pathway to generate NADPH and the generated NADPH is required for the maintenance of the antioxidant system [132]. Other tumor suppressor genes such as FOXO transcription factors, function by activating many antioxidant genes, such as Prdx3 and Prdx5 [133,134].

ROS also regulate mitogenic signals to promote cancer cell proliferation and also play a role in adapting to metabolic stress condition when a highly proliferative tumor tissue outstrips its blood supply [135,136]. Hypoxia-inducible factors (HIFs) are stabilized in the resulting hypoxic tissue [137,138]. Prolyl hydroxylases (PHDs) hydroxylate proline residues of HIF1α that are recognized by E3 ubiquitin ligase von Hippel-Landau (pVHL) protein [139]. This causes HIF1α to proteasome degradation [140]. Non-hydroxylated HIF1α is not recognized by pVHL and translocates to the nucleus [139,141]. In the nucleus, it dimerizes with HIF1β and then regulates the metabolic adaption to hypoxia and the expression of pro-oncogenes, such as vascular endothelial growth factor (VEGF) [142]. HIFs that are induced by ROS promote tumorigenesis in certain cancer cells [143]. Moreover, sirtuin proteins (SIRT3) upregulate the antioxidant system to prevent HIF activation [144,145].

In summary, two counter signaling mechanisms operate in cancerous cells to maintain tumor proliferation and tumor survival. One type of signaling is associated with excessive ROS generation that results from the highly active metabolism of cancerous cells and the other is to counter balance this excessive ROS and to prevent the cancerous cells from oxidative damage that involves the activation of the antioxidant defense system. Furthermore, increased ROS levels by endogenous sources are dangerous for cancerous cells, as well as for cancer development [109]. Thus, these antioxidant protective mechanisms can be targeted to kill cancer cells along with CSCs while protecting the normal cells and considered to be a better approach for cancer treatment and therapy [146,147].

6. Prdx3 and Carcinogenesis

As mentioned in Section 5, the cancer cells are endowed with extraordinarily high antioxidant capabilities; thus, it is not surprising that cancer cells contain upregulated levels of mitochondrial Prdxs [148]. Prdx3 is highly sensitive to oxidative stress, and is regulated by sirtuin1, a class III histone deacetylase (SIRT1). SIRT1 regulates the expression of Prdx3 by enhancing the formation of PGC-1α/FoxO3a transcriptional complex [149]. The signaling and regulation of Prdx3 varies, depending on the type of cancer and its interacting partners [150] (Figure 4). The debatable issue is whether these Prdxs function as cancer promoters or suppressors. Multiple lines of studies have shown that Prdx3 is upregulated in different types of cancerous cells [151–153]. The overexpression of Prdx3 is observed in colon cancer stem cells (CSCs) having elevated mitochondrial functions [150]. A positive correlation for upregulated expression of Prdx3 and CD133 is reported along with FOXM1 regulating the expression of Prdx3 and CD133 by binding to the promoter regions of both Prdx3 and CD133. Prdx3 knock out mice have shown a reduction in tumor volume and metastasis giving a clue for association of Prdx3 with FOXM1 associated pathways for cancer development [150]. FOXM1 is associated with cancer cell proliferation and the FOXM1/Prdx3 pathway plays a role in the survival of cells, so they can be specifically targeted to develop the efficient drugs. This study is significant in that both in vitro and in vivo results showed that Prdx3 is regulating the survival and metastasis of cancer stem cells through mitochondrial stabilization and the depletion of Prdx3 can lead to reduce tumor size and promote cell death by mitochondrial dysfunctions. The upregulated expression of Prdx3 is also observed in medulloblastoma (MB), along with the decreased expression of miR-383 [154]. RNA and protein levels of Prdx3 are both considerably reduced upon miR-383 restoration. This study provides clear evidence for the interaction of miR-383 with Prdx3 through miR-383 seed region in 3′ UTR of Prdx3 and reported the inverse relationship between miR-383 and Prdx3 in MB cells. However, this inverse relationship was not observed in MB samples [154,155]. This difference between Prdx3 and miR-383 relationship in MB cells and MB samples implies the need for in-vivo studies to further enhance the understanding of Prdx3 roles in cancer cell survival and metastasis. In addition to miR-383, the association of Prdx3 with miR-23b is also observed in prostate cancer (PCa). The PCa-cell line studies showed that Prdx3 is regulated by miR-23b in normal as well as in hypoxic conditions. miR-23b itself regulated by c-Myc in-turn regulates the expression of Prdx3 at both RNA and protein levels [156].

The association of Prdx3 with JunD is also observed in PCa, which indicates that JunD regulates cellular proliferation in PCa by regulating the expression of its associated genes, including Prdx3 with Myc family genes, as crucial downstream regulators [157]. These studies show that the involvement of Prdx3 in cancer is for more complicated as in PCa, the Prdx3 expression is regulated by miR-23b as well as by JunD. It is clear that Prdx3 is regulated by multiple factors and there is a dire need to analyze the co-relational regulation of Prdx3 by transcriptional factors and micro RNAs to identify the associated changes on the tumor microenvironment.

The involvement of Prdx3 with hypoxia is also established and it is interesting to note that the overexpression of Prdx3 suppressed the hypoxia mediated apoptosis of thymoma cells in vitro, but with regard to the role of von Hippel-Laundau protein (pVHL) in clear-cell renal cell carcinoma, the protein level of Prdx3 is downregulated and HIF-1α stabilization is induced by pVHL deficiency in conditions of normoxia or hypoxia; this leads to decrease in Prdx3 expression and is associated with cellular proliferation of clear cell renal cell carcinoma (CCRCC) [158]. Taken together, these studies demonstrate the important role of Prdx3 in hypoxia, which is associated with cancer development and the response to cancer therapies.

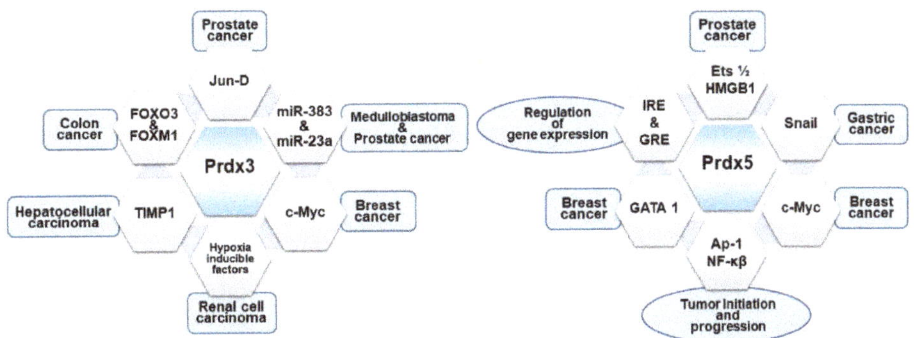

Figure 4. Microenvironment of mitochondrial Prdxs in cancer. The involvement of mitochondrial Prdxs in cancer depends on their interaction with different factors including transcription factors, different response elements, micro RNAs, and cancer-related genes. The microenvironment of Prdx3 involves its regulation by various transcription factors such as FOXO family transcription factors, Jun-D, and TIMP1 in colon, prostate, and hepatocellular carcinomas. These factors are associated with the upregulation of Prdx3 in the cancerous cells for the abovementioned cancer types. Prdx3 interacts with miR-383 and miR-23b in medulloblastoma and prostate cancer, further regulated by c-Myc in breast cancer and hypoxia-inducible factors in renal cell carcinoma. Similarly, the expression of Prdx5 in different malignancies depends on its regulation by a variety of interacting partners such as c-Myc and GATA1 in breast cancer, Snail in gastric cancer, and Ets1/2 and HMGB1 in prostate cancer. Additionally, Prdx5 interacts with transcription factors AP-1 and NF-κB, insulin and glucocorticoid response elements for the initiation of tumorigenesis and tumor progression.

The upregulated expression of Prdx3 is associated with an enhanced expression of ATP synthase and increased ATP production in hepatocellular carcinoma, and it plays a role in tumor growth and progression. However, at the same time, the downregulation of Prdx3 results in enhanced invasive properties of HepG2 cells through the downregulation of TIMP metallopeptidase inhibitor 1 (TIMP-1) and causes increased extracellular matrix (ECM) degradation [159–161]. This implies that the differential response of Prdx3 expression is regulated by diverse signaling pathways that are associated with cancer progression.

Breast cancer is one of the most common cancer among females with many subtypes and it is associated with a high mortality rate if not diagnosed at early stages [162]. Proteomic analysis of invasive ductal carcinoma of the breast with luminal B human epidermal growth factor receptor 2-positive (HER2-positive LB) and HER2-enriched (HE) subtypes have shown that Prdx3 with an upregulated expression of luminal B HER2 can serve as a promising bio-signature for LB subtype and it can also serve as potential biomarker for the diagnosis of early- and late-stage disease [163]. Enhanced expression of Prdx3 is also observed in MCF-7 cells [162,164,165]. In cervical cancer, single nucleotide polymorphism of Prdx3 leads to significant increased risk of cervical cancer and progression [166]. In addition, gene expression analysis has revealed the increase in the expression of Prdx3 in cervical cancer [166]. Endometrial cancer is among the common cancers in females that affect the genital tract. The expression of Prdx3 is upregulated in the endometrium of patients that suffer from this type of cancer as compared to that in normal endometrium [167]. A high expression of Prdx3 is associated with endometrial cancer and it has the potential to serve as a prognostic marker for endometrial cancer; thus, it can be targeted for the development of better therapeutic strategies [151,168]. The upregulated expression of Prdx3 is also observed in malignant mesothelioma (MM) cells and an overexpression of Prdx3 in MM cells maintains a redox set point that enables these cells to survive in conditions with elevated levels of mitochondrial ROS [169]. Any disturbance in this Prdx3 regulated redox set point impairs cellular proliferation by affecting the cell cycle dynamics operating between energy metabolism and mitochondrial network [170]. Taken together, the upregulated expression of Prdx3 is observed in different cancerous cell lines but the mechanistic details are lacking in how the upregulated expression affects the environment in cancerous cells.

Our survey of studies analyzing the expression of Prdx3 in human cancers clearly shows that Prdx3 is upregulated in many cancers and it is associated with cell proliferation. Hence, Prdx3 can be considered as pro-cell survival, either in healthy or diseased cells, and it can be targeted for cancer treatment. The cancerous cells have significantly high levels of ROS, irrespective of the upregulated expression of Prdxs as compared to the normal cell and these cancerous cells cannot respond to the increased ROS levels, like normal cells having many compensatory mechanisms to respond to ROS. Therefore, Prdxs inhibition can lead to a further increase in ROS and consequently can promote cell death of cancerous cells but not the normal cells having protective mechanisms [171]. However, there is evidence that downregulation of Prdx3 led to enhancing the tumor malignancies and invasiveness in certain cancers, so the targeting of Prdx3 for effective drug development requires further research and it will be beneficial to target the signaling pathway instead of a single factor.

7. Prdx5 and Carcinogenesis

Mitochondrial oxidants are produced in significantly large amounts in cancerous cells due to oncogenic transformation and metabolic reorganization [172]. Like Prdx3, the up- and downregulation of Prdx5 is also observed in many types of cancers (Figure 4). The expression of Prdx5 is regulated by different transcription factors, including AP-1, nuclear factor-κB (NF-κB), antioxidant response element (ARE), insulin response element (IRE), glucocorticoid response element (GRE), and c-Myc; further, c-Myc might directly regulate Prdx5 expression by interacting with putative responsive elements in the 5′-flanking region of the gene [54,173]. Despite these other transcription factors, such as nuclear respiratory factor 1 (NRF1) and nuclear respiratory factor 2 (NRF2; GABPA), which are associated with the mammalian cells, response to oxidative stress and mitochondrial biogenesis are also capable of indirectly regulating Prdx5 expression [53,174]. c-Myc not only directly regulates Prdx5 transcription, but also participates in the establishment of ROS homeostasis by selectively inducing the transcription of specific Prdxs when the function of one of the Prdxs is compromised [175]. In microenvironmental stress conditions, such as hypoxia, E-twenty-six transcription factor 1 and 2 (Ets1/2), and high-mobility-group protein B1 (HMGB1), mediate the upregulation Prdx5 in cancer cells, particularly in human prostate and epidermoid cancer cells exposed to H_2O_2 or hypoxia [176]. The interaction of Prdx5 with a variety of regulators complicates its functions in cell survival during normal and pathological conditions.

The expression of Prdx5 is closely related to the tumor size, depth, and lymphatic invasion in patients suffering from gastric cancer [177]. Moreover, the enhanced expression of Prdx5 leads to augmented carcinogenicity by increasing the proliferation and invasiveness of gastric cancer cells through the upregulation of Snail [177]. The treatment of Hodgkin's lymphomas is based on targeting ROS, but the increased expression of mitochondrial Prdxs leads to chemoresistance [178]. Increased levels of Prdx5 have been observed in aggressive Hodgkin's lymphomas [178]. In breast cancer, Prdx5 is upregulated in the mammary tissues and it is associated with poor prognosis. GATA1 transcription factor binds to the promoter region of Prdx5 in breast cancer cells and downregulates the transcription of Prdx5 [179]. The overexpression of Prdx5 protects cancerous cells from oxidative stress-induced apoptosis in a GATA1-regulated manner [165,179]. This implies that GATA1-regulated Prdx5 transcription can be targeted to treat breast cancer, but it requires further analysis involving the overexpression of GATA1 and its effects on tumor production and metastasis. Like Prdx3, the expression of Prdx5 is upregulated in endometrial cancer and this enhanced expression of Prdx5 in the endometrium of females with endometrial tumor can serve as a prognostic marker [167]. Mitochondrial Prdxs are overexpressed in ovarian cancer cells, and Prdx5 serves as a negative predictor of survival in patients suffering from ovarian cancer [180–182]. The upregulated expression of Prdx5 is also observed in malignant mesothelioma cells [169], while a reduction in Prdx5 expression has only been described in adrenocortical carcinoma [183].

In summary, the Prdx5 is upregulated in different cancers, except for adrenocortical carcinoma, but which factors are controlling the upregulation of Prdx5, the pathways associated with Prdx5 upregulation and cancer cell survival are lacking and need further investigation to demonstrate the exact mechanism of action of Prdx5 in different cancers and to develop strategies for effective drug designing.

8. Mitochondrial Prdxs and Chemoresistance

Cancerous cells are unique when compared with normal cells, in that they include elevated ROS levels as well as an increased level of antioxidants to counterbalance the ROS [111]. This distinctive feature of cancerous cells is attributable to the development of resistance in cancerous cells against chemo and radiotherapy, as these therapies are highly dependent on ROS-developed cytotoxicity [114]. A plethora of literature has described the association of elevated levels of Prdxs with chemo- or radioresistance to various drugs [184–188].

In breast cancer, the upregulated expression of Prdx3 is associated with the development of resistance to the drug doxorubicin [189]. Prdx3 regulates the apoptotic signaling pathway by controlling the release of cytochrome c from the mitochondria, along with establishing the linkage with leucine zipper kinase and IKB kinase [165]. Therefore, it will be a good strategy to develop drugs that target Prdx3 and mitochondrion specific electron suppliers, i.e., thioredoxin2 (Trx2), thioredoxin reductase2 (TrxR2), and sulfiredoxin (Srx), for response improvement of different chemotherapeutic agents, such as cisplatin, paclitaxel, and etoposide [189,190]. In many other cancers, such as breast cancer, ovarian cancer, and erythroleukemia, chemoresistance is developed by the upregulated expression of Prdx1, Prdx3, and Prdx6 [191]. In addition, there is evidence that Prdx5 is also involved in chemoresistance to adriamycin, bleomycin, vinblastine, and dacarbazine in patients of Hodgkin's lymphoma and in vitro lung carcinoma U1810 cell lines [192].

Chemoresistance is a complicated process that involves different factors and numerous modes of action affected by the tumor microenvironment as well as tumor biology [193,194]. Alterations in endogenous antioxidants play a determining role in the development of chemoresistance, as well as can serve as promising targets to design new drugs with better efficacy [195] (Figure 5).

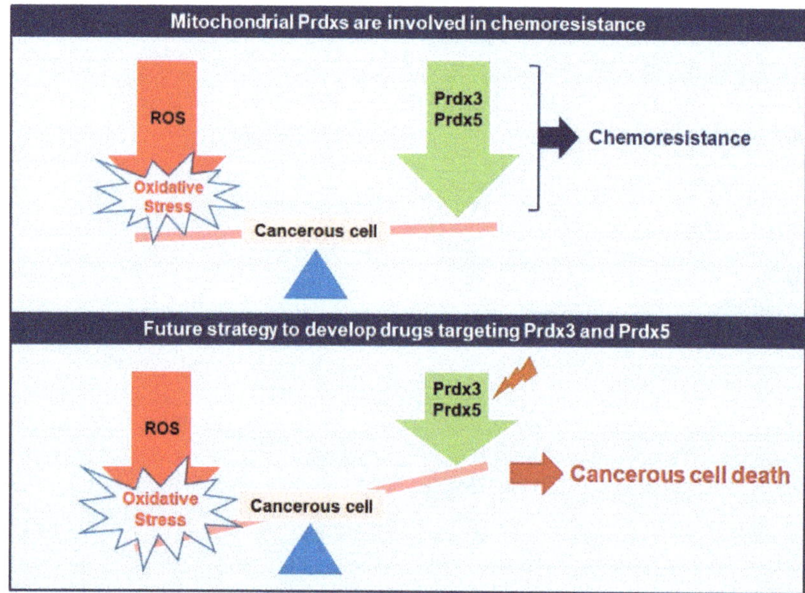

Figure 5. Mitochondrial Prdxs in development of Chemoresistance. The upregulated expression of mitochondrial Prdxs is associated with the development of drug resistance in a number of cancer types leading to complications in cancer treatment. These upregulated Prdxs can be specifically targeted in cancerous cells to develop new drugs against cancer with better efficacy and without causing any harm to the normal cells.

9. Concluding Remarks

Mitochondrial Prdxs, the multifunctional proteins of the cells, are well known for their physiological as well as pathological significance based on their interplay with ROS. The crosstalk between ROS and mitochondrial Prdxs is critical for the initiation and progression of various types of cancers. The targeting of mitochondrial Prdxs for developing drugs against cancer is a good strategy. However, because these Prdxs are associated with chemoresistance in certain cancers, it is conceivable that, instead of targeting mitochondrial Prdxs, it will be better to diagnose the signaling pathways and microenvironments of these Prdxs in particular cancer types and then develop a drug that can target the root, leading to the activation of these Prdxs in response to the elevated ROS levels.

10. Future Directions

Our survey of studies analyzing the involvement of mitochondrial Prdxs in human cancers shows that most of the conducted studies are based on the expression analysis of Prdx3 and Prdx5. It is depicted that mitochondrial Prdxs are upregulated in variety of cancer types and directly or indirectly regulated by transcription factors, microRNAs, and oncogenes. Further, the interaction of Prdx5 with response elements is also reported, but current studies for analyzing the roles of mitochondrial Prdxs and ROS in cancer need more in depth analyses, as described below

1. Most of the studies reporting the upregulation of mitochondrial Prdxs in human cancers are cell-line specific, it will be more advantageous to design in-vivo studies to explore the interaction of these Prdxs in cellular environment.
2. Mitochondrial Prdxs have the ability to function as molecular chaperons, enzyme activators, and can be involved in protein-protein interactions beyond their enzymatic peroxidase functions. Accordingly, the transgenic animal (mouse) models should be used to demonstrate the role

of mitochondrial Prdxs in oncogenic signaling. It will be beneficial to understand the exact mechanism of action of these Prdxs in cancer and to design effective drugs targeting a particular pathway associated with cancer survival and progression.

3. Although a plethora of studies describe the regulation of mitochondrial Prdxs by different transcription factors, oncogenes, and microRNAs in different types of cancer, but the exact mechanism of mitochondrial Prdxs in different types of cancers and their upstream and downstream regulators is lacking. Incorporation of variety of omics techniques i.e., transcriptomics, proteomics, and metabolomics into the in vitro and in vivo studies of mitochondrial Prdxs in cancer development can help to elucidate signaling mechanism in future studies.

4. More clinical investigations are needed to evaluate the differences in the expression of Prdx3 and Prdx5 between normal and diseased state. In addition, the expression of mitochondrial Prdxs during the early and late stage of cancer should be analyzed to demonstrate their role as anti-oncogenic or pro-oncogenic in different cellular context.

5. The upregulated expression of Prdx3 and Prdx5 is associated with the development of chemoresistance in different cancers and selective targeting of these mitochondrial Prdxs can lead to sensitization of cancer cells to chemotherapy. This fact should be investigated in detail to unveil the underlying mechanisms.

Author Contributions: T.I. and Y.K. wrote the manuscript and prepared figures. H.L. and D-S.L. revised and proofread the manuscript. T.I. and H-S.L. developed the central idea for the paper. H-S.L. critically analyzed it, and revised the manuscript to its final form.

Funding: This study was funded by the Ministry of Science, ICT, and Future Planning (MSIP) [grant no. 2015R1A4A1042271], the Republic of Korea.

Conflicts of Interest: The authors declare no conflict of interest.

Abbreviations

ROS	Reactive oxygen species
Prdxs	Peroxiredoxins
ATP	Adenosine triphosphate
SOD	Superoxide dismutase
Cys	Cysteine
Cp	Peroxidatic cysteine
Cr	Resolving cyotcine
ERK	Extracellular signal related kinase
JNK	Jun N-terminal kinase
MAPK	Mitogen activated protein kinase
PTP	Protein tyrosine phosphatases
PTEN	Phosphatase and tensin homologues
CSCs	Cancerous stem cells
NRF2	Nuclear factor erythroid 2 related factor
KEAP1	Kelch like ECH associated proteins
AREs	Antioxidant responsive elements
HIFs	Hypoxia inducible factors
PHDs	Prolyl hydroxylases
pVHL	von Hippel-Laundau protein
CCRCC	Clear cell renal cell carcinoma
FOXM1	Forkhead box protein 1
LNCaP	Lymph node carcinoma of prostate
TIMP1	TIMP metallopeptidase inhibitor 1
ECM	Extracellular matrix

HepG2	Human hepatocellular carcinoma/hepatoma cell lines
MCF-7	Mitichigan cancer foundation-7 cancerous cell lines
HER2	Human epithelial growth factor receptor 2
Trx2	Thioredoxin 2
TrxR2	Thioredoxin reductase 2
Srx	Sulfiredoxin
MM	Malignant mesothelioma
NF-κB	Nuclear factor κB
ARE	Antioxidant responsive element
IRE	Insulin responsive element
GRE	Glucocorticoid responsive element
Ets $\frac{1}{2}$	E twenty six transcription factor 1 and 2
HMGB1	High mobility group protein B1

References

1. Panchal, K.; Tiwari, A.K. Mitochondrial dynamics, a key executioner in neurodegenerative diseases. *Mitochondrion* **2018**. [CrossRef] [PubMed]

2. Austad, S.N. The Comparative Biology of Mitochondrial Function and the Rate of Aging. *Integr. Comp. Biol.* **2018**, *58*, 559–566. [CrossRef] [PubMed]

3. Moreira, O.C.; Estebanez, B.; Martinez-Florez, S.; de Paz, J.A.; Cuevas, M.J.; Gonzalez-Gallego, J. Mitochondrial Function and Mitophagy in the Elderly: Effects of Exercise. *Oxid. Med. Cell. Longev.* **2017**, *2017*, 2012798. [CrossRef] [PubMed]

4. Scorrano, L. Keeping mitochondria in shape: A matter of life and death. *Eur. J. Clin. Investig.* **2013**, *43*, 886–893. [CrossRef] [PubMed]

5. Horbay, R.; Bilyy, R. Mitochondrial dynamics during cell cycling. *Apoptosis* **2016**, *21*, 1327–1335. [CrossRef] [PubMed]

6. Jeong, S.Y.; Seol, D.W. The role of mitochondria in apoptosis. *BMB Rep.* **2008**, *41*, 11–22. [CrossRef] [PubMed]

7. Wang, Z.; Figueiredo-Pereira, C.; Oudot, C.; Vieira, H.L.; Brenner, C. Mitochondrion: A Common Organelle for Distinct Cell Deaths? *Int. Rev. Cell Mol. Biol.* **2017**, *331*, 245–287. [PubMed]

8. Parsons, M.J.; Green, D.R. Mitochondria in cell death. *Essays Biochem.* **2010**, *47*, 99–114. [CrossRef]

9. Fulda, S. Mitochondria, redox signaling and cell death in cancer. *Biol. Chem.* **2016**, *397*, 583. [CrossRef]

10. Chandel, N.S. Mitochondria as signaling organelles. *BMC Biol.* **2014**, *12*, 34. [CrossRef]

11. Chandel, N.S. Evolution of Mitochondria as Signaling Organelles. *Cell Metab.* **2015**, *22*, 204–206. [CrossRef] [PubMed]

12. Andreyev, A.Y.; Kushnareva, Y.E.; Starkov, A.A. Mitochondrial metabolism of reactive oxygen species. *Biochemistry (Moscow)* **2005**, *70*, 200–214. [CrossRef] [PubMed]

13. Turrens, J.F. Mitochondrial formation of reactive oxygen species. *J. Physiol.* **2003**, *552 Pt 2*, 335–344. [CrossRef]

14. Zhang, J.; Wang, X.; Vikash, V.; Ye, Q.; Wu, D.; Liu, Y.; Dong, W. ROS and ROS-Mediated Cellular Signaling. *Oxid. Med. Cell. Longev.* **2016**, *2016*, 4350965. [CrossRef] [PubMed]

15. Brieger, K.; Schiavone, S.; Miller, F.J., Jr.; Krause, K.H. Reactive oxygen species: From health to disease. *Swiss Med. Wkly.* **2012**, *142*, w13659. [CrossRef] [PubMed]

16. Roy, J.; Galano, J.M.; Durand, T.; Le Guennec, J.Y.; Lee, J.C. Physiological role of reactive oxygen species as promoters of natural defenses. *FASEB J.* **2017**, *31*, 3729–3745. [CrossRef] [PubMed]

17. Fridovich, I. The biology of oxygen radicals. *Science* **1978**, *201*, 875–880. [CrossRef] [PubMed]

18. Grivennikova, V.G.; Vinogradov, A.D. Mitochondrial production of reactive oxygen species. *Biochemistry (Moscow)* **2013**, *78*, 1490–1511. [CrossRef]

19. Wong, H.S.; Dighe, P.A.; Mezera, V.; Monternier, P.A.; Brand, M.D. Production of superoxide and hydrogen peroxide from specific mitochondrial sites under different bioenergetic conditions. *J. Biol. Chem.* **2017**, *292*, 16804–16809. [CrossRef]

20. Holzerova, E.; Prokisch, H. Mitochondria: Much ado about nothing? How dangerous is reactive oxygen species production? *Int. J. Biochem. Cell Biol.* **2015**, *63*, 16–20. [CrossRef]

21. Teppo, H.R.; Soini, Y.; Karihtala, P. Reactive Oxygen Species-Mediated Mechanisms of Action of Targeted Cancer Therapy. *Oxid. Med. Cell. Longev.* **2017**, *2017*, 1485283. [CrossRef] [PubMed]
22. Hybertson, B.M.; Gao, B.; Bose, S.K.; McCord, J.M. Oxidative stress in health and disease: The therapeutic potential of Nrf2 activation. *Mol. Asp. Med.* **2011**, *32*, 234–246. [CrossRef] [PubMed]
23. Halliwell, B. Free radicals and antioxidants: Updating a personal view. *Nutr. Rev.* **2012**, *70*, 257–265. [CrossRef] [PubMed]
24. Schieber, M.; Chandel, N.S. ROS function in redox signaling and oxidative stress. *Curr. Biol.* **2014**, *24*, R453–R462. [CrossRef] [PubMed]
25. Murphy, M.P. Antioxidants as therapies: Can we improve on nature? *Free Radic. Biol. Med.* **2014**, *66*, 20–23. [CrossRef] [PubMed]
26. Dandekar, A.; Mendez, R.; Zhang, K. Cross talk between ER stress, oxidative stress, and inflammation in health and disease. *Methods Mol. Biol.* **2015**, *1292*, 205–214. [PubMed]
27. Kattoor, A.J.; Pothineni, N.V.K.; Palagiri, D.; Mehta, J.L. Oxidative Stress in Atherosclerosis. *Curr. Atheroscler. Rep.* **2017**, *19*, 42. [CrossRef]
28. Oyewole, A.O.; Birch-Machin, M.A. Mitochondria-targeted antioxidants. *FASEB J.* **2015**, *29*, 4766–4771. [CrossRef]
29. Bjorklund, G.; Chirumbolo, S. Role of oxidative stress and antioxidants in daily nutrition and human health. *Nutrition* **2017**, *33*, 311–321. [CrossRef]
30. Cortassa, S.; O'Rourke, B.; Aon, M.A. Redox-optimized ROS balance and the relationship between mitochondrial respiration and ROS. *Biochim. Biophys. Acta* **2014**, *1837*, 287–295. [CrossRef]
31. Ji, Y.; Chae, S.; Lee, H.K.; Park, I.; Kim, C.; Ismail, T.; Kim, Y.; Park, J.W.; Kwon, O.S.; Kang, B.S.; et al. Peroxiredoxin5 Controls Vertebrate Ciliogenesis by Modulating Mitochondrial Reactive Oxygen Species. *Antioxid. Redox Signal.* **2019**, *30*, 1731–1745. [CrossRef] [PubMed]
32. Sies, H. Oxidative stress: A concept in redox biology and medicine. *Redox Biol.* **2015**, *4*, 180–183. [CrossRef] [PubMed]
33. Rhee, S.G. Overview on Peroxiredoxin. *Mol. Cells* **2016**, *39*, 5.
34. Rhee, S.G.; Kil, I.S. Multiple Functions and Regulation of Mammalian Peroxiredoxins. *Annu. Rev. Biochem.* **2017**, *86*, 749–775. [CrossRef] [PubMed]
35. Chae, S.; Lee, H.K.; Kim, Y.K.; Sim, H.J.; Ji, Y.; Kim, C.; Ismail, T.; Park, J.W.; Kwon, O.S.; Kang, B.S.; et al. Peroxiredoxin1, a novel regulator of pronephros development, influences retinoic acid and Wnt signaling by controlling ROS levels. *Sci. Rep.* **2017**, *7*, 8874. [CrossRef] [PubMed]
36. Karplus, P.A.; Hall, A. Structural survey of the peroxiredoxins. *Subcell. Biochem.* **2007**, *44*, 41–60. [PubMed]
37. Wood, Z.A.; Schroder, E.; Robin Harris, J.; Poole, L.B. Structure, mechanism and regulation of peroxiredoxins. *Trends Biochem. Sci.* **2003**, *28*, 32–40. [CrossRef]
38. Hall, A.; Nelson, K.; Poole, L.B.; Karplus, P.A. Structure-based insights into the catalytic power and conformational dexterity of peroxiredoxins. *Antioxid. Redox Signal.* **2011**, *15*, 795–815. [CrossRef]
39. Schroder, E.; Littlechild, J.A.; Lebedev, A.A.; Errington, N.; Vagin, A.A.; Isupov, M.N. Crystal structure of decameric 2-Cys peroxiredoxin from human erythrocytes at 1.7 A resolution. *Structure* **2000**, *8*, 605–615. [CrossRef]
40. Barranco-Medina, S.; Lazaro, J.J.; Dietz, K.J. The oligomeric conformation of peroxiredoxins links redox state to function. *FEBS Lett.* **2009**, *583*, 1809–1816. [CrossRef]
41. Kim, K.H.; Lee, W.; Kim, E.E. Crystal structures of human peroxiredoxin 6 in different oxidation states. *Biochem. Biophys. Res. Commun.* **2016**, *477*, 717–722. [CrossRef] [PubMed]
42. Poole, L.B.; Nelson, K.J. Distribution and Features of the Six Classes of Peroxiredoxins. *Mol. Cells* **2016**, *39*, 53–59. [PubMed]
43. Cao, Z.; Lindsay, J.G.; Isaacs, N.W. Mitochondrial peroxiredoxins. *Subcell. Biochem.* **2007**, *44*, 295–315. [PubMed]
44. Yamamoto, T.; Matsui, Y.; Natori, S.; Obinata, M. Cloning of a housekeeping-type gene (MER5) preferentially expressed in murine erythroleukemia cells. *Gene* **1989**, *80*, 337–343. [PubMed]

45. Nemoto, Y.; Yamamoto, T.; Takada, S.; Matsui, Y.; Obinata, M. Antisense RNA of the latent period gene (MER5) inhibits the differentiation of murine erythroleukemia cells. *Gene* **1990**, *91*, 261–265. [CrossRef]

46. Yang, H.Y.; Jeong, D.K.; Kim, S.H.; Chung, K.J.; Cho, E.J.; Yang, U.; Lee, S.R.; Lee, T.H. The role of peroxiredoxin III on late stage of proerythrocyte differentiation. *Biochem. Biophys. Res. Commun.* **2007**, *359*, 1030–1036. [CrossRef] [PubMed]

47. Tsuji, K.; Copeland, N.G.; Jenkins, N.A.; Obinata, M. Mammalian antioxidant protein complements alkylhydroperoxide reductase (ahpC) mutation in Escherichia coli. *Biochem. J.* **1995**, *307 Pt 2*, 377–381. [CrossRef]

48. Watabe, S.; Hiroi, T.; Yamamoto, Y.; Fujioka, Y.; Hasegawa, H.; Yago, N.; Takahashi, S.Y. SP-22 is a thioredoxin-dependent peroxide reductase in mitochondria. *Eur. J. Biochem.* **1997**, *249*, 52–60. [CrossRef]

49. Garrard, L.J.; Goodman, J.M. Two genes encode the major membrane-associated protein of methanol-induced peroxisomes from Candida boidinii. *J. Biol. Chem.* **1989**, *264*, 13929–13937.

50. Yamashita, H.; Avraham, S.; Jiang, S.; London, R.; Van Veldhoven, P.P.; Subramani, S.; Rogers, R.A.; Avraham, H. Characterization of human and murine PMP20 peroxisomal proteins that exhibit antioxidant activity in vitro. *J. Biol. Chem.* **1999**, *274*, 29897–29904. [CrossRef]

51. Knoops, B.; Clippe, A.; Bogard, C.; Arsalane, K.; Wattiez, R.; Hermans, C.; Duconseille, E.; Falmagne, P.; Bernard, A. Cloning and characterization of AOEB166, a novel mammalian antioxidant enzyme of the peroxiredoxin family. *J. Biol. Chem.* **1999**, *274*, 30451–30458. [CrossRef] [PubMed]

52. Seo, M.S.; Kang, S.W.; Kim, K.; Baines, I.C.; Lee, T.H.; Rhee, S.G. Identification of a new type of mammalian peroxiredoxin that forms an intramolecular disulfide as a reaction intermediate. *J. Biol. Chem.* **2000**, *275*, 20346–20354. [CrossRef] [PubMed]

53. Kropotov, A.; Usmanova, N.; Serikov, V.; Zhivotovsky, B.; Tomilin, N. Mitochondrial targeting of human peroxiredoxin V protein and regulation of PRDX5 gene expression by nuclear transcription factors controlling biogenesis of mitochondria. *FEBS J.* **2007**, *274*, 5804–5814. [CrossRef] [PubMed]

54. Knoops, B.; Goemaere, J.; Van der Eecken, V.; Declercq, J.P. Peroxiredoxin 5: Structure, mechanism, and function of the mammalian atypical 2-Cys peroxiredoxin. *Antioxid. Redox Signal.* **2011**, *15*, 817–829. [CrossRef] [PubMed]

55. Trujillo, M.; Ferrer-Sueta, G.; Thomson, L.; Flohe, L.; Radi, R. Kinetics of peroxiredoxins and their role in the decomposition of peroxynitrite. *Subcell. Biochem.* **2007**, *44*, 83–113. [PubMed]

56. Lambert, A.J.; Brand, M.D. Reactive oxygen species production by mitochondria. *Methods Mol. Biol.* **2009**, *554*, 165–181. [PubMed]

57. Kimura, S.; Waszczak, C.; Hunter, K.; Wrzaczek, M. Bound by Fate: The Role of Reactive Oxygen Species in Receptor-Like Kinase Signaling. *Plant Cell* **2017**, *29*, 638–654. [CrossRef]

58. Choudhury, F.K.; Rivero, R.M.; Blumwald, E.; Mittler, R. Reactive oxygen species, abiotic stress and stress combination. *Plant J.* **2017**, *90*, 856–867. [CrossRef]

59. Li, Z.; Xu, X.; Leng, X.; He, M.; Wang, J.; Cheng, S.; Wu, H. Roles of reactive oxygen species in cell signaling pathways and immune responses to viral infections. *Arch. Virol.* **2017**, *162*, 603–610. [CrossRef]

60. Finkel, T. From sulfenylation to sulfhydration: What a thiolate needs to tolerate. *Sci. Signal.* **2012**, *5*, pe10. [CrossRef]

61. Messens, J.; Collet, J.F. Thiol-disulfide exchange in signaling: Disulfide bonds as a switch. *Antioxid. Redox Signal.* **2013**, *18*, 1594–1596. [CrossRef] [PubMed]

62. Sun, M.A.; Zhang, Q.; Wang, Y.; Ge, W.; Guo, D. Prediction of redox-sensitive cysteines using sequential distance and other sequence-based features. *BMC Bioinform.* **2016**, *17*, 316. [CrossRef] [PubMed]

63. Deponte, M.; Lillig, C.H. Enzymatic control of cysteinyl thiol switches in proteins. *Biol. Chem.* **2015**, *396*, 401–413. [CrossRef] [PubMed]

64. Kil, I.S.; Ryu, K.W.; Lee, S.K.; Kim, J.Y.; Chu, S.Y.; Kim, J.H.; Park, S.; Rhee, S.G. Circadian Oscillation of Sulfiredoxin in the Mitochondria. *Mol. Cell* **2015**, *59*, 651–663. [CrossRef] [PubMed]

65. Mishanina, T.V.; Libiad, M.; Banerjee, R. Biogenesis of reactive sulfur species for signaling by hydrogen sulfide oxidation pathways. *Nat. Chem. Biol.* **2015**, *11*, 457–464. [CrossRef] [PubMed]

66. Murphy, M.P. Mitochondrial thiols in antioxidant protection and redox signaling: Distinct roles for glutathionylation and other thiol modifications. *Antioxid. Redox Signal.* **2012**, *16*, 476–495. [CrossRef] [PubMed]

67. Andreyev, A.Y.; Kushnareva, Y.E.; Murphy, A.N.; Starkov, A.A. Mitochondrial ROS Metabolism: 10 Years Later. *Biochemistry (Moscow)* **2015**, *80*, 517–531. [CrossRef]

68. Brand, M.D. Mitochondrial generation of superoxide and hydrogen peroxide as the source of mitochondrial redox signaling. *Free Radic. Biol. Med.* **2016**, *100*, 14–31. [CrossRef]

69. Winterbourn, C.C.; Hampton, M.B. Thiol chemistry and specificity in redox signaling. *Free Radic. Biol. Med.* **2008**, *45*, 549–561. [CrossRef]

70. Sharapov, M.G.; Ravin, V.K.; Novoselov, V.I. Peroxyredoxins as multifunctional enzymes. *Mol. Biol. (Mosk)* **2014**, *48*, 600–628. [CrossRef]

71. Kudryavtseva, A.V.; Krasnov, G.S.; Dmitriev, A.A.; Alekseev, B.Y.; Kardymon, O.L.; Sadritdinova, A.F.; Fedorova, M.S.; Pokrovsky, A.V.; Melnikova, N.V.; Kaprin, A.D.; et al. Mitochondrial dysfunction and oxidative stress in aging and cancer. *Oncotarget* **2016**, *7*, 44879–44905. [CrossRef] [PubMed]

72. Moloney, J.N.; Cotter, T.G. ROS signalling in the biology of cancer. *Semin. Cell Dev. Biol.* **2018**, *80*, 50–64. [CrossRef] [PubMed]

73. Gill, J.G.; Piskounova, E.; Morrison, S.J. Cancer, Oxidative Stress, and Metastasis. *Cold Spring Harb. Symp. Quant. Biol.* **2016**, *81*, 163–175. [CrossRef] [PubMed]

74. Zhao, Y.; Hu, X.; Liu, Y.; Dong, S.; Wen, Z.; He, W.; Zhang, S.; Huang, Q.; Shi, M. ROS signaling under metabolic stress: Cross-talk between AMPK and AKT pathway. *Mol Cancer* **2017**, *16*, 79. [CrossRef] [PubMed]

75. Chio, I.I.C.; Tuveson, D.A. ROS in Cancer: The Burning Question. *Trends Mol. Med.* **2017**, *23*, 411–429. [CrossRef]

76. Hanahan, D.; Weinberg, R.A. Hallmarks of cancer: The next generation. *Cell* **2011**, *144*, 646–674. [CrossRef] [PubMed]

77. Nagarajan, A.; Malvi, P.; Wajapeyee, N. Oncogene-directed alterations in cancer cell metabolism. *Trends Cancer* **2016**, *2*, 365–377. [CrossRef]

78. Pavlova, N.N.; Thompson, C.B. The Emerging Hallmarks of Cancer Metabolism. *Cell Metab.* **2016**, *23*, 27–47. [CrossRef]

79. Szatrowski, T.P.; Nathan, C.F. Production of large amounts of hydrogen peroxide by human tumor cells. *Cancer Res.* **1991**, *51*, 794–798.

80. Yuan, D.; Huang, S.; Berger, E.; Liu, L.; Gross, N.; Heinzmann, F.; Ringelhan, M.; Connor, T.O.; Stadler, M.; Meister, M.; et al. Kupffer Cell-Derived Tnf Triggers Cholangiocellular Tumorigenesis through JNK due to Chronic Mitochondrial Dysfunction and ROS. *Cancer Cell* **2017**, *31*, 771–789 e6. [CrossRef]

81. Weinberg, F.; Hamanaka, R.; Wheaton, W.W.; Weinberg, S.; Joseph, J.; Lopez, M.; Kalyanaraman, B.; Mutlu, G.M.; Budinger, G.R.; Chandel, N.S. Mitochondrial metabolism and ROS generation are essential for Kras-mediated tumorigenicity. *Proc. Natl. Acad. Sci. USA* **2010**, *107*, 8788–8793. [CrossRef] [PubMed]

82. Dhillon, A.S.; Hagan, S.; Rath, O.; Kolch, W. MAP kinase signalling pathways in cancer. *Oncogene* **2007**, *26*, 3279–3290. [CrossRef] [PubMed]

83. Faes, S.; Dormond, O. PI3K and AKT: Unfaithful Partners in Cancer. *Int. J. Mol. Sci.* **2015**, *16*, 21138–21152. [CrossRef] [PubMed]

84. Liu, X.; Xu, Y.; Zhou, Q.; Chen, M.; Zhang, Y.; Liang, H.; Zhao, J.; Zhong, W.; Wang, M. PI3K in cancer: Its structure, activation modes and role in shaping tumor microenvironment. *Future Oncol.* **2018**, *14*, 665–674. [CrossRef] [PubMed]

85. Dolcet, X.; Llobet, D.; Pallares, J.; Matias-Guiu, X. NF-kB in development and progression of human cancer. *Virchows Arch.* **2005**, *446*, 475–482. [CrossRef] [PubMed]

86. Tilborghs, S.; Corthouts, J.; Verhoeven, Y.; Arias, D.; Rolfo, C.; Trinh, X.B.; van Dam, P.A. The role of Nuclear Factor-kappa B signaling in human cervical cancer. *Crit. Rev. Oncol. Hematol.* **2017**, *120*, 141–150. [CrossRef] [PubMed]

87. De Sa Junior, P.L.; Camara, D.A.D.; Porcacchia, A.S.; Fonseca, P.M.M.; Jorge, S.D.; Araldi, R.P.; Ferreira, A.K. The Roles of ROS in Cancer Heterogeneity and Therapy. *Oxid. Med. Cell. Longev.* **2017**, *2017*, 2467940. [CrossRef] [PubMed]

88. Martindale, J.L.; Holbrook, N.J. Cellular response to oxidative stress: Signaling for suicide and survival. *J. Cell. Physiol.* **2002**, *192*, 1–15. [CrossRef]

89. Ranjan, P.; Anathy, V.; Burch, P.M.; Weirather, K.; Lambeth, J.D.; Heintz, N.H. Redox-dependent expression of cyclin D1 and cell proliferation by Nox1 in mouse lung epithelial cells. *Antioxid. Redox Signal.* **2006**, *8*, 1447–1459. [CrossRef]

90. Shimura, T.; Sasatani, M.; Kamiya, K.; Kawai, H.; Inaba, Y.; Kunugita, N. Mitochondrial reactive oxygen species perturb AKT/cyclin D1 cell cycle signaling via oxidative inactivation of PP2A in lowdose irradiated human fibroblasts. *Oncotarget* **2016**, *7*, 3559–3570. [CrossRef]

91. Wang, P.; Zeng, Y.; Liu, T.; Zhang, C.; Yu, P.W.; Hao, Y.X.; Luo, H.X.; Liu, G. Chloride intracellular channel 1 regulates colon cancer cell migration and invasion through ROS/ERK pathway. *World J. Gastroenterol.* **2014**, *20*, 2071–2078. [CrossRef] [PubMed]

92. Shi, Y.; Nikulenkov, F.; Zawacka-Pankau, J.; Li, H.; Gabdoulline, R.; Xu, J.; Eriksson, S.; Hedstrom, E.; Issaeva, N.; Kel, A.; et al. ROS-dependent activation of JNK converts p53 into an efficient inhibitor of oncogenes leading to robust apoptosis. *Cell Death Differ.* **2014**, *21*, 612–623. [CrossRef] [PubMed]

93. Zhao, W.; Lu, M.; Zhang, Q. Chloride intracellular channel 1 regulates migration and invasion in gastric cancer by triggering the ROS-mediated p38 MAPK signaling pathway. *Mol. Med. Rep.* **2015**, *12*, 8041–8047. [CrossRef] [PubMed]

94. Prasad, S.; Gupta, S.C.; Tyagi, A.K. Reactive oxygen species (ROS) and cancer: Role of antioxidative nutraceuticals. *Cancer Lett.* **2017**, *387*, 95–105. [CrossRef] [PubMed]

95. Leslie, N.R.; Bennett, D.; Lindsay, Y.E.; Stewart, H.; Gray, A.; Downes, C.P. Redox regulation of PI 3-kinase signalling via inactivation of PTEN. *EMBO J.* **2003**, *22*, 5501–5510. [CrossRef] [PubMed]

96. Xu, D.; Rovira, I.I.; Finkel, T. Oxidants painting the cysteine chapel: Redox regulation of PTPs. *Dev. Cell* **2002**, *2*, 251–252. [CrossRef]

97. Harris, I.S.; Blaser, H.; Moreno, J.; Treloar, A.E.; Gorrini, C.; Sasaki, M.; Mason, J.M.; Knobbe, C.B.; Rufini, A.; Halle, M.; et al. PTPN12 promotes resistance to oxidative stress and supports tumorigenesis by regulating FOXO signaling. *Oncogene* **2014**, *33*, 1047–1054. [CrossRef]

98. Qian, X.; Nie, X.; Yao, W.; Klinghammer, K.; Sudhoff, H.; Kaufmann, A.M.; Albers, A.E. Reactive oxygen species in cancer stem cells of head and neck squamous cancer. *Semin. Cancer Biol.* **2018**, *53*, 248–257. [CrossRef]

99. Ding, S.; Li, C.; Cheng, N.; Cui, X.; Xu, X.; Zhou, G. Redox Regulation in Cancer Stem Cells. *Oxid. Med. Cell. Longev.* **2015**, *2015*, 750798. [CrossRef]

100. Hamai, A.; Caneque, T.; Muller, S.; Mai, T.T.; Hienzsch, A.; Ginestier, C.; Charafe-Jauffret, E.; Codogno, P.; Mehrpour, M.; Rodriguez, R. An iron hand over cancer stem cells. *Autophagy* **2017**, *13*, 1465–1466. [CrossRef]

101. Zhang, B.B.; Wang, D.G.; Guo, F.F.; Xuan, C. Mitochondrial membrane potential and reactive oxygen species in cancer stem cells. *Fam. Cancer* **2015**, *14*, 19–23. [CrossRef] [PubMed]

102. Watson, J. Oxidants, antioxidants and the current incurability of metastatic cancers. *Open Biol.* **2013**, *3*, 120144. [CrossRef] [PubMed]

103. Negrini, S.; Gorgoulis, V.G.; Halazonetis, T.D. Genomic instability—An evolving hallmark of cancer. *Nat. Rev. Mol. Cell Biol.* **2010**, *11*, 220–228. [CrossRef] [PubMed]

104. Yeo, C.Q.X.; Alexander, I.; Lin, Z.; Lim, S.; Aning, O.A.; Kumar, R.; Sangthongpitag, K.; Pendharkar, V.; Ho, V.H.B.; Cheok, C.F. p53 Maintains Genomic Stability by Preventing Interference between Transcription and Replication. *Cell Rep.* **2016**, *15*, 132–146. [CrossRef] [PubMed]

105. Carroll, P.A.; Freie, B.W.; Mathsyaraja, H.; Eisenman, R.N. The MYC transcription factor network: Balancing metabolism, proliferation and oncogenesis. *Front. Med.* **2018**, *12*, 412–425. [CrossRef] [PubMed]

106. Gabay, M.; Li, Y.; Felsher, D.W. MYC activation is a hallmark of cancer initiation and maintenance. *Cold Spring Harb. Perspect. Med.* **2014**, *4*, a014241. [CrossRef] [PubMed]

107. Liou, G.Y.; Storz, P. Reactive oxygen species in cancer. *Free Radic. Res.* **2010**, *44*, 479–496. [CrossRef] [PubMed]

108. Galadari, S.; Rahman, A.; Pallichankandy, S.; Thayyullathil, F. Reactive oxygen species and cancer paradox: To promote or to suppress? *Free Radic. Biol. Med.* **2017**, *104*, 144–164. [CrossRef] [PubMed]

109. Calaf, G.M.; Aguayo, F.; Sergi, C.M.; Juarranz, A.; Roy, D. Antioxidants and Cancer: Theories, Techniques, and Trials in Preventing Cancer. *Oxid. Med. Cell. Longev.* **2018**, *2018*, 5363064. [CrossRef]

110. Mates, J.M.; Perez-Gomez, C.; Nunez de Castro, I. Antioxidant enzymes and human diseases. *Clin. Biochem.* **1999**, *32*, 595–603. [CrossRef]

111. Bazhin, A.V.; Philippov, P.P.; Karakhanova, S. Reactive Oxygen Species in Cancer Biology and Anticancer Therapy. *Oxid. Med. Cell. Longev.* **2016**, *2016*, 4197815. [CrossRef] [PubMed]

112. Schafer, Z.T.; Grassian, A.R.; Song, L.; Jiang, Z.; Gerhart-Hines, Z.; Irie, H.Y.; Gao, S.; Puigserver, P.; Brugge, J.S. Antioxidant and oncogene rescue of metabolic defects caused by loss of matrix attachment. *Nature* **2009**, *461*, 109–113. [CrossRef] [PubMed]

113. Dayem, A.A.; Choi, H.Y.; Kim, J.H.; Cho, S.G. Role of oxidative stress in stem, cancer, and cancer stem cells. *Cancers (Basel)* **2010**, *2*, 859–884. [CrossRef] [PubMed]

114. Tong, L.; Chuang, C.C.; Wu, S.; Zuo, L. Reactive oxygen species in redox cancer therapy. *Cancer Lett.* **2015**, *367*, 18–25. [CrossRef] [PubMed]

115. Yang, Y.; Karakhanova, S.; Hartwig, W.; D'Haese, J.G.; Philippov, P.P.; Werner, J.; Bazhin, A.V. Mitochondria and Mitochondrial ROS in Cancer: Novel Targets for Anticancer Therapy. *J. Cell. Physiol.* **2016**, *231*, 2570–2581. [CrossRef] [PubMed]

116. Krajka-Kuzniak, V.; Paluszczak, J.; Baer-Dubowska, W. The Nrf2-ARE signaling pathway: An update on its regulation and possible role in cancer prevention and treatment. *Pharm. Rep.* **2017**, *69*, 393–402. [CrossRef] [PubMed]

117. Deshmukh, P.; Unni, S.; Krishnappa, G.; Padmanabhan, B. The Keap1-Nrf2 pathway: Promising therapeutic target to counteract ROS-mediated damage in cancers and neurodegenerative diseases. *Biophys. Rev.* **2017**, *9*, 41–56. [CrossRef] [PubMed]

118. Pandey, P.; Singh, A.K.; Singh, M.; Tewari, M.; Shukla, H.S.; Gambhir, I.S. The see-saw of Keap1-Nrf2 pathway in cancer. *Crit. Rev. Oncol. Hematol.* **2017**, *116*, 89–98. [CrossRef]

119. Itoh, K.; Wakabayashi, N.; Katoh, Y.; Ishii, T.; Igarashi, K.; Engel, J.D.; Yamamoto, M. Keap1 represses nuclear activation of antioxidant responsive elements by Nrf2 through binding to the amino-terminal Neh2 domain. *Genes Dev.* **1999**, *13*, 76–86. [CrossRef]

120. Raghunath, A.; Nagarajan, R.; Sundarraj, K.; Panneerselvam, L.; Perumal, E. Genome-wide identification and analysis of Nrf2 binding sites—Antioxidant response elements in zebrafish. *Toxicol. Appl. Pharm.* **2018**, *360*, 236–248. [CrossRef]

121. Rojo de la Vega, M.; Chapman, E.; Zhang, D.D. NRF2 and the Hallmarks of Cancer. *Cancer Cell* **2018**, *34*, 21–43. [CrossRef] [PubMed]

122. Taguchi, K.; Yamamoto, M. The KEAP1-NRF2 System in Cancer. *Front. Oncol.* **2017**, *7*, 85. [CrossRef] [PubMed]

123. Nault, J.C.; Rebouissou, S.; Zucman Rossi, J. NRF2/KEAP1 and Wnt/beta-catenin in the multistep process of liver carcinogenesis in humans and rats. *Hepatology* **2015**, *62*, 677–679. [CrossRef] [PubMed]

124. Jeong, Y.; Hoang, N.T.; Lovejoy, A.; Stehr, H.; Newman, A.M.; Gentles, A.J.; Kong, W.; Truong, D.; Martin, S.; Chaudhuri, A.; et al. Role of KEAP1/NRF2 and TP53 Mutations in Lung Squamous Cell Carcinoma Development and Radiation Resistance. *Cancer Discov.* **2017**, *7*, 86–101. [CrossRef] [PubMed]

125. Budanov, A.V. The role of tumor suppressor p53 in the antioxidant defense and metabolism. *Subcell. Biochem.* **2014**, *85*, 337–358. [PubMed]

126. Bakalova, R.; Zhelev, Z.; Shibata, S.; Nikolova, B.; Aoki, I.; Higashi, T. Impressive Suppression of Colon Cancer Growth by Triple Combination SN38/EF24/Melatonin: "Oncogenic" Versus "Onco-Suppressive" Reactive Oxygen Species. *Anticancer Res.* **2017**, *37*, 5449–5458. [PubMed]

127. Ladelfa, M.F.; Toledo, M.F.; Laiseca, J.E.; Monte, M. Interaction of p53 with tumor suppressive and oncogenic signaling pathways to control cellular reactive oxygen species production. *Antioxid. Redox Signal.* **2011**, *15*, 1749–1761. [CrossRef] [PubMed]

128. Leonova, K.I.; Shneyder, J.; Antoch, M.P.; Toshkov, I.A.; Novototskaya, L.R.; Komarov, P.G.; Komarova, E.A.; Gudkov, A.V. A small molecule inhibitor of p53 stimulates amplification of hematopoietic stem cells but does not promote tumor development in mice. *Cell Cycle* **2010**, *9*, 1434–1443. [CrossRef] [PubMed]

129. Aubrey, B.J.; Kelly, G.L.; Janic, A.; Herold, M.J.; Strasser, A. How does p53 induce apoptosis and how does this relate to p53-mediated tumour suppression? *Cell Death Differ.* **2018**, *25*, 104–113. [CrossRef] [PubMed]

130. Bensaad, K.; Tsuruta, A.; Selak, M.A.; Vidal, M.N.; Nakano, K.; Bartrons, R.; Gottlieb, E.; Vousden, K.H. TIGAR, a p53-inducible regulator of glycolysis and apoptosis. *Cell* **2006**, *126*, 107–120. [CrossRef] [PubMed]

131. Ros, S.; Floter, J.; Kaymak, I.; Da Costa, C.; Houddane, A.; Dubuis, S.; Griffiths, B.; Mitter, R.; Walz, S.; Blake, S.; et al. 6-Phosphofructo-2-kinase/fructose-2,6-biphosphatase 4 is essential for p53-null cancer cells. *Oncogene* **2017**, *36*, 3287–3299. [CrossRef] [PubMed]

132. Kim, S.H.; Choi, S.I.; Won, K.Y.; Lim, S.J. Distinctive interrelation of p53 with SCO2, COX, and TIGAR in human gastric cancer. *Pathol. Res. Pr.* **2016**, *212*, 904–910. [CrossRef] [PubMed]

133. Coomans de Brachene, A.; Demoulin, J.B. FOXO transcription factors in cancer development and therapy. *Cell. Mol. Life Sci.* **2016**, *73*, 1159–1172. [CrossRef] [PubMed]

134. Link, W.; Fernandez-Marcos, P.J. FOXO transcription factors at the interface of metabolism and cancer. *Int. J. Cancer* **2017**, *141*, 2379–2391. [CrossRef] [PubMed]

135. Neuzillet, C.; Tijeras-Raballand, A.; de Mestier, L.; Cros, J.; Faivre, S.; Raymond, E. MEK in cancer and cancer therapy. *Pharm. Ther.* **2014**, *141*, 160–171. [CrossRef] [PubMed]

136. Zou, X.; Blank, M. Targeting p38 MAP kinase signaling in cancer through post-translational modifications. *Cancer Lett.* **2017**, *384*, 19–26. [CrossRef] [PubMed]

137. Chiu, D.K.; Tse, A.P.; Xu, I.M.; Cui, J.D.; Lai, R.K.; Li, L.L.; Koh, H.Y.; Tsang, F.H.; Wei, L.L.; Wong, C.M.; et al. Hypoxia inducible factor HIF-1 promotes myeloid-derived suppressor cells accumulation through ENTPD2/CD39L1 in hepatocellular carcinoma. *Nat. Commun.* **2017**, *8*, 517. [CrossRef]

138. Huang, Y.; Lin, D.; Taniguchi, C.M. Hypoxia inducible factor (HIF) in the tumor microenvironment: Friend or foe? *Sci. China Life Sci.* **2017**, *60*, 1114–1124. [CrossRef]

139. Pugh, C.W.; Ratcliffe, P.J. The von Hippel-Lindau tumor suppressor, hypoxia-inducible factor-1 (HIF-1) degradation, and cancer pathogenesis. *Semin. Cancer Biol.* **2003**, *13*, 83–89. [CrossRef]

140. Yang, M.; Su, H.; Soga, T.; Kranc, K.R.; Pollard, P.J. Prolyl hydroxylase domain enzymes: Important regulators of cancer metabolism. *Hypoxia* **2014**, *2*, 127–142.

141. Berra, E.; Benizri, E.; Ginouves, A.; Volmat, V.; Roux, D.; Pouyssegur, J. HIF prolyl-hydroxylase 2 is the key oxygen sensor setting low steady-state levels of HIF-1alpha in normoxia. *EMBO J.* **2003**, *22*, 4082–4090. [CrossRef] [PubMed]

142. Palazon, A.; Tyrakis, P.A.; Macias, D.; Velica, P.; Rundqvist, H.; Fitzpatrick, S.; Vojnovic, N.; Phan, A.T.; Loman, N.; Hedenfalk, I.; et al. An HIF-1alpha/VEGF-A Axis in Cytotoxic T Cells Regulates Tumor Progression. *Cancer Cell* **2017**, *32*, 669–683 e5. [CrossRef] [PubMed]

143. Guo, L.Y.; Zhu, P.; Jin, X.P. Association between the expression of HIF-1alpha and VEGF and prognostic implications in primary liver cancer. *Genet. Mol. Res.* **2016**, *15*. [CrossRef]

144. Cui, Y.; Qin, L.; Wu, J.; Qu, X.; Hou, C.; Sun, W.; Li, S.; Vaughan, A.T.; Li, J.J.; Liu, J. SIRT3 Enhances Glycolysis and Proliferation in SIRT3-Expressing Gastric Cancer Cells. *PLoS ONE* **2015**, *10*, e0129834. [CrossRef] [PubMed]

145. Torrens-Mas, M.; Oliver, J.; Roca, P.; Sastre-Serra, J. SIRT3: Oncogene and Tumor Suppressor in Cancer. *Cancers (Basel)* **2017**, *9*, 90. [CrossRef]

146. Athreya, K.; Xavier, M.F. Antioxidants in the Treatment of Cancer. *Nutr. Cancer* **2017**, *69*, 1099–1104. [CrossRef]

147. Bonner, M.Y.; Arbiser, J.L. The antioxidant paradox: What are antioxidants and how should they be used in a therapeutic context for cancer. *Future Med. Chem.* **2014**, *6*, 1413–1422. [CrossRef]

148. Hampton, M.B.; Vick, K.A.; Skoko, J.J.; Neumann, C.A. Peroxiredoxin Involvement in the Initiation and Progression of Human Cancer. *Antioxid. Redox Signal.* **2018**, *28*, 591–608. [CrossRef]

149. Olmos, Y.; Sanchez-Gomez, F.J.; Wild, B.; Garcia-Quintans, N.; Cabezudo, S.; Lamas, S.; Monsalve, M. SirT1 regulation of antioxidant genes is dependent on the formation of a FoxO3a/PGC-1alpha complex. *Antioxid. Redox Signal.* **2013**, *19*, 1507–1521. [CrossRef]

150. Song, I.S.; Jeong, Y.J.; Jeong, S.H.; Heo, H.J.; Kim, H.K.; Bae, K.B.; Park, Y.H.; Kim, S.U.; Kim, J.M.; Kim, N.; et al. FOXM1-Induced PRX3 Regulates Stemness and Survival of Colon Cancer Cells via Maintenance of Mitochondrial Function. *Gastroenterology* **2015**, *149*, 1006–1016 e9. [CrossRef]

151. Song, I.S.; Jeong, Y.J.; Seo, Y.J.; Byun, J.M.; Kim, Y.N.; Jeong, D.H.; Han, J.; Kim, K.T.; Jang, S.W. Peroxiredoxin 3 maintains the survival of endometrial cancer stem cells by regulating oxidative stress. *Oncotarget* **2017**, *8*, 92788–92800. [CrossRef] [PubMed]

152. Ummanni, R.; Barreto, F.; Venz, S.; Scharf, C.; Barett, C.; Mannsperger, H.A.; Brase, J.C.; Kuner, R.; Schlomm, T.; Sauter, G.; et al. Peroxiredoxins 3 and 4 are overexpressed in prostate cancer tissue and affect the proliferation of prostate cancer cells in vitro. *J. Proteome Res.* **2012**, *11*, 2452–2466. [CrossRef] [PubMed]

153. Whitaker, H.C.; Patel, D.; Howat, W.J.; Warren, A.Y.; Kay, J.D.; Sangan, T.; Marioni, J.C.; Mitchell, J.; Aldridge, S.; Luxton, H.J.; et al. Peroxiredoxin-3 is overexpressed in prostate cancer and promotes cancer cell survival by protecting cells from oxidative stress. *Br. J. Cancer* **2013**, *109*, 983–993. [CrossRef] [PubMed]

154. Li, K.K.; Pang, J.C.; Lau, K.M.; Zhou, L.; Mao, Y.; Wang, Y.; Poon, W.S.; Ng, H.K. MiR-383 is downregulated in medulloblastoma and targets peroxiredoxin 3 (PRDX3). *Brain Pathol.* **2013**, *23*, 413–425. [CrossRef] [PubMed]

155. Wang, X.M.; Zhang, S.F.; Cheng, Z.Q.; Peng, Q.Z.; Hu, J.T.; Gao, L.K.; Xu, J.; Jin, H.T.; Liu, H.Y. MicroRNA383 regulates expression of PRDX3 in human medulloblastomas. *Zhonghua Bing Li Xue Za Zhi* **2012**, *41*, 547–552. [PubMed]

156. He, H.C.; Zhu, J.G.; Chen, X.B.; Chen, S.M.; Han, Z.D.; Dai, Q.S.; Ling, X.H.; Fu, X.; Lin, Z.Y.; Deng, Y.H.; et al. MicroRNA-23b downregulates peroxiredoxin III in human prostate cancer. *FEBS Lett.* **2012**, *586*, 2451–2458. [CrossRef] [PubMed]

157. Elliott, B.; Millena, A.C.; Matyunina, L.; Zhang, M.; Zou, J.; Wang, G.; Zhang, Q.; Bowen, N.; Eaton, V.; Webb, G.; et al. Essential role of JunD in cell proliferation is mediated via MYC signaling in prostate cancer cells. *Cancer Lett.* **2019**, *448*, 155–167. [CrossRef]

158. Xi, H.; Gao, Y.H.; Han, D.Y.; Li, Q.Y.; Feng, L.J.; Zhang, W.; Ji, G.; Xiao, J.C.; Zhang, H.Z.; Wei, Q. Hypoxia inducible factor-1alpha suppresses Peroxiredoxin 3 expression to promote proliferation of CCRCC cells. *FEBS Lett.* **2014**, *588*, 3390–3394. [CrossRef] [PubMed]

159. Ismail, S.; Mayah, W.; Battia, H.E.; Gaballah, H.; Jiman-Fatani, A.; Hamouda, H.; Afifi, M.A.; Elmashad, N.; Saadany, S.E. Plasma nuclear factor kappa B and serum peroxiredoxin 3 in early diagnosis of hepatocellular carcinoma. *Asian Pac. J. Cancer Prev.* **2015**, *16*, 1657–1663. [CrossRef]

160. Liu, Z.; Hu, Y.; Liang, H.; Sun, Z.; Feng, S.; Deng, H. Silencing PRDX3 Inhibits Growth and Promotes Invasion and Extracellular Matrix Degradation in Hepatocellular Carcinoma Cells. *J. Proteome Res.* **2016**, *15*, 1506–1514. [CrossRef]

161. Shi, L.; Wu, L.L.; Yang, J.R.; Chen, X.F.; Zhang, Y.; Chen, Z.Q.; Liu, C.L.; Chi, S.Y.; Zheng, J.Y.; Huang, H.X.; et al. Serum peroxiredoxin3 is a useful biomarker for early diagnosis and assessemnt of prognosis of hepatocellular carcinoma in Chinese patients. *Asian Pac. J. Cancer Prev.* **2014**, *15*, 2979–2986. [CrossRef] [PubMed]

162. Liu, X.; Feng, R.; Du, L. The role of enoyl-CoA hydratase short chain 1 and peroxiredoxin 3 in PP2-induced apoptosis in human breast cancer MCF-7 cells. *FEBS Lett.* **2010**, *584*, 3185–3192. [CrossRef] [PubMed]

163. Pendharkar, N.; Gajbhiye, A.; Taunk, K.; RoyChoudhury, S.; Dhali, S.; Seal, S.; Mane, A.; Abhang, S.; Santra, M.K.; Chaudhury, K.; et al. Quantitative tissue proteomic investigation of invasive ductal carcinoma of breast with luminal B HER2 positive and HER2 enriched subtypes towards potential diagnostic and therapeutic biomarkers. *J. Proteom.* **2016**, *132*, 112–130. [CrossRef] [PubMed]

164. Kalinina, E.V.; Berezov, T.T.; Shtil, A.A.; Chernov, N.N.; Glazunova, V.A.; Novichkova, M.D.; Nurmuradov, N.K. Expression of peroxiredoxin 1, 2, 3, and 6 genes in cancer cells during drug resistance formation. *Bull. Exp. Biol. Med.* **2012**, *153*, 878–881. [CrossRef]

165. McDonald, C.; Muhlbauer, J.; Perlmutter, G.; Taparra, K.; Phelan, S.A. Peroxiredoxin proteins protect MCF-7 breast cancer cells from doxorubicin-induced toxicity. *Int. J. Oncol.* **2014**, *45*, 219–226. [CrossRef]

166. Safaeian, M.; Hildesheim, A.; Gonzalez, P.; Yu, K.; Porras, C.; Li, Q.; Rodriguez, A.C.; Sherman, M.E.; Schiffman, M.; Wacholder, S.; et al. Single nucleotide polymorphisms in the PRDX3 and RPS19 and risk of HPV persistence and cervical precancer/cancer. *PLoS ONE* **2012**, *7*, e33619. [CrossRef]

167. Byun, J.M.; Kim, S.S.; Kim, K.T.; Kang, M.S.; Jeong, D.H.; Lee, D.S.; Jung, E.J.; Kim, Y.N.; Han, J.; Song, I.S.; et al. Overexpression of peroxiredoxin-3 and -5 is a potential biomarker for prognosis in endometrial cancer. *Oncol. Lett.* **2018**, *15*, 5111–5118. [CrossRef]

168. Han, S.; Shen, H.; Jung, M.; Hahn, B.S.; Jin, B.K.; Kang, I.; Ha, J.; Choe, W. Expression and prognostic significance of human peroxiredoxin isoforms in endometrial cancer. *Oncol. Lett.* **2012**, *3*, 1275–1279. [CrossRef]

169. Kinnula, V.L.; Lehtonen, S.; Sormunen, R.; Kaarteenaho-Wiik, R.; Kang, S.W.; Rhee, S.G.; Soini, Y. Overexpression of peroxiredoxins I, II, III, V, and VI in malignant mesothelioma. *J. Pathol.* **2002**, *196*, 316–323. [CrossRef]

170. Cunniff, B.; Wozniak, A.N.; Sweeney, P.; DeCosta, K.; Heintz, N.H. Peroxiredoxin 3 levels regulate a mitochondrial redox setpoint in malignant mesothelioma cells. *Redox Biol.* **2014**, *3*, 79–87. [CrossRef]

171. Forshaw, T.E.; Holmila, R.; Nelson, K.J.; Lewis, J.E.; Kemp, M.L.; Tsang, A.W.; Poole, L.B.; Lowther, W.T.; Furdui, C.M. Peroxiredoxins in Cancer and Response to Radiation Therapies. *Antioxidants (Basel)* **2019**, *8*, 11. [CrossRef] [PubMed]

172. Idelchik, M.; Begley, U.; Begley, T.J.; Melendez, J.A. Mitochondrial ROS control of cancer. *Semin. Cancer Biol.* **2017**, *47*, 57–66. [CrossRef] [PubMed]

173. Nguyen-Nhu, N.T.; Berck, J.; Clippe, A.; Duconseille, E.; Cherif, H.; Boone, C.; Van der Eecken, V.; Bernard, A.; Banmeyer, I.; Knoops, B. Human peroxiredoxin 5 gene organization, initial characterization of its promoter and identification of alternative forms of mRNA. *Biochim. Biophys. Acta* **2007**, *1769*, 472–483. [CrossRef] [PubMed]

174. Usmanova, N.; Tomilin, N.; Zhivotovsky, B.; Kropotov, A. Transcription factor GABP/NRF-2 controlling biogenesis of mitochondria regulates basal expression of peroxiredoxin V but the mitochondrial function of peroxiredoxin V is dispensable in the dog. *Biochimie* **2011**, *93*, 306–313. [CrossRef] [PubMed]

175. Graves, J.A.; Metukuri, M.; Scott, D.; Rothermund, K.; Prochownik, E.V. Regulation of reactive oxygen species homeostasis by peroxiredoxins and c-Myc. *J. Biol. Chem.* **2009**, *284*, 6520–6529. [CrossRef]

176. Shiota, M.; Izumi, H.; Miyamoto, N.; Onitsuka, T.; Kashiwagi, E.; Kidani, A.; Hirano, G.; Takahashi, M.; Ono, M.; Kuwano, M.; et al. Ets regulates peroxiredoxin1 and 5 expressions through their interaction with the high-mobility group protein B1. *Cancer Sci.* **2008**, *99*, 1950–1959. [CrossRef]

177. Liu, F.; Zhang, Y.; Men, T.; Jiang, X.; Yang, C.; Li, H.; Wei, X.; Yan, D.; Feng, G.; Yang, J.; et al. Quantitative proteomic analysis of gastric cancer tissue reveals novel proteins in platelet-derived growth factor b signaling pathway. *Oncotarget* **2017**, *8*, 22059–22075. [CrossRef]

178. Bur, H.; Haapasaari, K.M.; Turpeenniemi-Hujanen, T.; Kuittinen, O.; Auvinen, P.; Marin, K.; Koivunen, P.; Sormunen, R.; Soini, Y.; Karihtala, P. Oxidative stress markers and mitochondrial antioxidant enzyme expression are increased in aggressive Hodgkin lymphomas. *Histopathology* **2014**, *65*, 319–327. [CrossRef]

179. Seo, M.J.; Liu, X.; Chang, M.; Park, J.H. GATA-binding protein 1 is a novel transcription regulator of peroxiredoxin 5 in human breast cancer cells. *Int. J. Oncol.* **2012**, *40*, 655–664.

180. Li, S.; Hu, X.; Ye, M.; Zhu, X. The prognostic values of the peroxiredoxins family in ovarian cancer. *Biosci. Rep.* **2018**, *38*. [CrossRef]

181. Sienko, J.; Gaj, P.; Czajkowski, K.; Nowis, D. Peroxiredoxin-5 is a negative survival predictor in ovarian cancer. *Ginekol. Polska* **2019**, *90*, 6. [CrossRef] [PubMed]

182. Zhang, W.; Zhou, Q.; Tao, X.; Shen, Z.; Luo, H.; Zhu, X. Expression and function of peroxiredoxins in gynecological malignancies. *Front. Biosci. (Landmark Ed.)* **2016**, *21*, 986–997. [PubMed]

183. Fernandez-Ranvier, G.G.; Weng, J.; Yeh, R.F.; Shibru, D.; Khafnashar, E.; Chung, K.W.; Hwang, J.; Duh, Q.Y.; Clark, O.H.; Kebebew, E. Candidate diagnostic markers and tumor suppressor genes for adrenocortical carcinoma by expression profile of genes on chromosome 11q13. *World J. Surg.* **2008**, *32*, 873–881. [CrossRef] [PubMed]

184. Chung, Y.M.; Yoo, Y.D.; Park, J.K.; Kim, Y.T.; Kim, H.J. Increased expression of peroxiredoxin II confers resistance to cisplatin. *Anticancer Res.* **2001**, *21*, 1129–1133. [PubMed]

185. Kwee, J.K. A paradoxical chemoresistance and tumor suppressive role of antioxidant in solid cancer cells: A strange case of Dr. Jekyll and Mr. Hyde. *Biomed. Res. Int.* **2014**, *2014*, 209845. [CrossRef] [PubMed]

186. Poschmann, G.; Grzendowski, M.; Stefanski, A.; Bruns, E.; Meyer, H.E.; Stuhler, K. Redox proteomics reveal stress responsive proteins linking peroxiredoxin-1 status in glioma to chemosensitivity and oxidative stress. *Biochim. Biophys. Acta* **2015**, *1854*, 624–631. [CrossRef] [PubMed]

187. Roininen, N.; Haapasaari, K.M.; Karihtala, P. The Role of Redox-Regulating Enzymes in Inoperable Breast Cancers Treated with Neoadjuvant Chemotherapy. *Oxid. Med. Cell. Longev.* **2017**, *2017*, 2908039. [CrossRef] [PubMed]

188. Wang, T.; Diaz, A.J.; Yen, Y. The role of peroxiredoxin II in chemoresistance of breast cancer cells. *Breast Cancer (Dove Med. Press)* **2014**, *6*, 73–80. [CrossRef] [PubMed]

189. Li, L.; Yu, A.Q. The functional role of peroxiredoxin 3 in reactive oxygen species, apoptosis, and chemoresistance of cancer cells. *J. Cancer Res. Clin. Oncol.* **2015**, *141*, 2071–2077. [CrossRef] [PubMed]

190. Song, I.S.; Kim, H.K.; Jeong, S.H.; Lee, S.R.; Kim, N.; Rhee, B.D.; Ko, K.S.; Han, J. Mitochondrial peroxiredoxin III is a potential target for cancer therapy. *Int. J. Mol. Sci.* **2011**, *12*, 7163–7185. [CrossRef] [PubMed]

191. Nicolussi, A.; D'Inzeo, S.; Capalbo, C.; Giannini, G.; Coppa, A. The role of peroxiredoxins in cancer. *Mol. Clin. Oncol.* **2017**, *6*, 139–153. [CrossRef] [PubMed]

192. Kropotov, A.; Gogvadze, V.; Shupliakov, O.; Tomilin, N.; Serikov, V.B.; Tomilin, N.V.; Zhivotovsky, B. Peroxiredoxin V is essential for protection against apoptosis in human lung carcinoma cells. *Exp. Cell Res.* **2006**, *312*, 2806–2815. [CrossRef] [PubMed]

193. Yeldag, G.; Rice, A.; Del Rio Hernandez, A. Chemoresistance and the Self-Maintaining Tumor Microenvironment. *Cancers (Basel)* **2018**, *10*, 471. [CrossRef] [PubMed]

194. Zhao, J. Cancer stem cells and chemoresistance: The smartest survives the raid. *Pharm. Ther.* **2016**, *160*, 145–158. [CrossRef] [PubMed]

195. Gorrini, C.; Harris, I.S.; Mak, T.W. Modulation of oxidative stress as an anticancer strategy. *Nat. Rev. Drug Discov.* **2013**, *12*, 931–947. [CrossRef] [PubMed]

MDPI
St. Alban-Anlage 66
4052 Basel
Switzerland
Tel. +41 61 683 77 34
Fax +41 61 302 89 18
www.mdpi.com

International Journal of Molecular Sciences Editorial Office
E-mail: ijms@mdpi.com
www.mdpi.com/journal/ijms

www.ingramcontent.com/pod-product-compliance
Lightning Source LLC
LaVergne TN
LVHW070054170726
843515LV00007B/1532